P9-CAZ-078

Worldwide Advances in Structural Concrete and Masonry

Proceedings of the CCMS Symposium held in conjunction with Structures Congress XIV

Sponsored by the Committee on Concrete and Masonry Structures of the Structural Division of the American Society of Civil Engineers

Chicago, Illinois
April 15-18, 1996

Edited by A.E. Schultz and S.L. McCabe

Property of Alumni Library
Wentworth Institute of Technology

Published by the
American Society of Civil Engineers
345 East 47th Street
New York, New York 10017-2398

624.183
.W6
1996

ADV 1489

ABSTRACT:

This proceedings contains the papers presented at the American Society of Civil Engineers (ASCE) Committee on Concrete and Masonry Structures (CCMS) Symposium on "Worldwide Advances in Structural Concrete and Masonry". The Symposium was held in conjunction with the Structures Congress XIV, Chicago, Illinois, USA, April 15-18, 1996. The proceedings serves as repository of selected advances made during the past decade in research and design pertaining to concrete and masonry structures. The content of the papers addresses wide-ranging issues in structural concrete and masonry including evaluation, finite element analysis, fracture mechanics, high performance concrete, performance-based design, prestressing, seismic design, shear design, shell structures, and stability analysis.

Library of Congress Cataloging-in-Publication Data

Worldwide advances in structural concrete and masonry : proceedings of the CCMS symposium held in conjunction with Structures Congress XIV / sponsored by the Committee on Concrete and Masonry Structures of the Structural Division of the American Society of Civil Engineers, Chicago, Illinois, April 15-18, 1996 ; edited by A.E. Schultz and S.L. McCabe.
p.   cm.
Includes indexes
ISBN 0-7844-0164-0
1. Concrete construction—Congresses. 2. Masonry—Congresses. I. Schultz, A. E. II. McCabe, S. L. III. Structures Congress (14th : 1996 : Chicago, Ill.) IV. American Society of Civil Engineers. Committee on Concrete and Masonry Structures.
TA680.W69 1996   96-6763
624.1'83—dc20   CIP

The Society is not responsible for any statements made or opinions expressed in its publications.

Photocopies. Authorization to photocopy material for internal or personal use under circumstances not falling within the fair use provisions of the Copyright Act is granted by ASCE to libraries and other users registered with the Copyright Clearance Center (CCC) Transactional Reporting Service, provided that the base fee of $4.00 per article plus $.25 per page is paid directly to CCC, 222 Rosewood, Drive, Danvers, MA 01923. The identification for ASCE Books is 0-7844-0164-0/96 $4.00 + $.25 per copy. Requests for special permission or bulk copying should be addressed to Permissions & Copyright Dept., ASCE.

Copyright (c) 1996 by the American Society of Civil Engineers,
All Rights Reserved.
Library of Congress Catalog Card No: 96-6763
ISBN 0-7844-0164-0
Manufactured in the United States of America.

PREFACE

These proceedings are a compendium of the papers presented at the American Society of Civil Engineers (ASCE) Committee on Concrete and Masonry Structures (CCMS) Symposium on "Worldwide Advances in Structural Concrete and Masonry" which comprised a twelve-session track of the ASCE Fourteenth Structures Congress held in Chicago, Illinois, USA, April 15-18, 1996. The knowledge contained in these papers is but a glimpse of the advances in structural concrete and masonry that have been made during the past decade.

As the ASCE Technical Administrative Committee that oversees the work of technical committees sponsored jointly by ASCE and the American Concrete Institute (ACI), the CCMS deemed it appropriate and timely to host an event which serves to disseminate innovations and advances in structural engineering research and practice pertaining to concrete and masonry structures.

The publication of these papers represents a monumental volunteer effort by the members of the CCMS, the members of the Structures Congress 1996 Steering Committee, and by the members of the ASCE Structural Division. All of the papers in these proceedings have been reviewed by the session organizers, and, as such, they are eligible for discussion in the *Journal of Structural Engineering* and are eligible for ASCE awards. A few of the papers included in the proceedings were not presented at the Symposium due to time constraints. However, these papers were reviewed using the same procedures and accepted by the same criteria as all other Symposium papers.

The CCMS Symposium organizers would like to acknowledge the support of the Symposium participants who contributed unselfishly to this endeavor through the submission and presentation of papers. Participants include members of the CCMS as well as invited speakers from the U.S.A., Canada, Germany, Great Britain, Italy, Japan, Mexico, New Zealand, Spain and the Netherlands. The CCMS also makes special mention of the past Chairman of the Committee, Daniel P. Abrams, who is also keynote speaker for the Symposium. It was he who first articulated the vision that fostered this undertaking, and who provided the guidance and advice that contributed to a successful Symposium.

A. E. Schultz and S. L. McCabe
Editors

CONTENTS

PLENARY SESSION
Chairs: A. E. Schultz, University of Minnesota, Minneapolis, MN, USA;
S. L. McCabe, University of Kansas, Lawrence, KS, USA

Session 401: RECENT DEVELOPMENTS IN PRESTRESSED MASONRY STRUCTURES
Chair: S. C. Anand, Clemson University, Clemson, SC, USA

Session 402: ADVANCES IN SEISMIC ANALYSIS AND DESIGN OF MASONRY STRUCTURES
Chairs: A. E. Schultz, University of Minnesota, Minneapolis, MN, USA;
D. P. Abrams, University of Illinois, Urbana, IL, USA

Session 403: NEW DEVELOPMENTS IN EVALUATION OF EXISTING MASONRY BUILDINGS
Chair: D. P. Abrams, University of Illinois, Urbana, IL, USA

v

Session 410: HIGH-STRENGTH CONCRETE - SHEAR RELATED ISSUES
Chairs: C. W. French, University of Minnesota, Minneapolis, MN, USA;
J. A. Ramirez, Purdue University, West Lafayette, IN, USA

Session 411: FRACTURE MECHANICS OF HIGH PERFORMANCE CONCRETE
Chairs: R. Gettu, University of Catalonia, Barcelona, SPAIN;
V. Gopalaratnam, University of Missouri, Columbia, MO, USA

Session 412: FAILURE ASSESSMENT OF HIGH PERFORMANCE CONCRETE FORMULATIONS
Chairs: K. J. Willam, University of Colorado, Boulder, CO, USA;
S. L. McCabe, University of Kansas, Lawrence, KS, USA

*Denotes papers that were not presented at the Symposium.

---

*Denotes papers that were not presented at the Symposium.

The Return of Masonry as a Structural Material

Daniel P. Abrams,[1] Member, ASCE

Abstract

Ancient applications where masonry has been used as a structural material are given ranging from Babylon to Renaissance Europe, China and the Americas. The evolution of structural masonry in the United States is then contrasted with these magnificent constructions which serves as a preface to a discussion on the return of masonry as a structural material. Development of rational masonry building code specifications is outlined, and present and future applications of structural masonry are presented.

Introduction

The use of masonry as a structural material is as old as masonry itself. Builders of early buildings, bridges and lifelines considered engineering parameters such as loads, spans, and stresses in their masonry designs. For several millennia, masonry remained the predominant building material in numerous far-separated cultures across the globe. However, with the advent of modern concrete, steel and timber structures at the end of the nineteenth century, the use of masonry as a structural material was all but forgotten.

For the last century, the structural skeleton of most multistory buildings has consisted of a frame of steel or reinforced concrete, and/or walls of concrete. Although masonry is predominant in many commercial, residential and institutional buildings because it conveys a message of strength and durability, it is commonly neglected as a structural component. Concrete masonry infill panels or clay masonry veneer are often prescribed for their good qualities for isolating thermal and

----

[1]Professor, University of Illinois at Urbana-Champaign, 1245 Newmark Civil Engineering Laboratory, 205 N. Mathews, Urbana, IL 61801

acoustical effects, or for their color and texture. However, the high compressive strength is not typically relied on to resist gravity or lateral forces.

Today, the use of masonry as a structural material is making a return particularly in the western states where structural systems of reinforced clay or concrete masonry are used exclusive of any other structural material. Buildings approaching thirty stories tall are being constructed with reinforced masonry walls as the sole resistance of lateral force. Building code requirements for structural masonry are now becoming as sophisticated as for other materials with the introduction of limit state concepts and performance based design.

Several examples are presented in this paper showing how masonry has been used as a structural material throughout the history of constructed works. The evolution of modern masonry construction in the United States is summarized to provide contrast with these ancient marvels of construction, and to serve as a preface to a discussion on how the structural use of masonry has made a return through research and rational building code requirements.

<u>The First Building Material</u>

The people of Jericho were building with brick more than 9000 years ago. Sumerian and Babylonian builders constructed city walls and palaces of sun-dried brick and covered them with more durable kiln-baked glazed brick. Mesopotamian architecture was necessarily based on the use of mud and clay brick because of the unavailability of stone and timber.

Ziggurats, a temple tower, were built with mud brick masonry from the 4th millennium BC to 600 BC. They rose in stepped stages to a small temple at the peak. The most famous ziggurat was the temple-tower of Etemenanki associated with the Tower of Babel (Fig. 1) which reached a height of 91 meters. The largest ruins are those of the Elamite ziggurat at Choga Zambil in Iran (13th century BC) which is 102 meters square at its base.

Figure 1 Tower of Babel

Stone masonry was used for construction of the Egyptian pyramids beginning c. 2500 BC. When constructed, the pyramid of Khufu (also known as the Great Pyramid) at Giza measured 147 meters high with a square base measuring 230 meters on each side, and remains today as one of the Seven Wonders of the World. The Pyramid of Khafre (Fig. 2), also at Giza, was constructed without the use of cranes, pulleys or lifting tackle. Stonework was squared and fitted, and no adhesive or mortar was used to join the stones. Beginning in the 10th century AD the Giza pyramids served as a source of building materials for the construction of Cairo, and as a result the pyramids were stripped of their smooth outer facing of limestone. Ancient examples of Cyclopean masonry, composed of large stone blocks laid without mortar, have been found throughout Europe and in China and Peru. During the same era, Egyptian houses were formed of mud-brick walls with columns made from bundles of reeds lashed together. The walls of stone buildings were generally thicker at the base and tapered.

Figure 2. Pyramid of Khafre

The Great Wall of China was constructed from 221 to 204 BC by linking earlier constructed walls of states along the northern frontier to thwart invasion from nomads to the north. The wall winds 2400 km from Gansu to the Yellow Sea, and is the longest human-made structure in the world. The width of the wall ranges from 4.6 to 9.1 meters at its base and tapers to nominally 3.7 meters at its top. Its height varies from 6 to 15 meters, and averages 7.6 meters. The wall is built of earth and stone, and faced with brick in the eastern parts.

As masonry dominated construction in Europe and Asia, it was also a popular material in the Americas. Early pyramids made of stone were built about 1200 BC at the Olmec site of LaVenta in Tabasco Mexico. Later monuments constructed by the Maya, Toltecs and Aztecs in central Mexico, the Yucatan, Guatemala, Honduras, El Salvador and Peru were based on the Olmec plan. The New World pyramids were four-sided, flat-topped polyhedrons with stepped sides. The larger American pyramids slope less than the Egyptian pyramids, but the smaller ones often have a steeper incline. One of the largest pyramids in Central America is the Pyramid of the Sun at Teotihuacan Mexico with a 66-meter height and constructed in the 2nd century AD.

Aztec, Mayan and other Indian cultures relied on masonry for housing and monuments. The capital of the Yucatan region of the classic Maya was Uxmal which had buildings faced with stone veneers. These 9th century buildings had unusual engineering refinements such as a slight outward lean that made them appear light and elegant. The Anasazi constructed multi-story pueblos in the American southwest from stone, mud and beams during the period of 1100-1300.

Greek and Roman Architecture

The Greeks and Romans developed masonry techniques that have continued in practice to the present day. Early Greek architecture (3000-700 BC) was characterized by the use of massive stone blocks for walls and by the use of corbeled masonry to make primitive forms of vaults and domes. They used limestone in Italy and Sicily, marble in the Greek islands and Asia minor, and limestone covered with marble on the Greek mainland. Later, they built primarily in marble.

The Romans inherited the architectural traditions of those lands they conquered, and added the construction skills of the Etruscans. Roman construction achievements reflected the use of a wider range of materials including concrete, terra cotta and fired clay bricks. Their refinements of the arch, vault and dome previously used by the Egyptians, Babylonians and Greeks (Fig. 3) gave rise to a number of building types of unprecedented size and complexity, which could not have been built with the Greek beam and column system. These included great aqueducts, coliseums, and palaces constructed of clay brick. Roman bridges

Figure 3   Arch and Vaulted Construction

consisted of one or more semicircular stone arches. The aqueduct at Pont du Gard at Nimes France, built circa 60 AD, has three tiers of arches rising 47.2 meters above the Gard river and spans a distance of 261 meters.

The continuity of Roman plans and construction techniques was strong with Byzantine architecture in the 4th century. Whereas slightly modified Roman basilican plans were used for Italian churches in the early 500's AD, huge domed churches were constructed in Constantinople, such as the Hagia Sophia (532-37), that were built on a scale far larger than anything achieved with the Western Roman Empire. During the middle ages, Islamic architects developed a rich variety of pointed, scalloped, horseshoe and S-curved arches for mosques and palaces.

Romanesque, Gothic and Renaissance Architecture

The primary characteristics of Romanesque architecture (mid-tenth to mid-twelfth century) were Roman in origin, however, large internal spaces were spanned by barrel vaults on thick, squat columns and piers, windows and doors had round-headed arches, and most of the major churches were laid out on the basilican plan, modified by the addition of the buttress, transept and tower. From the mid-twelfth to the 16th century, Gothic architecture was characterized by the use of the pointed arch, which minimized outward thrust and thus made possible lighter, thinner, window-filled walls, creating the lofty, spacious interiors of Gothic cathedrals.

In the 15th century, Renaissance architecture was influenced by the use of the round arch, the barrel vault, and the dome. Filippo Brunelleschi's design for the cathedral of Santa Maria degli Angeli in Florence (Fig.

Figure 4 Florence Domo

4) took the form of a domed octagon (39 meters in diameter and 91 meters high) with eight radiating chapels, a centralized plan that became the ideal among his contemporaries. His skills during construction of the dome (1420-61) of the cathedral marked the start of the Renaissance. By devising new scaffolding and hoists, by making the dome a self-supporting double shell, and by using new and lighter masonry, he was able to construct the dome without scaffolding.

Masonry was brought to the New World by European explorers for the construction of churches, fortifications and monuments. On his second trip, Columbus built the first catholic church in the Americas of stone masonry. However, this was far from the first application of masonry in the Americas as noted earlier with the construction of buildings and pyramids by the native Aztecs, Mayas and Anasazi cultures.

Early Applications of Modern Structural Masonry

Brick masonry construction was used for many of the buildings in Jamestown, Virginia as early as 1607. The oldest standing church in colonial America, St. Luke's Anglican Church was built of brick in 1632 in Wick County, Virginia. The popularity of brick construction in the American colonies was marked by a law passed in 1683 by the General Assembly of New Jersey to establish standard brick dimensions. Clay brick was made in Philadelphia as early as 1685, and by 1800, 80% of the buildings in that city were made of brick, and many are standing today.

Reinforced brick masonry was first used in 1825 with the Thames Tunnel in London. Two 50-foot diameter brick shafts were constructed 70 feet deep. The 30-inch thick walls were reinforced vertically with 1-inch diameter wrought iron bolts. Other applications of reinforced masonry followed by innovative builders with a sense for design, but having no basis for a rational design since flexural mechanics of reinforced sections had not been developed. However, these applications were the exception rather than the rule.   Unreinforced masonry continued to be the conventional option through the twentieth century.

Although present-day applications of masonry are usually limited to building structures, bridges, viaducts and sewers were commonly constructed of brick masonry through the early part of the twentieth century. The development of railroads precipitated a revival of semicircular-arch construction in cut stone. The Ballochmyle stone viaduct that crosses the Ayr River near Mauchline, Scotland has a semicircular arch spanning 55 meters. Opened in 1902, the longest stone-arch railroad viaduct is at Rockville, Pennsylvania and includes 48 arches with 21.3 meter spans for a total length of 1161 meters (Fig. 5). A viaduct of 222 arches and a total span of 3658 meters connects Venice with the Italian mainland. The longest span for a masonry arch in the world is 89.9 meters for the Syra Bridge at Plauen Germany completed in 1903. Massive piers constructed of limestone and granite for many of the bridges crossing rivers in urban centers were also constructed at this time, and are still standing today.   Some of these bridges such as the Queensboro Bridge crossing the East River in New York City are now being assessed for their seismic hazard potential which requires a detailed structural analysis of these masonry piers.

Figure 5   Railroad Viaduct at Rockville, Pennsylvania

The growth of the eastern cities in the United States after 1870 was marked by the construction of high-rise office and commercial buildings constructed with unreinforced brick masonry bearing walls. Building codes existed, but were based on crude empirical procedures for the structural design. One such design rule was to provide as many wythes in thickness as the number of stories above. Thus, the walls at the base of the fifteenth-story Monadnock Building (the highest and heaviest wall-bearing building in Chicago, built in 1891) are over four-feet thick.

Unreinforced brick bearing wall construction also remained popular for low-rise buildings up to three stories, many of which form the inner core of our urban areas today. These systems were typically designed to resist gravity forces, but because of their high lateral stiffness were also sound during high winds. Because little was known at the time about earthquake loads on buildings, these unreinforced brick buildings continued to be constructed even after the mass destruction in San Francisco after the 1906 earthquake. An ordinance was passed forbidding their construction in southern California after the 1933 Long Beach earthquake, however, many of these buildings remain today as earthquake hazards in middle and eastern America.

At the turn of the twentieth century, concrete block masonry became competitive with clay brick and timber as the Portland cement industry grew. Because concrete blocks could be made locally, freight charges were less than for other materials. Contractors quickly appreciated the enhanced speed of construction as one concrete block (12" x 9" x 32") could replace 28 common clay bricks. An estimated 1,500 manufacturers of concrete block emerged in just five years in the United States. Bricklayers resisted the movement to concrete block because of the heavy weight of a single unit (180 lbs.) until a lighter cored unit was developed.

The first specification for the manufacturing of concrete block was developed in 1908 by the National Association of Cement Users which later became the American Concrete Institute. Minimum compressive strength according to this specification was 1000 psi. By 1914, production of concrete block exceeded by three times that manufactured in 1901, and by 1919, 50 million units were produced. In 1941 total block production rose to 467 million units which exceeded the production of both brick and hollow-clay tile.

Despite the growth of the concrete block industry, clay-unit masonry remained popular for cladding and veneer. Multistory office and commercial buildings in urban centers at the turn of the century were often constructed of steel frames cladded with clay brick. Although the frames were assumed to carry all of the gravity and lateral loads, the unreinforced masonry cladding did stiffen the systems appreciably for lateral forces. Damage patterns to this effect were evident in a number of older buildings in Oakland following the 1989 Loma Prieta earthquake.

Rational Structural Design of Masonry

Rational, but simplistic, procedures started to emerge for the structural design of masonry as early as the first decade of the twentieth century when allowable compressive stresses were refined and standardized. However, nearly all masonry design procedures were still based on empirical rules because of the limited amount of engineering research done with structural masonry. As a result, other construction materials, namely reinforced concrete and structural steel, replaced masonry for structural applications in building and non-building structures. Although markets for brick and block masonry continued to increase, the utilization of masonry was largely limited to non-structural applications such as cladding, veneers, and infills. The age of enlightenment for masonry as a structural material had to wait for another half century.

Property of Alumni Library
Wentworth Institute of Technology

Much of the research that lead to building code recommendations for structural masonry was done at the Structural Clay Products Association and the Portland Cement Association. Each of these associations gave birth to the present-day Brick Institute of America (BIA) and the National Concrete Masonry Association (NCMA). In the early 1960's these and other masonry industry organizations began development of a data base of materials and assemblage performance through sponsored research and testing programs. In 1966, BIA published a set of recommendations for engineered brick masonry (BIA, 1966), and the NCMA followed four years later with a set of specifications for loadbearing concrete masonry (NCMA, 1970). A committee of the American Concrete Institute (ACI) on concrete masonry developed a report on concrete masonry construction (ACI, 1970) and a set of specifications on concrete masonry construction (ACI, 1976). These two documents served as the basis for a set of building code requirements for concrete masonry (ACI, 1979). However, by the end of the seventies, no single set of building code regulations existed for the design of both clay and concrete masonry.

The Masonry Society (TMS) was formed a few years before the First North American Masonry Conference was held at the University of Colorado in 1978. As a first activity of TMS, a masonry standard was developed which included both clay and concrete masonry. The TMS standard was used as the primary resource document for major changes in Chapter 24 of the 1985 Uniform Building Code (UBC) on masonry. This chapter has been further revised in 1988, 1991 and 1994 editions of the UBC (TMS, 1995), and now has evolved to include provisions on strength design of reinforced masonry walls, though it remains principally an allowable stress design document.

The ACI and the American Society of Civil Engineers (ASCE) created a joint committee on masonry (ACI-ASCE 530) in 1978 with a mission to develop a national consensus standard on masonry design comprising recommendations for both clay and concrete masonry. This committee produced a set of building code requirements and specifications for masonry construction in 1988. The committee was renamed the Masonry Standards Joint Committee (MSJC) when sponsorship of the committee was expanded to include The Masonry Society. Editions of the MSJC code have been published in 1992 and 1995 (TMS,ACI,ASCE, 1995) which are still based on allowable stresses. However, development of a limit states design code for masonry is underway at present for future editions of the code.

Provisions for structural design of masonry have been introduced to national guidelines for the seismic design, evaluation and rehabilitation of buildings. Chapter 12 of the *NEHRP Recommended Seismic Provisions for New Buildings* (FEMA, 1994) contains code-type language for the strength-design of newly constructed, unreinforced and reinforced masonry buildings. Appendix C of the *NEHRP Handbook for the Seismic Evaluation of Existing Buildings* (FEMA, 1992) is devoted to evaluation of unreinforced brick masonry buildings using strength-based concepts. The new FEMA/BSSC/ATC *Guidelines and Commentary for Seismic Rehabilitation*

Property of Ashburn Library
Wentworth Institute of Technology

of Buildings (ATC, 1995) prescribes analysis methods for evaluation of rehabilitated buildings comprised of masonry elements as well as elements constructed of steel, concrete or timber. For the first time, performance-based design concepts are being applied to masonry structural components.

Building code requirements for structural masonry will continue to be refined as more research is done to define behavior of masonry elements subjected to gravity and lateral forces. Much like code requirements for reinforced concrete in the 1950's, a great potential exists at present to improve design requirements for structural masonry. Such code development can be accomplished in a much shorter time than for concrete because of advances in the way research is conducted, the increased number of researchers and the enhanced capabilities of experimental facilities. With accelerated research programs, the return of masonry as a structural material can be accomplished in a relatively short period of a few decades.

Present Applications of Structural Masonry

The use of masonry as a structural material has been developing at a rapid pace for the last two decades in the western United States. This evolution is a result of the need to consider how masonry behaves structurally during seismic excitation. Unlike consideration of the structural system for static gravity loadings, portions of the building system (particularly the stiff masonry walls) cannot be simply neglected in the sense of being conservative. All of the structural elements whether supporting gravity loads or not, must remain stable while deflecting to sizable lateral drifts. Thus for the seismic case, masonry must be considered to act as a structural material no matter what the redundant system is.

As more and more has been learned about the structural characteristics of masonry, western engineers have become more aware of the fact that masonry could perhaps replace other construction materials for the design of new buildings whether in a high seismic zone or not. The use of reinforced masonry has come full front for the design of multistory buildings. The tallest structural masonry building is now just less than thirty stories (the Excaliber Hotel in Las Vegas) with many more structural masonry buildings in the design phase. Reinforced concrete masonry walls are the most typical structural component of these buildings, however, reinforced clay-unit masonry has also been used particularly in the northwest states. The advantage of these systems is that the beauty of the structural masonry can be displayed architecturally.

Despite the advances in the western U.S., masonry in eastern buildings is used predominantly for nonstructural applications. Architects, and the building owners they represent, prefer the look and texture of a masonry wall because it makes a statement of strength, durability and longevity. Yet ironically, the masonry structural properties are often neglected by the engineer. Whereas the veneer brick or infill block has sufficient compressive strength to resist gravity and lateral loads, a hidden, skeletal steel or concrete frame is designed to resist all of the load. For example, a building recently constructed on the Illinois engineering campus with decorative masonry arches was reinforced with a heavy steel wide-flange beam placed above the arch to span floor loads to steel columns.

Some structural masonry buildings have been constructed in the eastern U.S., but many with unreinforced masonry. Such systems are constructed by placing precast concrete planks on top of unreinforced concrete masonry walls. The heavy weight of the system provides the necessary resistance to lateral forces for buildings as tall as ten stories. Just in the present decade, some reinforced masonry has been introduced in the eastern U.S. by learning from the westerners on how to capture some of the market from tilt-up concrete construction. As a result of research in California and provisions first presented in the 1985 UBC, tall one-story buildings such as warehouses, discount stores, or gymnasiums are now economical with reinforced masonry. Strength concepts are used to design reinforced walls that are no longer restricted by code-limiting height-to-thickness ratios if a second-order analysis is done to account for the influence of lateral deflections on bending moments. With vertical reinforcement placed at the center, single wythe walls of hollow concrete or clay units are now possible at heights exceeding forty feet. This design concept survived its most stringent test to date with the 1994 Northridge earthquake that excited, with nearly no structural damage, a large number of post 1985 buildings of this type (TMS, 1994).

Future Applications of Masonry Construction

Over the next two decades the structural use of masonry will continue to rise as reinforced masonry buildings become common place across the country. As limit states design codes for masonry become more mature, such as is now taking place with Chapter 21 of the 1994 Uniform Building Code and the next code cycle for the 1998 Masonry Standards Joint Committee Code, more and more engineers will adopt an attitude that strong and serviceable designs can be obtained with masonry. In addition, design offices will be populated with an increasing number of young engineers educated at universities that have introduced masonry courses into their engineering curricula.

In the future, the structural use of masonry will extend beyond simply buildings. Retaining and sound barrier walls constructed of masonry are presently being introduced. Research is currently being done to evaluate the use of unbonded masonry pavers for airport and highway pavements. Perhaps in the years to come, transportation and hydraulic structures may adopt reinforced masonry in lieu of conventional reinforced concrete for small applications. To meet these applications, prestressed masonry will become much more popular as American engineers learn of practices abroad, and as design procedures are introduced in U.S. codes for both pretensioned and post-tensioned walls.

Concluding Remarks

A brief history of masonry constructed works has been presented to illustrate that masonry was indeed the first structural material. More masonry structures have been constructed over time than any other, yet masonry fails to be the structural material of preference for present-day building construction. The strength of masonry is comparable to that of high-performance materials, but its use has not been exploited fully for modern-day structural applications.

The time since publication of the first building code requirements for structural masonry ten years ago is quite short relative to that of codes for other materials, and nearly insignificant relative to the history of masonry. If the duration of all masonry construction were represented across a time line of 24 hours, these code requirements would be introduced only in the last 90 seconds. With further advances in construction technology, a potential exists for future structural applications of masonry to rival the wondrous achievements of ancient civilizations.

Appendix 1: References

1. *Building Code Requirements for Concrete Masonry Structures*, American Concrete Institute (ACI 531-79), 1979.

2. *Building Code Requirements for Masonry Structures* (TMS 402-95, ACI 530-95/ASCE 5-95) and *Specifications for Masonry Structures* (TMS 602-95, ACI 530.1-95/ASCE 6-95), 1995.

3. *Commentary on Chapter 21 of the Uniform Building Code*, The Masonry Society, Boulder, Colorado, 1995.

4. *Concrete Masonry Structures - Design and Construction*, American Concrete Institute Committee 531, 1970.

5. *Guidelines and Commentary for Seismic Rehabilitation of Buildings*, FEMA-BSSC-ASCE-ATC, 75% draft, Applied Technology Council, Redwood City, CA (ATC 33), October 1995.

6. *NEHRP Handbook for the Seismic Evaluation of Existing Buildings*, FEMA 178, Appendix C: Evaluation of Unreinforced Masonry Bearing Wall Buildings, June 1992.

7. *NEHRP Recommended Seismic Provisions for New Buildings*, Chapter 12, 1994.

8. *Performance of Masonry Buildings in the Northridge Earthquake*, The Masonry Society, Boulder, Colorado, 1994.

9. *Recommended Practice for Engineered Brick Masonry*, Brick Institute of America, 1966.

10. *Specifications for Concrete Masonry Construction*, American Concrete Institute (ACI 531.1-76), 1976.

11. *Specifications for Loadbearing Concrete Masonry*, National Concrete Masonry Association, 1970.

Innovative Development of Prestressed Masonry
Eur Ing G Shaw CEng FICE FIStructE MConsE

1.0   Asbstract

The paper describes the Author's experience in the innovative development of structural pre-stressed masonry through designs, research and application over a thirty five year period. The cross fertilisation of ideas and developments, from other materials, and the unique   advantages this revealed for masonry are described. The techniques are illustrated with examples of applications. The paper concludes with a description of the most recent developments in prefabrication techniques which place masonry on a new horizon for exciting further developments and applications.

2.0   Introduction

It is the cost and ease of construction which guides the good engineer when considering development of his ideas through project application. "Keep it simple" has long been the guidance for true innovation and if the proposals are to be attractive to the client they must be cost effective in relation to capital cost and maintenance.

The projects which punctuate the developments in this paper clearly indicate the successful application by the Author of simple techniques applied to structural masonry in its development over the past 35 years. These examples are a small number selected from the hundreds of pre-stressed applications over these years. The developments in this case originate from basic principles of engineering, extended through analysis, application, design and research to solve structural problems. Observation of stability and strength gained from geometry of layout of bonded masonry units ran parallel with the authors experience in the analysis

----

Gerard Shaw, Curtins Consulting Engineers plc
18 Thanet Street, Kings Cross, London WC1H 9QL

13

and design of prestressed concrete.   During this period the principles of prestressed concrete and the structural efficiency of geometric sections such as pre-stressed hollow sections were examined.

It was noted that sections with a high Z/A ratio (ie section modulus divided by the sectional area) have greater efficiency in bending resistance and have an improved radius of gyration and hence resistance to buckling. The Authors early designs of masonry structures in the 1960s had been basically gravity structures with compressive loading on the external cladding and internal walls. As the buildings became taller and of lighter construction the problems relating to lateral loading became more critical. The problem for masonry was particularly serious when lightweight roofs were adopted for tall single storey open plan structures where compressive loads on the walls was minimal and lateral load due to wind was significant.

The design criteria for these buildings was not axial compressive strength but the flexural tensile strength of the wall section. This is fundamentally important since the flexural tensile strength of masonry is only approximately 1/30 of its compressive strength. The governing structural design factor at first, became to limit the tensile stress and this was achieved by increasing the section modulus of the structural element by using Fin Wall and Diaphragm wall sections, (ie T sections and hollow box sections) see ref 1 and 2.

As is well known the moment of resistance MR = f x Z (ie bending stress times the section modulus. Thus an increase of Z results in a decrease in the bending stress f and hence the bending tensile stress. The normal masonry wall has a low Z/A ratio and is not structurally efficient compared with box, Tee and L sections see Ref 5. The use of section geometry seemed an obvious way forward but in a number of situations section sizes became cumbersome since the compressive stress of the masonry was not being fully developed and the governing factor was its very low bending tensile strength. Masonry being strong in compression and weak in tension is similar to concrete and, whilst construction methods are different, the basic methods of overcoming the weakness are related.

3.0   Material Displacement in Pre-stressed Masonry

The ease by which a material can be displaced from the neutral axis, to give a high Z/A ratio, varies depending upon method of manufacture and/or construction. Masonry is constructed from small preformed standardised units, and to make economic geometric forms it is necessary to arrange them into larger and more efficient sectional shapes.

4.0   Masonry Strengthening

Changes in section geometry can compensate for the low tensile strength and or tensile stresses can be carried by steel reinforcement or eliminated by prestressing. See ref 4 and 7. In the UK however reinforced masonry has not usually been shown to be cost effective. In addition to construction and supervision difficulties with reinforced sections there is the disadvantage that over 50% of the cross-section is structurally wasted in the tensile zone where it makes little contribution in resisting bending. Prestressing on the other hand utilises the whole cross-section and, too, can reduce the amount of microcracking by maintaining compression in the section.

The location and magnitude of the prestress within the section is critical to the dimensions of the required cross-section. For sections subjected to compression and single axis bending the use of prestress applied eccentricity to counteract the applied bending moment is generally the most economic. For sections subjected to reversal of bending moments, such as an element resisting wind loading concentric prestress is often more suitable.

The increase in bending resistance of an eccentricly prestressed hollow masonry section from that of a similar cross section of solid, plain masonry is in the region of 400 to 500 times.

5.0 Section Analysis

In the early stages of these developments the basic analysis of post-tensioned masonry sections were based upon maintaining zero tensile stress in the cross section at ultimate load for all conditions. However, for later designs improved efficiency was gained by allowing a cracked section in some situations at ultimate load but maintaining a zero tensile stress approach for serviceability limit state.

The following selected projects punctuated the developments in prestressed masonry using the above approach.

6.0 Masonry Development Through Project Applications

6.1 Pre-Stressed Cavity Construction

In the late 60s and early 70s the Author designed a number of buildings using tall prestressed cantilevered cavity walls to resist wind loading, these included many school buildings, Libraries and churches. See ref 4 and 5.

6.2 Oaktree Lane Community Centre

In 1978 in the Oaktree Lane Community Centre, the author incorporated the first use of post-tensioned diaphragm walls, (hollow box sections) as the perimeter walls to the main sports hall. See ref 8.

Post-tensioning was used to overcome the diagonal tensile stresses produced in the masonry panels during mining subsidence and wind loading. The site was situated in an active coal mining area with both past workings and future extraction to consider. Three seams were due to be worked within the first five year period and the calculated subsidence for the first wave indicated a maximum of 1,080mm (3ft-7ft) and a differential subsidence across the site which would leave the building in a tilted state after the first wave had passed through. The effects of other workings, within the five year plan were less severe and predicted to have a righting effect on this initial tilt.

At Oaktree Lane Community Centre prestressed propped cantilever diaphragm walls was adopted. The solution allowed a "brittle" material to accommodate distortions as the subsidence wave passed under the building.

In order to limit the strain applied from the mining subsidence a sand slip plane was provided below the foundation, and a compression ring of loose filling around the perimeter of the foundation. The masonry panels were jointed into short lengths and the heads of the panels were linked together by a continuous reinforced concrete beam. This was used both as the capping for the post-tensioned rods and a tie to link the panels together. It is believed that this project provided the first application of post-tensioning to a diaphragm wall and, too, to solve a mining subsidence problem.

6.3   **Warrington Citadel**

The post tensioned masonry on this project was restricted to the main hall which was approximately 25m (82ft) long by 15m (49ft) wide and 8.5m (27ft 10ins) high. The top of the wall was not propped due to the difficulty of accommodating the propping through the clear storey window which ran around the high level of the wall. It was decided that the economic solution was to design a post-tensioned diaphragm as a free cantilever. See photograph 1.

A combination of clay brickwork outer leaf and concrete block inner leaf was adopted in spite of the differential movement characteristics of the two materials. The use of two materials was due to the aesthetic requirement which over-ruled other requirements relating to movement.

This combination of materials added further difficulties to the design in that the likely long-term changes in stress within the masonry section were more unpredictable. It was anticipated that after prestress some load transfer would be likely from the shrinking concrete blockwork into the clay brickwork on the other face. It was anticipated that stress losses would occur in the concrete and stress gain in the clay masonry. The client the Salvation Army noted significant savings over their previous steel structures.

Photograph 1

6.4   Braintree Ambulance Port

The ambulance port forms the main entrance for both ambulances and the general public into the new hospital rehabilitation department. The intention was to provide an open structure on all four sides. See photograph 2. The structural design had to solve the problems of overall stability and accidental damage from impact.

Photograph 2

The solution for the main support consisted of six hollow brick columns 440m sq (1ft 5¼ins) with a central void of 235mm sq (9¼ins). Through this void two number high yield 16mm (approx ½ins) diameter bars were used to post-tension each column. The post-tensioning was used not only to increase the moment of resistance but also to convert the brickwork from a brittle to a ductile behaviour and to increase its impact resistance. The design allowed for a column to be accidentally removed and the remaining columns to act as anchorage using the post-tensioning rods as ties in the opposite corners to prevent total roof collapse.

6.5 Osborne Memorial Halls, Boscombe

For this 600 seat Hall a framework of small post-tensioned channel sections was used to support the roof and to resist lateral wind loading, see ref 6 and 7 and photograph 3. On this project the solution demonstrated that post-tensioned masonry does not require high strength bricks and mortar, the bricks having a characteristic compressive strength of 19N/mm² (2,700 lbs/in²) in a 1:1:6 mortar mix.

Photograph 3

Because of site limitations there was a restriction on any projections and it was essential to use the minimum channel cross-section. This was achieved by propping the cantilever channel, against a braced steel roof deck to reduce the wall bending moments, and calculating the required size based upon a cracked section at ultimate load and zero tension at serviceability limit state.

6.6 Prefabricated Post Tensioned Brickwork

In 1984 the author produced details for prefabricated post-tensioned masonry in the Masonry Detailers Manual for Granda Publications, and numerous opportunities to develop the technique further have since arisen. In 1990 the Author engineered and verified the buildability and handling suitability of prefabricated prestressed masonry by the use of both design and prototype construction for prefabricated stack bonded masonry for outer cladding for an office building in London.

Further use of prototype construction was used to verify the prefabricated prestressed brick cladding for the Power Gen headquarters in Coventry. Good practical details confirmed by prototype construction is a way of debugging and testing the buildability of new techniques and has certainly been of value in developing prefabricated masonry as a cost effect alternative for use in fast track construction.

6.7   A Break Through For Horozontal Masonry

The difficulty of constructing horizontal prestressed masonry has meant that until recently the application of post tensioned members has only proved economic when used vertically.

In 1984 the author had realised that prefabrication of horizontal members constructed vertically could prove economic see Ref 9. Calculations followed by research has shown that such solutions are viable, practical and easily constructed and by 1994 the development was used for foot bridge construction.

6.7.1   Tring Bridges

Introduction

In 1994 two 7m (22ft 11½ins) long bridge decks designed by the author were constructed on end in Tring, Hertfordshire and tipped over to cross a stream in a nature park. The bridges are the first prestressed brick box girder decks and their performance is being monitored by the University of Plymouth.

The first bridge used parafil rope as the pre-stressing tendons. The second bridge has steel tendons which will allow comparisons between the two materials. The construction of these two bridges results from a 5 year search by the author for a suitable site to progress the application of structural masonry into a new era.

Design

The design intention was to maximise durability and minimise maintenance and part of this aim was to incorporate as little reinforcement as possible within the decks.

The design of the decks was developed from a preliminary sized brick bonding arrangement of adequate webs for shear and stiffness and the largest practical voids to minimise weight and accommodate tendons.   The importance of durability of the decks was, at the forefront of the design priorities and a two-directional surface camber was used to shed water, reduce algae growth and improve frost resistance.   A camber of 20mm (¾ins) across the deck and 75mm (3ins) along the length of the deck was considered adequate. (see figures 1 & 2). This two directional curvature, using standard bricks, creates a foothold key on the walking surface.

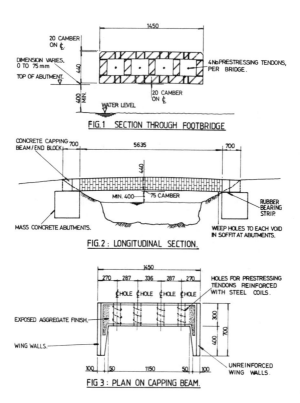

FIG.1   SECTION THROUGH FOOTBRIDGE

FIG.2 : LONGITUDINAL SECTION.

FIG 3 : PLAN ON CAPPING BEAM.

Figs 1, 2 & 3

In order to make the walking surface more aesthetically pleasing a diamond pattern of red engineering bricks was incorporated into the main background of blue engineering bricks. Engineering bricks and 1:¼:3 mix mortar was specified to satisfy the need for a durable deck.

Design calculations were carried out to check the bending and shear stresses, the required post-tensioning forces and the likely deflections for all critical stages of construction, erection and use. The tendons were designed as unbonded, straight tendons and used the camber to provide the required eccentricity for the pre-stress at mid span. The end-block anchorages/capping beams (see figure 3) incorporate wing walls to retain the top of the earth embankment when in the final position and to elevate the pre-stressing anchorages to provide access when standing on end during construction. Apart from the four pre-stressing tendons the only reinforcement in the decks consists of steel coils around the anchorages in the end block to prevent bursting.

Construction

The basis of the construction method adopted is the "Shaw System" which consists of initially constructing the masonry sections vertically, then pre-stressing them prior to moving the sections into their final horizontal position (see photograph 4). The actual construction sequence adopted in the case of the two bridges was to cast the mass concrete abutment foundations whilst prefabricating the bridge end blocks. For each bridge one end block was then stood on end on the abutment foundation and the box sections constructed vertically in a similar fashion to a four flue chimney stack. The stack was built to an outward two-directional curve using timber profile boards prior to curing, pre-stressing and tipping over onto the abutments..

Vertical profiles of timber/ply were used on the face side of the brick tower to control the longitudinal camber from top to bottom. The back side of these profiles was straight and continuous so that it could be easily "plumbed" as the work progressed.

The camber on the Tring bridges was designed to match the required eccentricity of the pre-stressing force,. The cavities/flues had to be kept clean and 200mm (7¾ins) squares of ply were lowered into each void each morning and raised and cleaned off each night. The underside of the bridge deck followed the same camber as the upper deck surface, to prevent the cross-joints becoming too big. This required two batten profile boards equal to the bridge width one concave and one convex. The method of construction proved simple in spite of instrumentation for monitoring purposes.

Photograph 4: Laying steel tendon Bridge on abutments

After vertically constructing the brick decks pull strings were dropped though each flue and through the lower anchorage holes to assist in pulling through the pre-stressing tendons. The upper end blocks were then bedded in place. In the case of the "Parafil" bridge this was prior to installing the tendons, whilst in the case of the "Macalloy" bridge the tendons were placed prior to fixing the upper end block. This was due to the need to accommodate the wiring to the foil sensor strain gauges fixed to the pre-stressing tendons. The wiring for the monitoring devices was threaded through the end blocks as they were bedded onto the brickwork.

After construction the decks were allowed to cure for a further two weeks prior to the pre-stress being applied. The stressing was then carried out using a hydraulic jack up to a final load in the tendons of 20 tons per tendon which 2 days later was topped up to compensate for the early losses of pre-stress.

On completion of topping up the stress, the decks were attached to a crane and the scaffolding, which had propped the decks during construction, was removed. The decks were then hung from one crane whilst a second crane pulled out the lower end of the deck to allow it to be lowered down on the rubber bearing strips on the abutments.

Instrumentation and monitoring

The construction of the footbridges provided the opportunity to study the behaviour of these new forms of construction.

In order to monitor strains across the section, 22 vibrating wire gauges were installed within the voids, near one end of the parafil tendon bridge and one temperature sensor located in each of the four voids. These gauges are connected to a data-logger which sends data regularly to the University via a telecom link. Demec studs are provided in a regular pattern along the side elevations, of both bridges for data collection during test loading.

Foil sensor strain gauges are installed at the centre of each of the four macalloy pre-stressing tendons in order to measure strain under both ambient and test load conditions. The bridges are both working bridges and have been providing data under ambient conditions for the past year, the results will be published in a later paper.

5.7   Conclusions

Modern structural masonry is a 'new' structural materials (not merely an attractive cladding) and pre-stressed hollow masonry can achieve strengths of 400 to 500 time those of plain 'plate' section walls. This 'new' material has in the past 35 years proved itself cost effective, as a structural framework, in competition with other structural materials, for many building applications.

With appropriate details masonry rates highly for buildability being within the normal capabilities of most small, medium and large contractors. Experience has shown that the material when used properly, has excellent durability and with further development will have a long and lasting future.

Though structural masonry has advanced considerably in the last 35 years it is obvious that there is still much potential awaiting development. The present weakness in the material requires innovative thinking in order to allow maximum use of its available strength. The use of pre-stressing is likely to be in the forefront of such development.

The opportunities already opened up for horizontal masonry members is clearly indicated by the prestressed prefabricated masonry solutions in this paper. The research and present designs on the drawing board, indicate more ambitious and elegant possibilities but the brick industry needs to work more closely if full advantage of development is to be achieved.

References

1.   Curtin and Shaw 'The Development and Design of Brick Diaphragm Walls' Ibmac, Washington, (1979)

2.   Curtin, Shaw, Beck and Bray 'Fin Wall Construction in tall single storey buildings' Ibmac, Rome, (1982)

3.   Curtin, Shaw, Beck and Parkinson 'Structural Masonry Detailing' Granada Publications (1984)

4.   Curtin, Shaw and Beck 'Design of Reinforced and Prestressed Masonry' Thomas Telford (1988)

5.   Curtin, Shaw, Beck and Bray 'Structural Masonry Designers Manual' 2nd edition BSP Professional Books (1988)

6.   Shaw, Othick and Priestly 'The Osborne Memorial Halls at Boscombe' The Brick Development Association Engineers File Note No 6 (1986)

7.   Shaw, Curtin, Priestley and Othick 'Prestressed Channel Section Masonry Walls' The Structural Engineer Vol 66 No 7 (1988)

8.   Shaw Practical Application of Post tensioned and Reinforced Masonry ICE Symposium 1986

9.   Shaw Inovative Development in Masonry ISE/BRE Building The Future Conference 1992

VSL's Experience with Post-Tensioned Masonry

Hans Rudolf GANZ
Member ASCE

ABSTRACT

The paper presents a new post-tensioning system for masonry structures which has been introduced in Switzerland in 1988. Since then different structures have been post-tensioned with the system. After a brief description of the system a summary of references is given. Design basis and typical assumptions used for the design of these structures are discussed. Select projects and details are presented.

The paper also describes innovative methods for the strengthening of lateral load resisting frames and walls of structures using the post-tensioning technique. The examples include small houses with small monostrand tendons up to large structures using multistrand tendons.

INTRODUCTION

In Switzerland masonry construction is extremely popular, however, traditionally only as an unreinforced material. With the increasing awareness of the risk involved with earthquakes the Swiss Loading Code (1989) has recently increased minimum lateral load requirements and limited the maximum height of unreinforced masonry buildings to 12 m (40ft) in certain areas. In addition, the new Swiss Masonry Standard (1992) does not allow to consider the tensile strength in the design of masonry structures. These provisions can limit the use of plain masonry significantly. Different options to avoid such limitaions include the provision of reinforced concrete tie beams and columns, structural steel stiffeners, etc. However, these options quite often are unattractive, provide problems with heat insulation, and are not economical.

Typical Swiss Masonry units are 120 to 250 mm wide, 190 mm high, and 300 mm long (5 to 10 in / 8 in/ 12 in). Net gross sectional area is typical in the order of 40 to 50 % with lots of small holes. Grouted masonry is not known as a construction practice. Hence, placement and corrosion protection of reinforcing steel in these blocks is difficult.

With these circumstances in mind, a new post-tensioning system for masonry structures has been introduced in Switzerland. It uses high strength prestressing strand. Hence, only a small number of tendons has to be placed in a typical structure. The tendons are continuous from anchorage to anchorage and thus, avoid multiple laps of bars which are a potential weakness in the structure. The tendons can be placed in traditional masonry units with relatively small holes of approximately 50 mm (2 in) diameter. The use of individually greased and sheathed monostrand provides a reliable corrosion protection to the tendon. Finally, the applied prestress enhances the cracking resistance of masonry and therefore, avoids cracking problems typical in lightly loaded walls, as well as brittle behaviour.

The VSL Post-Tensioning System for Masonry

The VSL Post-Tensioning System for Masonry is an unbonded system. It utilizes monostrands, i.e. high strength steel strands that are greased and coated with extruded plastic for maximum corrosion protection. A solid and durable duct around the monostrand tendon provides a third layer of protection.

A typical tendon for post-tensioned masonry is illustrated in Fig. 1. At the lower end of the tendon a self-activating dead-end anchorage is placed in a cast-in-place concrete slab or beam. The stressing anchorage is located at the upper end of the tendon. It may be placed either in a prefabricated concrete block laid on top of the masonry wall or in a cast-in-place concrete member. Both anchorages are filled with a special grease for corrosion protection of the prestressing steel which is bare inside the anchorage. Low relaxation 7-wire strand of 0.6 in diameter with a guaranteed ultimate strength of 265 kN (58.4 kips) are used. The tendons are placed inside a galvanized steel or plastic duct of 28 mm (1 1/8 in) diameter. The VSL System includes accessoires such as pre-assembled chairs at the dead-end anchorage and caps on top of each duct segment for temporary protection.

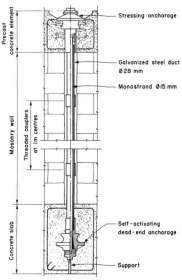

Fig. 1. : VSL Tendon for Post-Tensioned Masonry

Although the application of post-tensioning is not restricted to a certain compressive strength of masonry, its benefits can be best exploited if a minimum strength of 8 MPa (1,200 psi) based on gross cross sectional area is specified and a cement mortar is used. The common application of the system with tendons running vertically in the centre of a single leaf wall is greatly facilitated if masonry units with specially formed cores are used. Fig. 2 illustrates two examples of clay and calcium silicate units which can be used laying along or across a wall to form wall thickness of 180 or 250 mm (7 or 10 in), respectively.

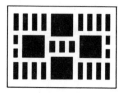
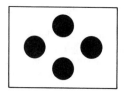

Clay unit   Calcium Silicate unit
180 x 250 x 190 mm   180 x 250 x 190 mm

Fig. 2 . : Two Examples of Clay and Calcium Silicate Units

First step in the construction of a post-tensioned masonry wall is the placement of the dead-end anchorage in the formwork of the cast-in-place concrete member. After the concrete has been poured, wall construction can commence. Duct segments are threaded to the anchorage or previously placed segments according to the progress of wall construction. At this stage, ducts only serve as a void former for the tendon to be installed after construction. This procedure allows the units to be laid easily because they need only be threaded a small number of lifts, if any, over the duct segment. When the final wall height is reached, the final duct segment is cut to the required length and a prefabricated concrete element, containing the stressing anchorage and a sleeve to tightly fit over the duct, is placed on top of the wall. After the masonry reaches the specified strength, typically in the order of 7 days, prestressing may commence. First, the monostrand tendons are fed through the stressing anchorage and duct into the self-activating dead-end anchorage. The tendons are subsequently stressed with a light hydraulic jack to 75% of their tensile strength, i.e. 200 kN (43.8 kips), and locked off.

DESIGN

Swiss Code

There is no particular code for reinforced or prestressed masonry in Switzerland. Nevertheless, this construction method has been used successfully and is now becoming common practice.

The design of post-tensioned masonry is based on the principles set out in the Swiss Masonry Standard SIA 177 (1992). This standard applies to the design of masonry, in general, and uses serviceability and ultimate limit states. For ultimate limit state design factored load effects are compared with the design resistance of a member or structure, calculated with factored material strength. Material factors are 2.0 for the masonry compressive strength, and 1.2 for the reinforcing and prestressing steel as well as for friction in the bed joints. These factors convert into ϕ - factors of 0.5 and 0.83 according to North American practice. Since, in general, ductile failures prevail the above material factors come close to a member/cross section ϕ - factor of 0.8 according to ACI notation. Applied to unbonded tendons the designer checks the flexural resistance of a post-tensioned masonry section with the tendon force « effective prestress/1.2 » and a rectangular masonry stress block at a stress « masonry strength/2.0 ».

Second Order Effects

There is at least one significant difference between prestress loads and an applied load. Applied loads can lead to buckling of walls. Post-tensioning tendons laterally restrained/guided inside the masonry wall, however, cannot cause buckling since the line of action of prestress is not changing relative to the centroid of the section when walls deflect. Hence, restrained/guided prestress does not cause second order effects. This is not the case for unrestrained tendons which are free to move inside the wall cross section. For such a case, the effect of prestress comes close to those of applied loads. Hence, for restrained tendons buckling must be checked for applied loads only. For unrestrained tendons, however, buckling shall be checked for the superimposed effects of the applied load and the prestress force.

Anchorage of tendons

Tendons must be anchored by mechanical anchorages in concrete elements or in masonry, see Fig. 4. Anchorage by bond must be used in concrete elements only since there is not enough experience with transfer and development of prestressing steel in grout. A limitation of bearing stresses in masonry beneath anchorages or anchorage blocks to 50 percent of masonry strength under maximum permissible jacking force provides generally accepted safety levels.

Loss of prestress

Loss of prestress in masonry still seems to be a concern to many researchers and designers. Losses exceeding 40 percent of initial prestress have been reported and assumed in design of concrete masonry post-tensioned with prestressing bars. With proper material selection such excessive losses can be avoided, see Table 1. The key to low losses such as those common in prestressed concrete is in the use of high strength prestressing steel rather than medium to low strength bar. Since high strength prestressing steel such as 7-wire 1860 MPa (270 ksi) strand is stressed to high strains, typically 7×10^{-3} mm/mm, the shortening of masonry due to creep and shrinkage in the order of 0.7×10^{-3} mm/mm creates only a small relative loss. The relative loss for 690 MPa (100 ksi) bar stressed initially to the same percentage of GUTS (Guaranteed Ultimate Tensile Strength), however, is 270 percent of the value of the strand above. In addition, relaxation losses are well controlled and don't exceed 2.5 percent in 1000h stressed to 70 percent GUTS for 7-wire strand which is quite different for medium strength bar. Above considerations are the reason for limiting prestressing steel specification in masonry to those commonly used in prestressed concrete. More detailed design considerations have been published elsewhere, VSL (1990).

Table 1 : Loss of Prestress

	Clay Brick Masonry		Concrete Block Masonry	
	Assumed values	Associated Losses	Assumed Values	Associated Losses
Shrinkage	0 (mm/m	0	-　0.4　x　10-3 (mm/mm)	7 %
Creep	1.0	4 %	2.0	8 %
Relaxation	3 %	3 %	3 %	3 %
Total		7 %		18 %

Notes :　　Initial stress in masonry based on gross area : 2MPa (280 psi)
　　　　　Initial stress in strand : 1250 MPa (175 ksi)
　　　　　Modulus of masonry : 8 GPa (1,120 ksi)
　　　　　Modulus strand : 195 GPa (27,300 ksi)

DURABILITY

High strength prestressing steel is more sensitive to corrosion than reinforcing steel. Its protection in concrete structures relies mainly on a sufficiently thick and dense concrete cover. For bonded tendons cement grouts are injected into ducts to provide a highly alkaline environment around the prestressing steel for a secondary barrier. For unbonded tendons grease and plastic sheathing take this role.

Masonry units and grout do not provide the same quality of protection as a thick concrete cover. Therefore, the secondary barrier takes a much more important role in masonry construction. It is one of the designer's prime obligations to assure permanent corrosion protection of post-tensioning tendons in masonry construction.

APPLICATIONS

• General

The system as described above has been introduced in 1988. After a slow start with two projects in 1988, none in 1989 and one in 1990, applications now become more regular and larger. By the end of 1995 some 30 projects have been carried out in Switzerland. In addition, the system has been used in research projects on post-tensioned masonry, and with adaptations to 0.5 in strands, for one new construction and several retrofit projects in the US, VSL (1990).

• Individual examples for new construction

Movie Theatre, Wattwill :

All the perimeter walls and one inside wall of the movie theatre were designed in post-tensioned masonry. The walls were laterally supported on roof level and built into a reinforced concrete beam and slab at the bottom. Design lateral wind load was 1.1 kPa (22 psf). The subsequently applied factor on these wind loads for ultimate design was 1.5.

The plan of the building and a cross section of a post-tensioned wall are illustrated in Fig. 3. The masonry walls are up to 26.5 m (87ft) long, 5.15 m (17 ft) high, and are 180 mm (7 in) thick. Calcium silicate units were used in running bond. The dead-end anchorages were cast into an in-place reinforced concrete beam and slab. The walls were vertically post-tensioned with 33 tendons at a spacing between 1.7 and 2.2 m (5 ½ to 7 ft). The stressing anchorages were placed into a 350 mm (14 in) deep cast-in-place reinforced concrete tie beam on top of the wall. All the bursting forces were covered with the reinforced concrete members and therefore, no bed joint reinforcement was provided in the masonry.

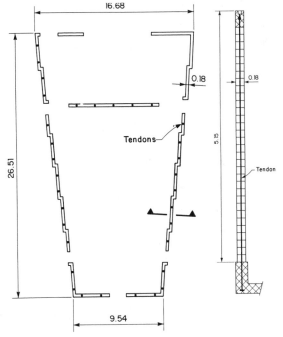

Fig. 3 : Plan of a Movie Theatre and a Cross Section of a Post-Tensioned Wall

Industrial Centre, Altendorf :

Adjacent to an existing building of the industrial centre a new warehouse was to be added. Two walls of the warehouse facing the existing building were designed in post-tensioned masonry. The lower part of these walls were in contact with the adjacent building while the top was completely exposed. The lateral wind load for design was assumed as 0.3 kPa (6 psf) and 1.0 kPa (20 psf) for the lower and top part of the wall, respectively. The subsequently applied load factor for wind was 1.5.

The plan of the warehouse and a section through the post-tensioned wall is shown in Fig. 4. The walls are 43.35 m (140 ft) and 9.5 m (31 ft) long. The lower 5.6 m (18 ft) of the 15 m (50 ft) tall walls were cast in concrete. At elevation + 8.0 m a cast-in-place reinforced concrete tie beam was introduced and laterally supported at the concrete columns of the adjacent building. The masonry walls were constructed in 250 mm (10 in) thick calcium silicate units. The dead-end anchorages were cast 0.5 m (1.5 ft) into the concrete wall and properly lapped with the reinforcement in this area. A total of 71 tendons were installed at spacings of 0.57 m (1' - 10'') and 0.95 m (3' - 2'') depending on actual wind exposure conditions. The stressing anchorages were placed in precast elements on top of the wall with layers of bed joint reinforcement in the three joints below the elements.

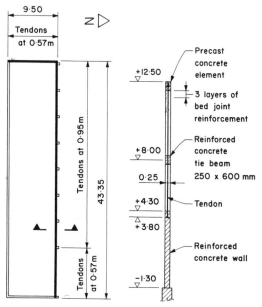

Fig. 4: Plan of Warehouse and a Section through the Post-tensioned Wall

• Individual examples of strengthening applications

Los Gatos Brick Castle

One of the structures retrofitted after the 1989 Loma Prieta Earthquake. The building is locally known as the Los Gatos Brick Castle and approximately 100 years old. It consists of unreinforced clay brick masonry exterior walls, 200 mm (8 in) thick, timber floors, partitions and roof, and a stone rubble foundation. The plan of the building is fairly irregular with lots of window and door openings, as well as a particularly small total wall length on the South side, Fig.5

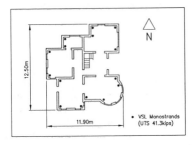

Fig. 5 : Floor Plan of Brick Castle with Tendon Location.

During the Loma Prieta Earthquake the building had lost parts of the masonry parapets, suffered considerable structural and non-structural cracking and as a consequence was red-tagged, i.e. habitants were no longer allowed to use the building. Structural cracking consisted mainly of diagonal shear cracks on the South side of the building, and vertical separation cracks along window and door openings.

Structural repair work included grouting of cracks in the masonry walls, reconstruction of damaged or lost parapets and the addition of continuous structural steel chords and anchors along all the floor-wall connections to properly tie the walls to the floor diaphragms. At roof level a reinforced concrete tie beam was added.

Preliminary in-situ testing indicated that the masonry was strong enough to allow the use of post-tensioning to increase the shear strength of the building. Consequently, a total of 15 vertical monostrand tendons were introduced in the load-bearing walls, anchored in the stone rubble foundation and the reinforced concrete tie beam at roof level. The layout of the tendons is illustrated in Fig. 5. Tendons were placed in small slots cut into the wall corners, thus tendon eccentricities were minimized and taken by cross walls. For longterm corrosion protection, individually greased and sheated monostrands were used. For fire protection and added mechanical protection, the tendons were placed inside steel tubes and the slots were grouted.

Repair and strengthening work started in Fall 1990 and was completed first half of 1991. Installation and stressing of the monostrand tendons did not reveal any problems.

Holy Cross Church, Santa Cruz

The 104 years old structure was severely damaged during the Loma Prieta Earthquake and as consequence, the bell tower had to be removed and the church was closed. The church and tower consist of unreinforced brick masonry walls built on a stone rubble foundation. Timber trusses span from buttresses across the church and provide support for the roof. Tower and walls of the church were severely cracked both in shear and flexure. However, masonry and foundation were generally found to be in good condition.

Retrofit of the church included grouting of cracks in the masonry, reconstruction of parts of the bell tower in steel, timber, and cladding to reduce weight and seismic effects, addition of reinforced concrete beams on top of the butresses and a new roof with in-plane steel trusses.

In-situ testing indicated masonry strengths between 4 MPa and roughtly 8 MPa (600 to 1200 psi) based on gross area. Hence, introduction of post-tensioning forces into the structure was feasible. A total of 26 tendons were introduced into the structure to enhance its shear and flexural resistance. Fig. 6 illustrates the layout of the tendons. Tendons in the end walls and towers consist of seven strands diameter 0.5 in. Tendons in the buttresses include twelve strands, in general. The tendons are detailed similar to ground anchors, i.e. use greased and sheated strand in the free unbonded length in the masonry and bare strand bonded to the ground and rubble foundation by cement grout. The tendons are stressed to 910 kN (200 kips) and 1560 kN (350 kips), respectively, from the top end and anchored with bearing plates cast into reinforced concrete beams. The 100 mm (4 in) and 175 mm (7 in) holes for the tendons were core drilled through the walls from the top. Since the wall thickness varies over the height and the tendons follow the centre line of the wall section, those holes had to be drilled at an angle.

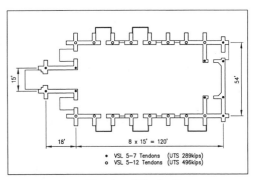

Fig. 6 : Plan of Holy Cross Church and Tendon Layout

Buttresses are laterally supported by the roof. To provide adequate stiffness and strength a new steel truss was added in the plane of the roof. Connections between post-tensioned buttresses and steel truss were designed to yield at a chosen load level. While securing stability for overturning of the buttresses for transverse loads this technique helps to limit the maximum forces introduced into the buttresses and hence, failure of the masonry buttresses can be positively avoided. In addition, this technique allows to safely estimate the maximum loads transferred from the buttresses through the roof to the end walls. Thus, a safe design of the end walls for shear can be achieved.

Development of the first retrofit concepts started in 1990. Refined investigations including non-linear analyses as well as review of the concepts went well into 1991. The retrofit work was done in 1992 and on October 17, 1992, after successful completion of all tendon stressing, the new steeple was installed on the bell tower.

CONCLUSIONS

Post-tensioned masonry has successfully been applied in Switzerland over the last 7 years. Slowly, this new construction method is gaining wider acceptance. There is a large potential for post-tensioned masonry for industrial buildings, warehouses, shopping centres, and alike with typical wall heights in the order of 6 m (20 ft). Such potential definitely exists also in North America for a variety of structures where traditionally tilt-up concrete is used.

The strengthening of masonry structures with post-tensioning makes optimum use of existing materials. Thus, major structural alterations such as the introduction of structural steel frames or the addition of new foundations can be avoided, in general. Basically, many different types of structures can and have been strengthened with the proposed technique. However, the method is best suited for larger structures made of reasonably strong masonry.

REFERENCES

[1] Swiss Standard SIA 160, « Einwirkungen auf Tragwerke (Loadings for Structures) », Schweizerischer Ingenieur - und Architekten - Verein, Zürich, 1989,   103 pp.

[2] Swiss Standard SIA 177/2, « Bemessung von Mauerkswäden (Design of Masonry Walls) », Schweizerischer Ingenieur - und Architekten - Verein, Zürich, 1992, 32 pp.

[3] VSL International Ltd « Post-Tensioned Masonry Structures », Berne, 1990, 35 pp.

Testing of Prestressed Clay-Brick Walls

Gary L. Krause[1], Ravi Devalapura[2], and Maher K. Tadros[3]

Abstract

   A new concept for the construction of clay-brick walls has been developed. The use of prestressing in a veneer type wall has been shown to provide improved resistance to cracking under service level loads, while eliminating the need for intermittent wall tie connections to the support structure. The prestressing concept involves the use of high-strength threaded bars, two-core bricks which index in a running bond pattern, and direct tension indicating washers for measuring the prestressing force. The prestressing of each bar is accomplished by use of a simple torque wrench. Several panels were constructed at the structures laboratory at the University of Nebraska-Lincoln and tested for lateral load capacity, cracking capacity, and prestress losses. The results showed that the panels can be designed using standard prestressed concrete assumptions, to resist significant lateral wind pressures without cracking. This would significantly improve the performance of veneer systems. The tests also showed that prestress losses are not very significant for this type of system.

Introduction

   Clay-Brick masonry walls are used quite frequently as veneer walls in the construction of moderate size buildings. The system of supporting these masonry walls by attaching the to a gridwork of light gage steel stud members has been used for decades. The stiffness variation between the bricks and steel studs in this system can lead to maintenance problems such as water entry and cracking of the masonry.

---

[1] Assistant Professor, Department of Civil Engineering, University of Nebraska-Lincoln, Member, ASCE

[2] Engineering and Research Manager, Composite Technology Corp., Ames, Iowa

[3] Cheryl Prewett Professor, Department of Civil Engineering, University of Nebraska-Lincoln, Member, ASCE

The introduction of water can lead to safety concerns due to the potential corrosion of the wall support ties.

This research project was undertaken to determine if prestressing techniques could be applied to the masonry wall in order to allow for the separation of the masonry wall from the steel stud gridwork. The objective of the research was to develop a simple prestressing system that could be applied at the site, which would eliminate the need for attachment of the masonry wall to the steel studs, by providing the wall with enough flexural strength to span from floor to floor. The introduction of a prestress would also prevent the wall from cracking due to service load stresses caused by wind loading. Complete details of the testing can be found in Devalapura (1995).

System Development

The system developed for this research project consisted of new two-core brick units, high strength threaded rods, a sponge system for clearing brick cores, and direct tension indicating (DTI)washers. The two-core brick units were designed to provide cores that were large enough to permit the use of prestressing bars up to approximately 1.5 inch (38 mm) in diameter. Figure 1 shows a sketch of the two core brick unit with dimensions. The cores were also designed to line up vertically (index) when laid in a running bond pattern.

High strength threaded rods 120 ksi(828 MPa) were chosen for the purpose of prestressing the clay brick walls. Rods with a higher strength would also be acceptable, as long as they meet the minimum requirements of the MSJC (1995) provisions. The rods used in this research were continuously threaded, but the threads are only needed at the ends.

In the construction of the walls, mortar will inevitably fall into the open cores of the brick units. In order to prevent this from occurring in the cores used for prestressing rod placement, a sponge attached to a wire rod was placed in the necessary brick core of the first course, and then continuously pulled up through the cores of subsequent courses(see Figure 2). This provided for an obstruction free core into which the prestressing bars could be place and later grouted.

The prestressing force was applied by using a hand held torque wrench to turn the nut placed at the top of the wall panel. A DTI washer was placed under the nut in order to measure the prestressing force applied to the wall. Tests were performed in the laboratory, and strain gages were placed on the bars to verify that the DTI washers were providing an accurate measure of the prestressing force. Figure 3 shows a sketch of a typical wall panel specimen.

Laboratory Testing

The lab testing was performed in three stages. Stage I testing consisted of out of plane bending tests of six wall panel specimens. Figure 4 shows the laboratory test set-up. Stage II testing consisted of out of plane load testing of eight wall panel specimens. The Stage III testing consisted of the monitoring of prestress losses in nine wall panel specimens.

The testing in stage I was performed in order to determine the feasibility of the proposed system, and to help develop parameters for the stage II testing. Six wall panel were constructed and tested in the laboratory. Three panels were prestressed without any grouting of the cores in order to help determine the behavior of an unbonded prestressed system. These specimens were labeled TPIUG1, TPIUG2, and TPIUG3 respectively. The other three specimens were grouted after the prestressing was applied. These specimens were labeled TPIG1, TPIG2, and TPIG3 respectively. The prestresing was applied to all panels at seven days after construction. The prestressing force was the same for all six panels, and was 19 kips (85.5 kN) for a 5/8 inch (15.9 mm) diameter, 120 ksi (828 MPa) bar. The 28-day prism strength of these panels was 2845 psi (196. MPa). The load versus deflection curves for the stage I specimens are shown in Figure 5. The results indicated that a higher load capacity could be obtained by grouting the prestressing rod.

The testing in stage II was conducted in order to determine the effect of various parameters on the behavior of the prestressed panels. All eight panels were grouted in the core where the prestressing bar was placed. The same size and strength bar was used as in stage I, and the 28-day prism strength of the panels was 3430 psi (23.7 MPa). The parameters that were investigated were, age at prestressing, use of a thick plate or a standard shelf angle at the anchorage locations, and the amount of prestressing force. Table 1 shows how these parameters were varied in the stage II test specimens. The load versus deflection curves for the stage II specimens are shown in Figure 6.

The stage III testing was conducted in order to determine what magnitude of prestress losses could be expected in this new system. The eight panels tested are described in Table 2. The 28-day prism strength for these panels was 3600 psi (24.8 MPa) The strain in the prestressing bars was monitored using strain gages, and the strain in the masonry wall was monitored using a target and dial gage system. Three specimens (PLP-2, PLA-2, and PLA-4) were placed outside after an initial 30 days in the testing lab. The remaining specimens were kept inside the lab for the full 180 days of monitoring. The results of these test indicated an average loss of prestressing force of about 9% for the outside panels, and a net gain of 15% for the inside panels. However, it was clear from the large variations in the test data that the actual prestress loss/gain in a constructed wall could differ from these average values.

Discussion and Conclusions

Figure 7 shows a comparison of all of the stage I and II results with a theoretical analysis of the sections using standard prestressed concrete procedures. Only the specimens with the 19 kip (85.5 kN) prestressing force are shown in the figure. The figure shows clearly that the test specimens behavior was very consistent with the behavior predicted by the Moment Curvature Method. Two other analytical methods, the Bi-Linear Method (UBC 1994) and the Horton Tadros Method (1990), gave conservative predictions of the panel behavior. Table 3 shows the predicted and average measured moment strength at cracking and ultimate. These values also show that the behavior of the test specimens was easily predicted using standard procedures.

The results of the stage II tests indicated that the prestressing force could be applied to the wall as early a one day after construction without altering the behavior or strength of the system. The use of a standard shelf angle at the anchorage zone also did not appear to significantly affect the behavior of the panels. The use of a smaller prestressing force led to behavior that was consistent with the predicted behavior, and was as simple to implement using the DTI washers.

The objectives of this project were met, and several conclusions can be made:
1. It was feasible to introduce prestressing to a standard size clay brick masonry wall.
2. The use of DTI washers provided a simple yet accurate way of determining the prestressing force in the bar.
3. The prestressed wall system behaved in a manner that was predictable using standard structural theory and methods.
4. The losses/gains in this system appeared to be well within the bounds of standard prestressed concrete systems.
With the development of appropriate connection details, it is thought that this system can be used in veneer type wall applications as an acceptable alternative/replacement to the steel stud system.

Acknowledgments

The research project in progress at the University of Nebraska-Lincoln, is a collaborative research effort sponsored by the following organizations: United States Army Corps of Engineers (CPAR Program), Center for Infrastructure Research at the University of Nebraska-Lincoln, Post-Tensioning Institute, Brick Institute of America, Masonry Contractors Association of Eastern Nebraska, and the Nebraska Masonry Institute. The authors would like to thank Steve Sweeney, Ervell Staab, Don Littler, Karl Anderson, and Gary Davis for their contributions to the project. The authors extend their thanks to Dr. Amin Einea, CIR, University of Nebraska who helped with the data-acquisition system during the test, and Say-Gunn Low, graduate research assistant for his help during the test. Part of the steel used in the project was donated by Williams Form Engineering, Corporation, MI. This paper represents the opinions of the authors and does not necessarily reflect the opinion of the sponsors.

References

Devalapura R. (1995), "Development of Prestressed Clay Brick Masonry Walls," Doctoral Dissertation presented to the University of Nebraska-Lincoln Graduate College, Lincoln, Nebraska, August, 1995.

Horton, R.T. and Tadros, M.K. (1990), "Deflection of Reinforced Masonry Members," *ACI Structural Journal*, Vol. 87, No. 4, July-August 1990, pp. 453-463.

MSJC Provisions (1995), Prestressed Masonry Design Criteria, Standards Joint Committee, Draft No. 6, April 1995.

UBC (1994), Uniform Building Code, International Conference of Building Officials, Whittier California, 1994.

Table 1 Details of Stage II Panels

Panel Designation	Dimension (in.)	Prestressing Force (lb)	End Bearing Condition
PLP-1	36 x 48	19,000	12 x 3 5/8 x 1 in. Plate
PLP-2	36 x 48	19,000	12 x 3 5/8 x 1 in. Plate
PLA-1	36 x 48	19,000	4 x 4 x 3/8 in. - 36 in. long Angle
PLA-2	36 x 48	19,000	4 x 4 x 3/8 in. - 36 in. long Angle
PLA-3	36 x 48	19,000	4 x 4 x 3/8 in. - 36 in. long Angle
PLA-4	36 x 48	16,000	4 x 4 x 3/8 in. - 36 in. long Angle
PLA-5	36 x 48	16,000	4 x 4 x 3/8 in. - 36 in. long Angle
PLA-6	36 x 48	16,000	4 x 4 x 3/8 in. - 36 in. long Angle

Table 2 Details of Stage III Panels

Panel Designation	Dimension (in.)	Prestressing Force (lb)	End Bearing Condition
PLP-1	36 x 48	19,000	12 x 3 5/8 x 1 in. Plate
PLP-2	36 x 48	19,000	12 x 3 5/8 x 1 in. Plate
PLA-1	36 x 48	19,000	4 x 4 x 3/8 in. - 36 in. long Angle
PLA-2	36 x 48	19,000	4 x 4 x 3/8 in. - 36 in. long Angle
PLA-3	36 x 48	19,000	4 x 4 x 3/8 in. - 36 in. long Angle
PLA-4	36 x 48	16,000	4 x 4 x 3/8 in. - 36 in. long Angle
PLA-5	36 x 48	16,000	4 x 4 x 3/8 in. - 36 in. long Angle
PLA-6	36 x 48	16,000	4 x 4 x 3/8 in. - 36 in. long Angle

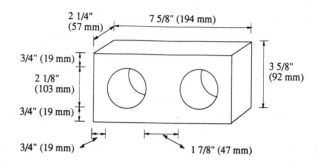

Fig. 1 Two-Cored Brick

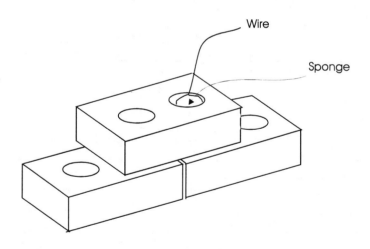

Fig.2 Blocking the Core With Wet Sponge

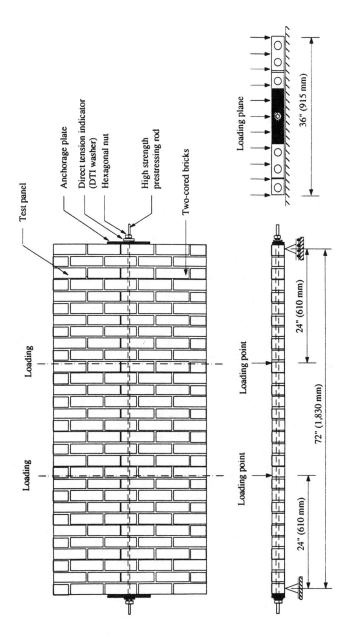

Fig. 3 Typical Detail of a Prestressed Test Specimen For Out-of-Plane Loading

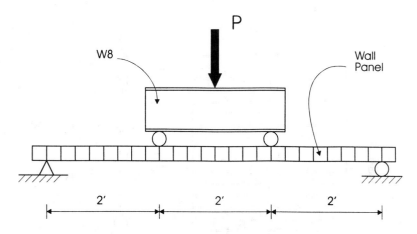

Fig. 4 Panel Test Set-Up

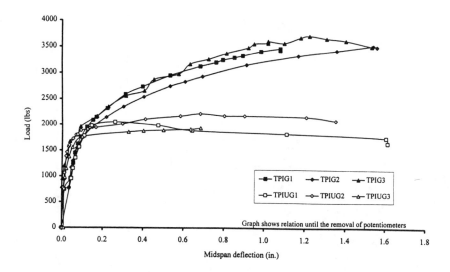

Fig. 5 Load vs. Midspan Deflection for Stage I Specimens

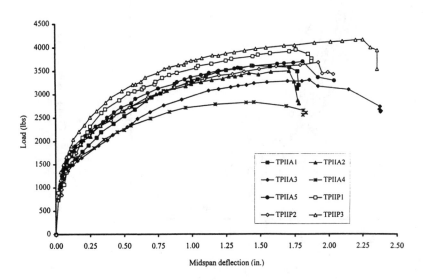

Fig. 6 Load vs. Midspan Deflection for Stage II Specimens

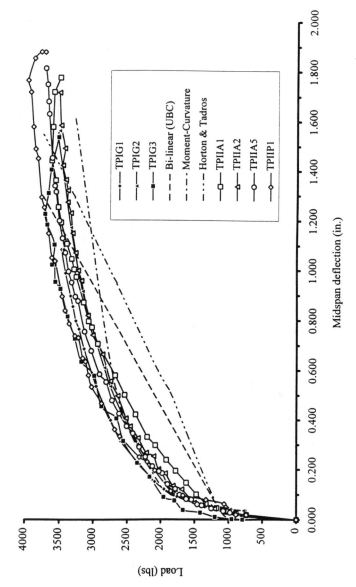

Fig. 7 Comparison Between Theoretical and Measured Load vs. Deflection Curves

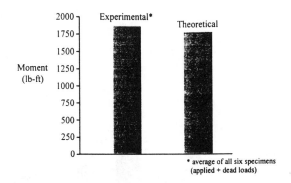

Fig. 8 Stage I Cracking Moments

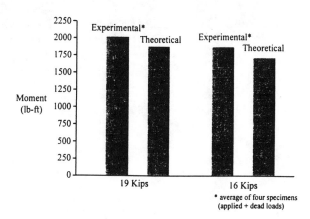

Fig. 9 Stage II Cracking Moments

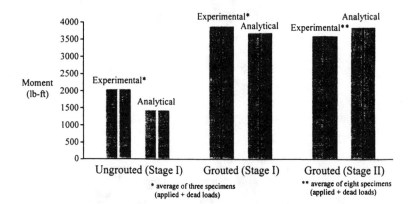

Fig. 10 Ultimate Moments Stage I and Stage II

Prestress Loss Due to Creep in Post-Tensioned Clay Masonry

Subhash C. Anand*, F. ASCE and Naresh Bhatia**

Abstract

Numerical schemes using the finite element technique as well as method of superposition are utilized in this paper to compute total load strains (which include creep strains) in stack-bond and running-bond clay brick laboratory wall specimens. It is shown that the resulting values of strains compare favorably with those calculated using the two-phase composite model and measured experimentally. The magnitude of creep strains in walls after 300 days is 50% more than the instantaneous elastic strains. Accordingly, post-tensioned masonry walls necessitate application of additional prestress, especially during the first two months of their initial prestressing. Creep strains in a masonry pier specimen are calculated by utilizing the commercially available finite element package - ABAQUS. The results indicate that the creep strains in this case are approximately one-half of the instantaneous elastic strains. Nevertheless, they still represent a substantial loss of prestress in post-tensioned masonry piers and must be accounted for in their design.

Introduction

It is well known that brick masonry, when subjected to sustained loads, not only experiences instantaneous deformation but continues to deform with time. This phenomenon of additional long term deformation due to load, the major part of which occurs within a year, is termed creep. This creep deformation in masonry walls and piers could be two to four times the magnitude of instantaneous deformation (Lenczner 1970,1986).

----

* Prof., Dept. Of Civil Engr., Clemson Univ., Clemson, SC 29634
** Engr., Seelye Stevenson Value and Knecht, Inc. 225 Park Ave., NY, NY 10019
   (formerly Grad. Stud., Dept. Of Civil Engr., Clemson Univ., Clemson, SC 29634)

Although masonry is strong in compression, it is quite weak in tension. Consequently, masonry cannot support loads which cause tensile stresses, such as lateral loads on masonry walls due to wind, and vertical loads in beams. However, if initial compressive stresses are introduced in these elements by prestressing, which counteract tensile stresses caused due to loads, such members can be employed effectively in masonry construction. Consequently, prestressing can be used economically in beams, non-load bearing and load bearing walls, eccentrically loaded masonry columns, masonry sound barrier walls, etc.

A recent study was conducted at the North Carolina State University (Schultz and Scolforo 1991, 1992) to investigate the state-of-the-art of prestressed masonry in the U.S. and the world, and to suggest possible ways for its adoption in the U.S. industry. One of the concerns in the development of prestressed clay brick masonry found in this study, is the loss of prestress in masonry from various sources. As stated by the authors of the above cited study..."One of the large sources of prestress losses in masonry is creep deformation under sustained loads". It is evident from the above statement that accurate estimation of creep in clay brick masonry is of utmost importance if the development and practice of prestressed masonry is to be promoted.

Much of the development in prestressed masonry has occurred in the United Kingdom during the past thirty years (Shaw 1986, 1995). This development resulted from the need to build lighter and slender load bearing masonry structures subjected to lateral loads. New systems for post-tensioning have also been developed in the continental Europe, especially in Switzerland and have been reported by Ganz (1993, 1995). The last reference describes successful use of post-tensioned masonry in new structures in Switzerland and rehabilitation of damaged buildings and churches due to earthquakes in California. Only a very limited amount of prestressed masonry research has been undertaken in the U.S. The only prestressed masonry research currently in progress is at the University of Nebraska-Lincoln by Krause et al. (1995), which is focused towards masonry veneer walls.

The pioneering work on creep in masonry was conducted in the United Kingdom by David Lenczner (1970). Besides conducting creep tests on brick panels and piers, he analyzed the existing creep data of other investigators and proposed preliminary empirical equations using logarithmic functions of elastic strain and brickwork cube strength as parameters to predict creep. He later modified these empirical equations by introducing the square root of brick strength as the major parameter in defining specific creep in masonry. Improved creep-time functions using rheological and Maxwell models were proposed by him and his colleagues to estimate the loss of prestress in post-tensioned masonry (1986). All of the creep equations presented by Lenczner are empirical and are based on the square root of brick strength with no specific consideration of mortar properties.

A very different approach to predict creep in masonry has been taken by Shrive and his colleagues (1981) in Canada and Brooks (1986) in England. They defined the masonry creep properties in terms of those of the units and mortar by connecting the bricks and mortar layers in series and parallels. No Poisson's effects were considered and the materials were assumed to be isotropic and linear elastic. The authors proposed a two-phase composite model for stack-bond brick masonry and later extended this model to include the vertical and horizontal mortar joints in order to model running bond masonry walls, as well as piers.

Analytical and computational procedures to estimate creep in composite masonry walls based on the method of superposition using specific creep curves for brick and block masonry using the finite element technique have been developed by Anand and Rahman (1991) and Anand and Bhatia (1994, 1995). The use of the superposition principle for creep has been derived from the literature on concrete and other time dependent materials. The developed technique is very versatile as it can take into account any geometric and loading configuration without difficulty and also account for stress redistribution in a step-by-step time incremental solution. The superposition technique was utilized to compute long term load strains in stack-bond and running-bond laboratory wall specimens. The results from these analyses were compared with the laboratory tests and the two-phase model, and were found to agree to these very favorably. No analyses to predict creep strains in piers using either the two-phase model or the method of superposition have been reported in the literature.

The purpose of this paper is to summarize the results of the two previous papers by the authors (Anand and Bhatia 1994, 1995) which define the total load strains (including creep strains) in stack-bond and running-bond laboratory wall specimens, and to estimate creep strains in masonry piers using a technique built into the commercial finite element package - ABAQUS (Ref. 1). Although the loss of prestress in a masonry wall or pier involves stress relaxation, which changes the amount of stress acting in masonry with time, the analyses presented in this paper do not include the effect of stress changes. The primary objective of the paper is to demonstrate that the calculated creep strains in piers are substantially smaller than those in the walls as has been observed experimentally. It is also shown that the creep strains are large enough to cause substantial loss of prestress in post-tensioned masonry walls and piers.

Stack-Bond Laboratory Wall Specimen

The laboratory specimen of Brooks (1986) consisted of 13 layers of calcium silicate bricks of size 219x102x65mm in stack bond with 12mm thick mortar joints, as shown in Fig. 1. The elastic moduli for brick and mortar were given as 17,100 N/mm^2, and 12,700 N/mm^2, respectively. The corresponding values for the Poisson's ratio are assumed as 0.25 and 0.20. The specimen was subjected to a uniform load intensity of

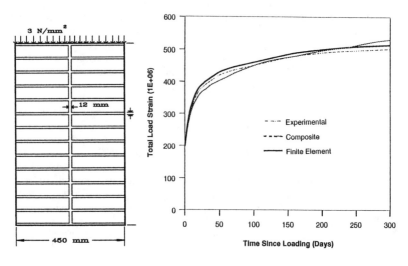

Fig. 1 Stack Bond Specimen

Fig. 2 Load Strain for Stack Bond Specimen

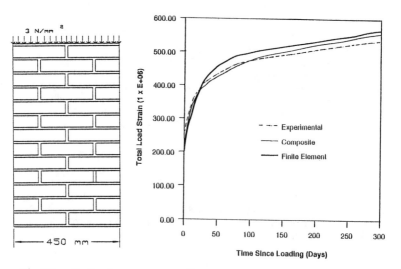

Fig. 3 Running Bond
Specimen

Fig. 4 Load Strain for Running Bond Specimen

$3N/mm^2$ for which long term vertical strains were recorded. The details of the experimental set-up are available in Brooks (1986) and the measured axial strains are plotted in Fig. 2.

The two-phase composite model, as developed by Brooks (1986), leads to an expression for the effective modulus, E_{wy}, for the stack-bond laboratory wall specimen at any time after loading as

$$\frac{1}{E_{wy}} = \frac{1}{1.152E_{by} + 0.032E_m} + \frac{0.168}{E_m} \tag{1}$$

Eq. (1) is an expression for strain in the laboratory wall specimen at any time per unit of stress, in which E_{by} and E_m are effective moduli of the brick and mortar prism given below in Eq. (2) where strains are measured experimentally. The effective modulus of brick or mortar (i.e., E_{by} or E_m) can be given by

$$E = \frac{\sigma}{\epsilon} = \frac{\sigma}{\epsilon_o + \epsilon_c} \tag{2}$$

where ϵ_o = instantaneous elastic strain and ϵ_c= creep strain. Eq. (2) can also be written as

$$E = \frac{1}{\dfrac{\epsilon_o}{\sigma} + \dfrac{\epsilon_c}{\sigma}} = \frac{1}{e_o + c_s} \tag{3}$$

in which e_o = specific elastic strain and c_s = specific creep strain at any time. The load strains (total strains) in the specimen at any time are obtained by using $\epsilon_{wy} = \sigma_{wy}/E_{wy}$ in which $1/E_{wy}$ is calculated from Eq. (1). These load strains are plotted in Fig. 2.

The laboratory specimen was also analyzed using the method of superposition in conjunction with the finite element technique, the details of which are available in the paper by Anand and Bhatia (1994). These results along with those obtained experimentally are also plotted in Fig. 2. It is seen that both the two-phase model and the finite element model yield results which are very close to the experimental data. The final load strain after 300 days is approximately equal to $2^1/_2$ times the magnitude of the instantaneous elastic strain.

Running-Bond Laboratory Wall Specimen

The running-bond laboratory wall specimen has the same dimensions as the stack-bond specimen and is shown in Fig. 3. As before, the specimen was subjected to a uniform load intensity of $3N/mm^2$ for which long term load strains were recorded. The details of data collection are given by Brooks (1986) and the results are plotted in Fig. 4.

Because of the different volumes of mortar in alternating layers in a running-bond wall specimen, two different expressions must be developed for the effective modulus of the specimen, one for a layer with one mortar joint and the other for a layer with two mortar joints. These expressions (Bhatia, 1994) are

$$\frac{1}{E_{wy}} = \frac{1}{1.179E_{by} + 0.033E_m} + \frac{0.182}{E_m} \tag{4}$$

for a layer having one vertical mortar joint, and

$$\frac{1}{E_{wy}} = \frac{1}{1.121E_{by} + 0.063E_m} + \frac{0.152}{E_m} \tag{5}$$

for a layer with two mortar joints. The meanings of various symbols in these equations have been defined earlier. The average of Eqs. 4 and 5 gives the strain in the specimen per unit of stress (i.e., specific strain) at any time. This average is multiplied by the applied stress of $3N/mm^2$ to yield the total vertical load strain in the specimen at any time which is plotted in Fig. 4.

This wall specimen was also analyzed using the method of superposition along with the finite element technique. The procedure is similar to that described by Anand and Bhatia (1994, 1995) in which the vertical and horizontal mortar joints are modeled individually. The resulting computed strains are also plotted in Fig. 4. It can be seen from this figure that the total load strains after 300 days are between $2^1/_2$ times to $2^3/_4$ times larger than the instantaneous elastic strains. In addition, the measured as well as calculated values predicted by either of the two models are acceptably close to one another.

Creep Predictions in a Masonry Pier

It has been observed and reported in the literature that creep strains in masonry piers are much smaller than the corresponding creep strains in masonry walls at the same level of stress, i.e. the specific creep for a masonry pier is much smaller than that for a wall of the same materials (Lenczner 1986, Brooks 1988). It is shown in this section that this observed phenomenon can also be predicted with the help of a numerical finite element model.

As in the case of laboratory wall specimens made of 13 layers of calcium silicate bricks presented in the previous two sections, a 13 layer pier specimen of calcium silicate bricks, shown in Fig. 5, is considered in the analysis. The alternating brick layers are selected as shown to assure symmetry of the pier about the two axes. The elastic moduli selected for the brick and mortar are the same as used earlier for the stack-bond laboratory specimen.

The finite element analysis of the pier required a 3-dimensional model, and none was available at Clemson University which had used the previously described superposition technique for creep analysis. It was decided, therefore, to utilize the commercially available ABAQUS (1989) finite element package for 3-D creep analysis. The creep properties of the constituent materials (i.e., bricks and mortar) are defined differently in this package as will be described later. Because of the double symmetry in the pier specimen, only one quarter of the specimen cross-section was modeled. As modeling mortar joints using individual elements would require an extremely large number of 3-D elements, a smearing technique was employed to model the mortar joints. The mortar joints in the pier are replaced by an equivalent volume of mortar elements which are lumped at discrete convenient locations. In the horizontal cross-section, this requires calculating the brick and mortar areas separately and distributing these symmetrically at discrete locations across the cross-section. In the pier under consideration, brick occupies 86.95% and mortar 13.05% of the cross-sectional area. The resulting smeared cross-section of the finite element model is shown in Fig. 6 and consists of a total of 100 elements with mortar elements interspersed between brick elements. Only half of the pier height was considered in the vertical direction due to symmetry. The smearing for mortar in the vertical direction was performed similarly and is shown in Fig. 7 which resulted in six brick layers each 65 mm high, two mortar layers each 42 mm high and one brick layer 32.5 mm high. The thickness for each mortar layer of 42 mm was obtained by lumping together seven mortar layers of 12 mm each and dividing the resulting mortar thickness into two layers. This simplification was carried out in order to reduce the number of elements in the analysis and is completely justified as we are interested only in the overall creep behavior of the pier. Each layer was modeled with 100 solid elements with brick layers containing interspersed mortar elements and mortar layers containing only mortar elements. The complete finite element model consisted of 900 solid cube elements, 1210 nodes and 3630 degrees of freedom.

An ABAQUS input file was created for creep analysis using the built in creep subroutine of ABAQUS based on a specified time law. According to this time law, the rate of creep strain at any time is defined by

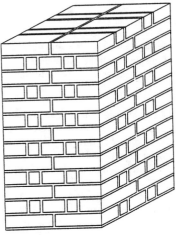

Fig. 5 Brick Masonry Pier

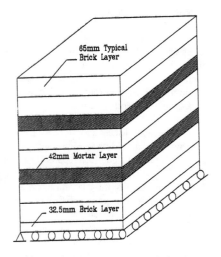

Fig. 7 Mortar Smearing Along Pier Height

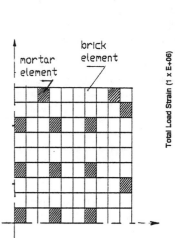

Fig. 6 Smeared $^{1}/_{4}$ Cross-Section

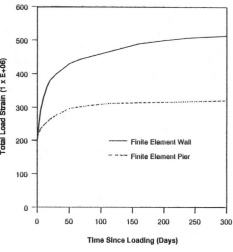

Fig. 8 Load Strain in Wall and Pier

$$d\bar{\epsilon}/dt = Aq^n t^m \qquad\qquad (6)$$

where A, n and m are constants to be defined for a particular creep behavior, q is the equivalent stress, ϵ is the von Mises equivalent strain and t is time. Assuming the creep strain rate to vary linearly with stress, the value of n can be taken as 1. The values of A and m for brick and mortar can be found using the corresponding specific creep strain curves and reducing Eq. 6 to one of the standard curve-fitting functions suggested by Lenczner (1986). Details of this procedure have been given by Bhatia (1994).

Creep analysis in ABAQUS can be carried out for a maximum of 10 time steps and the time step size is automatically selected by the software depending upon the input values of the constants A, n and m. As stated earlier, n was taken as unity and the values of A and m were obtained for different time scale units, i.e., one day, 5 days and 40 days, and were utilized in the creep analysis resulting in total strain values for 5.87 days, 46.8 days and 235 days, respectively. Some corrections to incorporate the effects of the various time step sizes were introduced, details of which are again available in Bhatia (1994). The resulting load strains are plotted in Fig. 8.

The previously calculated load strains for the stack-bond specimen calculated by the finite element method are also plotted in Fig. 8 for comparison with the strains in a pier. The initial elastic strain at zero days is almost the same in the wall and pier, thus, insuring the validity of the finite element idealization for the pier. The long term creep strains in the wall, on the other hand, are almost three times larger than the corresponding pier strains. A similar phenomenon has also been observed in the experimental results by Brooks and Abdullah (1988). This behavior is attributed to the smaller volume to exposed surface area ratio for a wall when compared to the corresponding ratio for a pier. The smaller volume of the wall allows for a shorter path for the moisture to travel to the surface, thus resulting in larger creep strains. Also, the larger exposed surface area of bricks and mortar allows more area for mortar diffusion in the wall than in a pier.

Results and Conclusions

It is obvious from the results presented in the previous sections that creep strains in clay masonry walls subjected to a uniform constant stress are approximately 150% larger than the instantaneous strains. Most of the creep occurs in the walls during the first two months during which time there is a tremendous relaxation of the prestress in post-tensioned masonry walls. Accordingly, adequate provisions must be made in the field for prestressed masonry to be reloaded to its desired level of prestress.

As the rate of increase in creep strains decreases substantially after the first 60 days of loading, and comes almost to a halt in a years time, the prestressing magnitude will eventually remain constant without any further need of additional post-tensioning.

The magnitude of creep strains in masonry piers, on the other hand, are much smaller than in the walls, and are less than the value of instantaneous elastic strains. Accordingly, some of the prestress in masonry piers will always be sustained even if no additional post-tensioning is applied in due course of time. Nevertheless, it is highly desirable to have load cells installed in the prestressing steel to monitor the level of prestress in masonry piers as a function of time.

It should be noted that in the analyses presented in this paper, strains due to long term effects other than creep, such as moisture expansion and/or shrinkage, creep in prestressing steel, as well as slippage in post-tensioning mechanisms, have been ignored. The magnitudes of these movements must also be included in an appropriate design of prestressed masonry.

References

1.   'ABAQUS - A finite element software.' (1989). Hibbitt, Carlsson and Sorensen, Inc., Pawtucket, RI.

2.   *ACI 530-92/ASCE 5-92/TMS 402-92*. (1992). "Building Code Requirements for Masonry Structures," ASCE, NY, NY.

3.   Anand, S.C. and Rahman, M.A. (1991). "Numerical modeling of creep in composite masonry walls." *J. Struct. Engrg.*, ASCE, 117(7), 2149-2165.

4.   Anand, S.C. and Bhatia, N. (1994). " A finite element model to estimate creep in masonry." *Proc., 10th Int. Brick/Block Masonry Conf.*, Calgary, Canada, 787-801.

5.   Anand, S.C. and Bhatia, N. (1995). "Modeling creep in masonry based on creep of units and mortar." *Proc. Structures Congress XIII*, Boston, MA, ASCE, 397-400.

6.   Bhatia, N. (1994). "Investigation of creep strains in brick masonry wall and pier specimens." *MS Thesis*, Clemson Univ., Clemson. SC.

7.   Brooks, J.J. (1986). "Composite models for predicting long-term movements in brickwork walls." *Proc., Brit. Cer. Soc., No. 1*, Stoke-on-Trent, UK, 20-23.

8. Brooks, J.J. (1986). "Time-dependent behaviour of calcium silicate and fletton clay brickwork walls." *Proc., Brit. Cer. Soc., No. 1*, Stoke-on-Trent, UK, 17-19.

9. Brooks, J.J. and Abdullah, C.S. (1988). "Composite model prediction of the geometry effect on creep and shrinkage of clay brickwork." *Proc. 8th Int. Brick/Block Masonry Conf.*, Dublin, Ireland, 342-349.

10. Ganz, H.R.(1993). "Strengthening of masonry structures with post-tensioning." *Proc. 6th North American Masonry Conf.* (6NAMC), Philadelphia, PA, 645-655.

11. Ganz, H.R. (1996). "VSL's experience with post-tensioned masonry." *Proc. Structures Congress XIV*, ASCE, Chicago, IL (12pp.).

12. Krause, G.L., Devalapura, R., and Tadros, M.K. (1996). "Testing of prestressed clay-brick walls." *Proc. Stuctures Congress XIV*, ASCE, Chicago, IL (12pp.).

13. Lenczner, D. (1970). "Creep in brickwork." *Proc., 2nd. Int. Brick Masonry Conf.*, Stoke-on-Trent, UK, 44-49.

14. Lenczner, D. (1986). "Creep and prestress losses in brick masonry." *The Structural Engineer*, 64 (3), 57-62.

15. Schultz, A.E. and Scolforo, M.J. (1991). "An overview of prestressed masonry." *The Mas.Soc.Jour.*, 10 (1), 6-21.

16. Schultz, A.E. and Scolforo, M.J. (1992). "Engineering design provisions for prestressed masonry. Part 2: Steel stresses and other considerations." *The Mas.Soc.Jour.*, 10 (2), 48-64.

17. Scolforo, M. and Ganz, H.R. (1995). "Prestressed masonry provisions in ACI 530/ASCE 5/TMS 402 - Building Code Requirements for Masonry Structures - draft." *Private Communication.*

18. Shaw, G. (1986). "Practical application of post-tensioned and reinforced masonry." *Practical Design of Masonry Structures*, Thomas Telford, Ltd. London, 197-212.

19. Shaw, G. (1996). "Innovative development of pre-stressed masonry." *Proc. Structures Congress XIV*, ASCE, Chicago, IL (12pp.).

20.   Shrive, N.G. and England, G.L. (1981). "Elastic, creep and shrinkage behaviour of masonry." *Int. J. Masonry Const.*, 1 (3), 103-109.

ANALYSIS OF MASONRY STRUCTURES WITH
ELASTIC/VISCOPLASTIC MODELS

By Teymour Manzouri[1], Student M. ASCE, P. Benson Shing [2], M. ASCE, and Bernard Amadei[2], M. ASCE

Abstract

Two elastic/viscoplastic constitutive models are presented for the modeling of masonry structures. One is a two-dimensional continuum model based on the Drucker-Prager failure criterion for modeling masonry units. The other is an interface model for modeling mortar joints. The latter allows mixed-mode fracture and shear dilatancy. In these models, the Perzyna-type viscoplasticity is employed. The ability of the continuum model in diffusing failure and removing the mesh-size sensitivity problem is demonstrated with numerical examples. The models are validated with experimental results obtained from an unreinforced masonry wall.

INTRODUCTION

Masonry structures can exhibit a brittle fracture behavior when subjected to strong earthquake loads. Fracture is manifested in the form of cracks propagating through the mortar joints and masonry units, and it often leads to a compressive as well as tensile softening phenomenon. The modeling of the fracture behavior in masonry structures is a most challenging problem. First, the fracture process is very much a time dependent phenomenon. Brittle materials may exhibit different crack patterns and strengths when subjected to different rates of loading. Secondly, modeling of compressive or tensile softening with

---

[1]Research Asst., Dept. of Civ., Environ., & Archit. Engrg., Univ. of Colorado, Boulder, CO 80309-0428

[2]Prof., Dept. of of Civ., Environ., & Archit. Engrg. Univ. of Colorado, Boulder, CO 80309-0428

a continuum F.E. model is often plagued with the mesh-size sensitivity problem (Bažant 1976; Needleman 1988). However, mesh-size dependency can be reduced by means of a localization limiter, in the form of nonlocal damage formulations (Bažant et al. 1987), fracture-energy based plasticity formulations (Promono and Willam 1989), second-order gradient plasticity models (de Borst et al. 1992), Cosserat models (Dietsche and Willam 1992), or viscoplastic models (Needleman 1988; Loret and Prevost 1989). Viscoplasticity has been viewed as an efficient regularization tool in both static and dynamic problems.

Two elastic/viscoplastic finite element models are presented in this paper for the analysis of the masonry structures. One is a continuum model for masonry units and the other is an interface model for masonry joints. The performance of these models is demonstrated by numerical examples.

FORMULATION OF ELASTIC/VISCOPLASTIC MODELS

In this study, masonry units are modeled with continuum elements, while mortar joints are modeled with elastic/viscoplastic interface elements. In either case, the classical theory of viscoplasticity is adopted, in which the total strain rate is decomposed to an elastic part and a viscoplastic part as follows.

$$\dot{\epsilon}_{ij} = \dot{\epsilon}_{ij}^e + \dot{\epsilon}_{ij}^{vp} \tag{1}$$

The stress rate, $\dot{\sigma}_{ij}$, is related to the elastic strain rate via the following constitutive relation:

$$\dot{\sigma}_{ij} = D_{ijkl}\dot{\epsilon}_{kl}^e \tag{2}$$

in which D_{ijkl} represents a fourth-order elasticity tensor. The viscoplastic strain can be expressed as (Perzyna 1966):

$$\beta_{ij} = \dot{\epsilon}_{ij}^{vp} = \gamma < \phi(F) > \bar{m}_{ij} \tag{3}$$

in which γ is a fluidity parameter, ϕ is a flow function, F is a yield function, and $\bar{m}_{ij}$ is a vector defining the direction of viscoplastic flow and is expressed as:

$$\bar{m}_{ij} = \frac{m_{ij}}{||m_{ij}||} \tag{4}$$

where

$$m_{ij} = \frac{\partial Q}{\partial \sigma_{ij}} \tag{5}$$

in which Q is the viscoplastic potential and $|| \cdot ||$ signifies an Euclidean norm. As in plasticity, the function F can be governed by a hardening and/or softening evolution law. The viscoplastic flow function $\phi(F)$ is in general a non-dimensionalized function and the bracket $< . >$ is defined to have the following properties.

$$\begin{aligned} &< \phi(F) >= 0 &&\text{when} \quad F \leq 0 \\ &< \phi(F) >= \phi(F) &&\text{when} \quad F > 0 \end{aligned} \tag{6}$$

Continuum Model

Yield and Failure Criteria. - The adopted model is based on the classical theory of viscoplasticity, in which a yield criterion has to be defined. For cementitious materials, a yield surface and a failure surface are usually defined. These criteria should account for (a) the relatively low tensile strength and relatively high compressive strength of the material, (b) the pronounced dependency of the strength of the material on the level of hydrostatic pressure, and (c) the increase in volume under plastic deformation, which is termed dilatancy.

In this study, the yield and failure surfaces are represented by the generalized Drucker-Prager (Drucker and Prager 1952) criterion, which can be expressed in terms of the deviatoric stress and hydrostatic stress as follows.

$$F(\boldsymbol{\sigma}, \chi) = \sqrt{J_2} + \alpha(\chi)I_1 - \kappa(\chi) = 0 \qquad (7)$$

in which I_1 is the first invariant of the stress tensor and J_2 is the second invariant of the deviatoric stress, and parameters α and κ are material parameters, both expressed as functions of an *equivalent viscoplastic strain* χ. Figure 1 illustrates the generalized Drucker-Prager criterion in the $\sqrt{J_2} - I_1$ space.

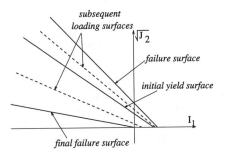

Figure 1: Generalized Drucker-Prager Criterion.

Once the yield function is specified, one can define the flow function ϕ used in Eq. 3. In this study, the flow function is defined as follows.

$$\phi(F) = (\frac{F}{F_0})^n \qquad (8)$$

in which F_0 is a nonzero positive function of the stress state to make the function ϕ non-dimensionalized, and n is a material constant.

Plastic Potential and Flow Rule. - Based on experimental evidence, cementitious materials show a pronounced non-associativity in their flow rule. Hence, the following function is adopted as the plastic potential.

$$Q = \sqrt{J_2} + \mu(\chi)I_1 - \kappa^* \tag{9}$$

in which μ is a dilatancy parameter and κ^* is a constant which is eliminated in the derivatives of the plastic potential and does not need to be calibrated.

Evolution of Loading Surface. - In the Drucker-Prager criterion defined by Eq. 7, parameters κ and α are assumed to be functions of equivalent viscoplastic strain χ. These functions are defined in such a way that the model can be readily calibrated with the uniaxial stress-strain relation of the material. The functions adopted here are as follows.

$$\begin{aligned}
\kappa &= A\chi^2 + B\chi + C \quad &\text{for} \quad 0 < \chi < \chi_2 \\
\kappa &= De^{-E\chi} \quad &\text{for} \quad \chi > \chi_2
\end{aligned} \tag{10}$$

and

$$\begin{aligned}
\alpha &= \bar{A}\chi^2 + \bar{B}\chi + \bar{C} \quad &\text{for} \quad 0 < \chi < \chi_2 \\
\alpha &= \bar{D}e^{-\bar{E}\chi} + \alpha_r \quad &\text{for} \quad \chi > \chi_2
\end{aligned} \tag{11}$$

in which χ_2 is the point at which the functions change from a quadratic form to an exponential form. Parameters A, B, C, D, E, $\bar{A}$, $\bar{B}$, $\bar{C}$, $\bar{D}$, and $\bar{E}$ are calibrated to capture the uniaxial tension and compression behavior of the material. Parameter α_r represents the capability of the material to sustain hydrostatic compression once the resistance to uniaxial tension and compression is depleted.

Evolution of Plastic Potential. - The direction of viscoplastic flow is defined by the plastic potential (Eq. 9). Experimental results indicate that plastic dilatancy is generally reduced as the damage level increases. This property is governed by μ in Eq. 9. Hence, μ can be called the dilatancy parameter. A simple exponential function is adopted for μ to describe the aforementioned dilatancy property.

$$\mu = \mu_0 e^{-\eta\chi} + \mu_r \tag{12}$$

in which μ_0, η, and μ_r are parameters that have to be calibrated.

Evolution of Fluidity γ. - In brittle materials, once fracture or damage propagates, the material will no longer exhibit a viscous behavior due to the loss of internal bonds. Hence, it is realistic to reduce the viscosity of the model once damage initiates. For this reason, the fluidity is assumed to increase linearly from an initial value γ_{in} to an ultimate value γ_{ult} as a function of χ. The change in fluidity starts at $\chi = \chi_1$ and the ultimate value for the fluidity parameter is assumed to be reached at $\chi = 10\chi_1$. Parameter χ_1 is a value of equivalent viscoplastic strain corresponding to the maximum values of functions α and κ defined in Eqs. 11 and 10.

Equivalent Viscoplastic Strain χ. - For cementitious materials, isotropic strain hardening/softening laws suffer from a major drawback that damage develops uniformly in all directions. Another problem associated with this type of model can arise when the equivalent viscoplastic strain is defined in a conventional way as follows.

$$\dot{\chi} = \sqrt{\dot{\epsilon}_{ij}^{vp}\dot{\epsilon}_{ij}^{vp}} \qquad \chi = \int_0^t d\chi = \int_0^t \sqrt{\dot{\epsilon}_{ij}^{vp}\dot{\epsilon}_{ij}^{vp}}\,dt \qquad (13)$$

in which ϵ_{ij}^{vp} is the viscoplastic strain tensor and the repeated indices imply summation. The difficulty with such a definition is that if the model is calibrated with uniaxial compression, then for failure in uniaxial tension, it will develop the same level of cumulative viscoplastic strain as in uniaxial compression. This problem is shown in Figure 2a, in which ϵ_c and ϵ_t are the total strains associated with compression and tension failures, respectively. This is evidently not realistic.

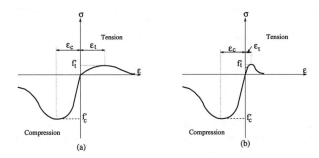

Figure 2: Uniaxial Behavior of the Model: a) Conventional χ; b) Modified χ.

To address the above problem, we assume that the viscoplastic tensile strain will induce more damage than the viscoplastic compressive strain, and define damage parameter χ accordingly. The principal viscoplastic strain rates $\dot{\epsilon}_i^{vp}$ (with $i = 1, 2, 3$) are evaluated with a given $\dot{\epsilon}_{ij}^{vp}$. These principal strain increments are then checked for the sign. If any of them is positive, it is multiplied by a factor ξ to obtain a damage measure ω, i.e,

$$\dot{\omega}_i = \xi \cdot \dot{\epsilon}_i^{vp} \qquad \text{for} \quad i = 1, 2, 3 \qquad (14)$$

where

$$\xi = \begin{cases} (k-1)\zeta + 1.0 & \text{for} \quad \sqrt{J_2} > \frac{1}{\sqrt{3}}I_1 \quad \text{and} \quad I_1 > 0 \\ & \text{or} \quad \sqrt{J_2} > -\frac{1}{\sqrt{3}}I_1 \quad \text{and} \quad I_1 < 0 \\ k & \text{for} \quad \sqrt{J_2} \leq \frac{1}{\sqrt{3}}I_1 \quad \text{and} \quad I_1 > 0 \end{cases} \qquad (15)$$

$$\zeta = \frac{I_1 + \sqrt{3J_2}}{2\sqrt{3J_2}} \qquad (16)$$

in which k is a positive material property defined as the ratio of the total strain associated with the uniaxial compression failure to the total strain associated with the uniaxial tension failure. The equivalent viscoplastic strain is then defined as

$$\dot{\chi} = \sqrt{\dot{\omega}_i \dot{\omega}_i} \qquad\qquad \chi = \int_0^t d\chi = \int_0^t \sqrt{\dot{\omega}_i \dot{\omega}_i} dt \qquad (17)$$

The behavior of the modified model is shown in Figure 2b.

Interface Model

The dilatant interface constitutive model proposed by Lotfi and Shing (1994) is adopted in this study. In the present work, the inviscid interface model is extended to a rate-dependent model. The primary objectives of this extension are to simulate the rate-dependent property of joints in masonry, concrete, and rocks, and to achieve a higher computational efficiency.

Element Formulation. - In the interface model, the tractions are related to the relative displacements between the two faces of an interface. The viscoplastic constitutive relations have the following form.

$$\begin{aligned}
\dot{\mathbf{d}} &= \dot{\mathbf{d}}^e + \dot{\mathbf{d}}^{vp} \\
\dot{\mathbf{d}}^{vp} &= \gamma < \phi(F) > \bar{\mathbf{m}} \\
\dot{\boldsymbol{\sigma}} &= \mathbf{D}(\dot{\mathbf{d}} - \dot{\mathbf{d}}^{vp}) \\
\bar{\mathbf{m}} &= \frac{\mathbf{m}}{\|\mathbf{m}\|}
\end{aligned} \qquad (18)$$

in which $\mathbf{d} = [d_n \ d_t]$ is the relative displacement vector as defined in Figure 3a, γ is the fluidity parameter, $\bar{\mathbf{m}} = \frac{\mathbf{m}}{\|\mathbf{m}\|}$ is the normalized flow vector with $\mathbf{m} = \frac{\partial Q}{\partial \boldsymbol{\sigma}}$, with Q being the plastic potential.

Yield (Slip) Surface and Viscoplastic Flow Function. - A three parameter hyperbolic yield criterion proposed by Lotfi and Shing (1994) is used. The yield function is expressed as follows.

$$F(\boldsymbol{\sigma}, \mathbf{q}) = (\sigma - s) - \frac{r}{\mu^2} + \sqrt{(\frac{r}{\mu^2})^2 + (\frac{\tau}{\mu})^2} = 0 \qquad (19)$$

in which $\mathbf{q}^t = \{s, r, \mu\}$ represents the internal variables, with r being the curvature of the hyperbola at the vertex, s the tensile strength, and μ the slope of the asymptote to the hyperbola, as shown in the Figure 3b. The flow function is defined as follows.

$$\phi(F) = (\frac{F}{F_0})^n \qquad (20)$$

in which F_0 is defined as

$$F_0 = s + \frac{r}{\mu^2} \qquad (21)$$

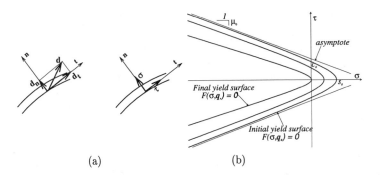

Figure 3: Interface Model: (a) Relative Displacements and Tractions; (b) Yield (Slip) Surface.

and n is a material constant.

Plastic Potential and Flow Rule. - Shear dilatancy and the change in dilatancy properties with the level of damage are modeled by the following plastic potential.

$$Q(\boldsymbol{\sigma}, \mathbf{q}) = \bar{\eta}\tau^2 + (r - r_r)(\sigma - s) \tag{22}$$

in which $\bar{\eta}$ is a parameter which scales the dilatancy and r_r represents the residual value of r. The details of this model and the evolution of $\mathbf{q}$ are explained by Lotfi and Shing (1994).

NUMERICAL EXAMPLES

The performance of the elastic/viscoplastic continuum model in removing the mesh-size dependency problem is illustrated by shear wall examples. A square shear wall is discretized into a 6x6 mesh and a 12x12 mesh, respectively. The wall is subjected to a constant vertical load and a monotonically increasing lateral load. Two lateral displacement rates are used, which are 25.4 mm/sec and 254 mm/sec. The lateral load-vs.-lateral displacement curves obtained for the lower rate with the two different meshes are shown in Figure 4a and those with the higher rate in Figure 4b. It can be observed that the mesh-size dependency diminishes as the displacement rate increases for a given viscosity level. The damage patterns obtained from these examples are shown in Figures 5 and 6. It can be observed that the higher the displacement rate is, the wider is the damage zone. With the same displacement rate, the damage zones obtained with the two different meshes are almost identical as far as the yielding and crushing are concerned, but tensile softening still tends to localize in a single row of elements.

The accuracy of the model in analyzing masonry structures is examined by test results obtained from an unreinforced masonry shear wall (Manzouri 1995). The wall is analyzed under in-plane vertical and horizontal loads with the aforementioned viscoplastic models. The continuum model is calibrated for masonry units based on the results of prism tests. The interface model is calibrated for masonry bed joints based on the direct shear test results obtained from these joints. The load-deformation results obtained from both the analysis and the experiment are shown in Figure 7. The rate of applied displacement in the analysis is 25.4 mm/sec, which is considerably higher than the displacement rate applied in the experiment (0.1 mm/sec). The fluidity parameter used for the continuum model is 1.0/sec, while that for the interface model is 1.7/sec. With these fluidity values, the aforementioned displacement rate can be considered close to static. The results of the analyses indicate an excellent performance of the models in capturing the failure mechanisms, as shown by the respective damage patterns in Figure 8.

CONCLUSIONS

Two elastic/viscoplastic finite element models are presented for the analysis of masonry structures. One is for the modeling of the masonry units and the other for the mortar joints. It has been demonstrated that the viscoplastic continuum formulation helps to mitigate the problem of mesh-size dependency of numerical solutions and prevent the localization of damage into a narrow band. Nevertheless, the numerical results show that the problem is not completely eliminated.

ACKNOWLEDGMENT

The study presented in this paper is supported by the National Science Foundation under Grant No. MSM-9017149. However, opinions expressed in this paper are those of the writers, and do not necessarily represent those of the sponsor.

APPENDIX I. REFERENCES

Bažant, Z. P., Lin, F. B. and Pijaudier-Cabot, G. (1987). "Yield limit degradation: Nonlocal continuum model with local strain." *International Conference on Computational Plasticity, Models, Software and Applications,* Barcelona, Spain.

Bažant, Z. P. (1976). "Instability, ductility and size effect in strain-softening concrete." *Journal of the Engineering Mechanics Devision, ASCE,* 102(EM2), 331-334.

de Borst, R., Muhlhaus, H. B. and Pamin, J. (1992). "A gradient continnum model for mode-I fracture in concrete and rock." *First International Conference on Fracture Mechanics of Concrete Structures,* 251-259.

Dietsche, A. and Willam, K. J. (1992). "Localization analysis of elasto-plastic cosserat continua." Wu, J. W. and Valanis, K. C., editors, *Damage Mechanics and Localization, ASME,* AMD-142, MD-34, 109 - 123.

Drucker, D. C. and Prager, W. (1952). "Soil mechanics and plasticity analysis or limit design." *Qurt. Applied Mathematics,* 10(2), 157-165.

Loret, B. and Prevost, J. H. (1986). "Dynamic strain localization in elasto-(visco)-plastic solids." *Computer Methods in Applied Mechanics and Engineering,* 83, 247-294. North Holand.

Lotfi, H. and Shing, P. B. (1994). "An interface model applied to fracture of masonry structures." *Journal of Structural Engineering, ASCE,* 120(1), 63-80.

Manzouri, T. (1995). *Nonlinear finite element analysis and experimental evaluation of retrofitting techniques for unreinforced masonry structures.* PhD Thesis, Dept. of Civil, Environmental and Architectural Engineering, University of Colorado, Boulder, CO (In preparation).

Needleman, A. (1988). "Material rate dependence and mesh sensitivity in localization problems." *Computer Methods in Applied Mechanics and Engineering,* 67, 69-85.

Perzyna, P. (1966). "Fundamental problems in viscoplasticity." *Advances in Applied Mechanics,* 9, 244-368.

Promono, E. and Willam, K. J. (1989). "Fracture energy-based plasticity formulation of plain concrete." *ASCE Journal of Engineering Mechanics,* 115(6), 1183-1204.

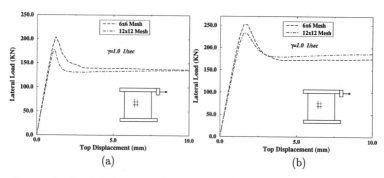

Figure 4:   Load-Deformation Curves under Different Displacement Rates: a) Rate=25.4 mm/sec; b) Rate=254 mm/sec.

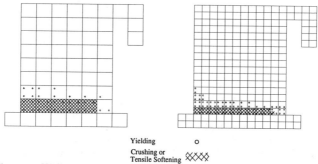

Figure 5: Wall Damage at 4.0 mm Displacement with Rate=25.4 mm/sec.

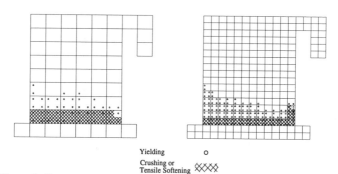

Figure 6: Damage pattern at 4.0 mm Displacement with Rate=254 mm/sec.

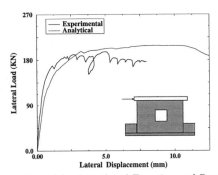

Figure 7: Comparison of Analytical and Experimental Results for Wall #4.

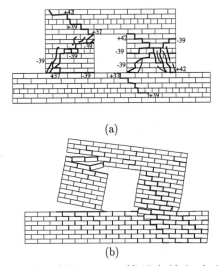

Figure 8: Failure Modes: a) Experiment (Cyclic); b) Analysis (Monotonic).

A Finite Element Damage Model for the Evaluation
and Rehabilitation of Brick Masonry Shear Walls

Luigi Gambarotta[1] and Sergio Lagomarsino[2]

Abstract

Constitutive equations for brick masonry, taking into
account both de-cohesion and slipping in mortar joints
and failure in bricks, are applied in the finite element
analysis of shear walls. The simulation of experimental
results, obtained from a large scale perforated wall,
shows the capability of the model to identify the lateral
strength, the post-peak response and the hysteretic
behavior. The procedure, implemented in a standard FE
code, is also applied in evaluating both the dynamic
response of shear walls and the effectiveness of different techniques of structural rehabilitation.

Introduction

The idea of a scanty performance of the masonry
building under seismic actions is, in some cases, the result of applying simplified methods of analysis. When
considering the case of in-plane loaded shear walls,
these procedures may underestimate the lateral strength
of the wall (Abrams and Costley, 1994) and in any case
they neglect the benefits of the dissipative behavior of
the wall itself. The need comes out of improving the
theoretical models for the shear wall analysis. To
achieve this goal two distinctive aspects must be taken
into account: the two dimensional shape of the wall and
the composite structure of the brick masonry.

All of this requires both a two-dimensional finite
element modeling of the wall and *ad hoc* constitutive

---
[1] Professor, Faculty of Engineering, Univ. of Messina,
Contrada Sperone 31, 98166 S.Agata (ME), Italy
[2] Researcher, Istituto di Scienza delle Costruzioni,
Univ. of Genoa, via Montallegro 1, 16145 Genoa, Italy

equations for the masonry. Thus, a proper treatment of the material model has to be carried out which is able to properly describe the quasi-brittle behavior of brick units and mortar and the de-cohesion and frictional sliding in the brick-mortar interfaces.

With the aim of modeling large scale shear walls, rather than describing separately the brick units and the mortar joints, it is preferable to take on continuum models entailing a lower degree of freedom. This approach is pursued in the present paper in which the constitutive equations proposed by the Authors (Gambarotta et al., 1994) are assumed. These ones are obtained by homogenizing a two-phase layered model in which the mechanisms of inelasticity due to both stiffness degradation in the brick unit layer and frictional sliding coupled with damage in the mortar bed joints are taken into account. Through this constitutive model a finite element procedure has been developed and applied, as an example, in the simulation of the lateral response of a large scale wall experimented by Calvi and Magenes (1994). To point out the possibilities provided by the numerical procedure also the dynamic response to seismic excitations of the wall has been evaluated, allowing comparisons with the results from the cyclic quasi-static analysis. Finally, the model has been applied for evaluating the effectiveness of different techniques of structural rehabilitation of the walls.

Damage model for brick masonry

The brick masonry response to in-plane loads is represented through the constitutive equations proposed by Gambarotta et al. (1994), summarized as follows.

The continuum model for brick masonry is obtained as being a stratified medium consisting of bed joint layers and layers representing brick and mortar head joints. By homogenizing mortar bed joints and brick layers it follows that the mean strain $\boldsymbol{\varepsilon} = \{\varepsilon_1\ \varepsilon_2\ \gamma\}^{\cdot}$, ε_2 being the vertical strain normal to the bed joints, is given as follows:

$$\boldsymbol{\varepsilon} = \mathbf{K}\boldsymbol{\sigma} + \boldsymbol{\varepsilon}_m^* + \boldsymbol{\varepsilon}_b^* \quad , \tag{1}$$

where $\mathbf{K}$ is the elastic orthotropic compliance matrix of the masonry, which depends on the elastic moduli of mortar (E_m, v_m) and brick (E_b, v_b) and on the volume fraction η_m of mortar joints, $\boldsymbol{\sigma} = \{\sigma_1\ \sigma_2\ \tau\}^{\cdot}$ is the mean stress, $\boldsymbol{\varepsilon}_m = \{0\ \varepsilon_m\ \gamma_m\}^{\cdot}$ and $\boldsymbol{\varepsilon}_b = \{0\ \varepsilon_b\ \gamma_b\}^{\cdot}$ are the inelastic strains in mortar and bricks respectively.

The mechanisms of inelasticity in bed mortar joints induce the extension ε_m and the sliding γ_m, which are assumed linearly dependent on the damage variable α_m, as follows

$$\varepsilon_m = c_{mn}\alpha_m \, \mathcal{H}(\sigma_2) \, \sigma_2 \quad , \qquad \gamma_m = c_{mt}\alpha_m (\tau - f) \quad , \tag{2}$$

where σ_2 and τ are the resolved stresses on the mortar bed joints, f represents the friction in the mortar brick interface, $\mathcal{H}$ is the Heaviside function taking into account the unilateral response of the interface and, finally, c_{mn} and c_{mt} are the extensional and tangential compliance parameters characterizing the mortar joints.

Brick damage is described by considering the effects of the vertical compressive and shear stresses on the inelastic strain components, which are assumed as follows:

$$\varepsilon_b = c_{bn}\alpha_b \, \mathcal{H}(-\sigma_2) \, \sigma_2 \quad , \qquad \gamma_b = c_{bt}\alpha_b \, \tau \quad , \tag{3}$$

where α_b represents the damage in bricks, c_{bn} and c_{bt} are the compressive and tangential compliance parameters of the brick.

Equations (1-3) allow the mean strain to be obtained once the internal variables α_m, γ_m and α_b are evaluated for a given stress state. To this end, the evolution equations must be defined in terms of the corresponding variables, i.e. the damage energy release rate in joint mortars

$$Y_m = \frac{1}{2}c_{mn} \, \mathcal{H}(\sigma_2) \, \sigma_2^2 + \frac{1}{2}c_{mt} \, (\tau - f)^2 \quad , \tag{4}$$

the friction f and the damage energy release rate in bricks

$$Y_b = \frac{1}{2}c_{bn} \, \mathcal{H}(-\sigma_2) \, \sigma_2^2 + \frac{1}{2}c_{bt} \, \tau^2 \quad . \tag{5}$$

The evolution equations are formulated on the basis of three conditions which must be satisfied during the loading process. The damage evolution in mortar joints and bricks is defined by imposing the corresponding damage energy release rate to be less than or equal to a proper toughness function (assumed dependent on the damage variable), that is

$$\phi_{dm} = Y_m - R_m(\alpha_m) \le 0 \quad , \tag{6}$$

$$\phi_{db} = Y_b - R_b(\alpha_b) \le 0 \quad , \tag{7}$$

where R_m and R_b are the joint and brick toughness functions. Thus, when the limit condition $\phi_{dm} = 0$ ($\phi_{db} = 0$) is attained in the infinitesimal load step, the joint damage rate $\dot{\alpha}_m$ (brick damage rate $\dot{\alpha}_b$) is assumed to take place. Moreover, the variable f has to satisfy the friction limit condition

$$\phi_s = \left| f \right| + \mu \sigma_2 \leq 0 \quad , \tag{8}$$

involving the friction coefficient μ to which the non-associated flow rule

$$\dot{\gamma}_m = v \dot{\lambda} \quad , \quad \dot{\lambda} \geq 0 \quad , \tag{9}$$

is related $(v=f/\left|f\right|)$.

The above assumptions imply an incremental formulation of the constitutive equations which may exhibit unstable response to applied stress rate according to the structure of the functions R_m and R_b. In fact, the stable response is obtained when $R'=dR/d\alpha>0$, until the maximum toughness R_{mc} (R_{bc}) is reached; after this state the response turns out to be unstable. Considering this transition as a limit strength condition obtained for $\alpha_m=1$ ($\alpha_b=1$), the following limit strength domain can be obtained, referring to the proportional loading path, as the envelope of the following limit domains:

• bed mortar joints

$$\left| \tau \right| + \mu \sigma_2 = \tau_m \quad , \qquad \text{if } \sigma_2 < 0 \tag{10}$$

$$\sigma_2^2 + \rho_m \tau^2 = 2R_{mc}/c_{mn} = \sigma_m^2 = \rho_m \tau_m^2 \quad , \qquad \text{if } \sigma_2 \geq 0 \tag{11}$$

• bricks

$$\sigma_2^2 + \rho_b \tau^2 = 2R_{bc}/c_{bn} = \sigma_b^2 = \rho_b \tau_b^2 \quad , \qquad \text{if } \sigma_2 \leq 0 \tag{12}$$

$$\left| \tau \right| = \tau_b \quad , \qquad \text{if } \sigma_2 > 0 \tag{13}$$

where σ_m and τ_m are the tensile and the shear strengths, respectively, of the mortar joints and σ_b and τ_b are the compressive strength of the brick masonry and the shear strength of bricks.

Good simulations of the brick masonry response have been obtained by the Authors (Gambarotta and Lagomarsino, 1994) assuming the toughness functions in the following form

$$R(\alpha) = \begin{cases} R_c \alpha & 0 < \alpha < 1 \quad , \\ R_c \alpha^{-\beta} & 1 \leq \alpha \quad , \end{cases} \tag{14}$$

where R_c is the maximum value ($\beta>0$).

The evaluation of the material parameters can be carried out by means of uniaxial tests on bricks (elastic moduli E_b, v_b and tensile strength τ_b), mortar (elastic moduli E_m, v_m), masonry (compressive strength σ_b, failure strain c_{bn} and softening β_m) and tests on the mortar joints (direct shear strength τ_m, friction coefficient μ, failure shear strain c_{mt}, softening β_m, direct tensile strength σ_m).

Finite element modeling of large scale walls

The constitutive equations described here have been applied in an incremental finite element procedure for the in-plane analysis of brick masonry walls based on four node isoparametric elements. The incremental solution is obtained by means of an iterative *initial stress* method whose fundamental peculiarity is the algorithm for the finite load step integration of the constitutive equations. In fact, this task turns out to be very complex due to the interaction of the three inelastic deformation mechanisms involved in the present model. The finite increments of the stress and internal variables corresponding to finite strain increments are obtained by means of a predictor-corrector procedure, similar to that proposed by Simo *et al.* (1988), in which the correction phase is carried out by means of a Newton-Raphson algorithm involving the limit functions of the active mechanisms at each iteration. Furthermore, in order to take into account size effects, rectangular finite elements are chosen with dimensions such that, for each point of integration, a representative volume element of the brick masonry is considered as having width equal to the mean distance between the head joints, and height equal to one course of bricks and mortar.

To provide a useful tool for structural analysis, the finite load step algorithm has been implemented in the code ANSYS (1992) by defining a proper quadrilateral masonry element. So making, both static and dynamic analyses can be carried out and the analyzer is allowed to sum up different elements for modeling the masonry wall in a more complex structural context.

In order to obtain information on the capability of the constitutive model and the related finite element procedure for application to brick masonry, several experimental tests on brick masonry shear walls have been simulated. Among these ones the simulation of the large scale shear wall tests carried out by Calvi and Magenes (1994) is here considered as an example (see figure 1b).

In this case the masonry wall (6 m width, 6.43 m height) was built using 250×120×55 mm brick units, implying 250 mm wall depth, and 10 mm mortar joint thickness. Bricks, mortar and their assemblages have been tested and characterized by Binda *et al.* (1994), allowing the identification of the model parameters (E_b=2160 Mpa, ν_b=0.28, σ_b=5.9 MPa, τ_b=2 MPa, $1/c_{bn}$=32000 MPa, β_b=0.4, E_m=533 MPa, G_m=87 Mpa, ν_m=0.18, η_m=0.154, σ_m=0.05 MPa, τ_m=1.8 MPa, μ=0.577, $1/c_{mt}$=195 MPa, β_m=0.8. The finite element model is formed by 185×130 mm four node isoparametric elements.

In the experimental tests, the wall was first subjected to the uniformly distributed vertical force P=21.2 kN/m, applied at the first and second floor level, and subsequently to horizontal displacements (w_1, w_2), cyclically applied to each floor level in order to obtain equal applied story forces.

In figure 1, the theoretical results are summarized in terms of the hysteretic response and of the distribution of the mortar joint damage α_m and vertical stress σ_2. By comparing these results with the experimental ones provided by Calvi and Magenes (1994), it turns out that the model is able to identify both the horizontal strength of the wall and its dissipative behavior.

During the loading process, the first damage appears for 0.1% drift and is localized in the mortar bed joints in the architrave of the first floor, where the vertical stress σ_2 attains low values. For higher horizontal displacements, the base sections of the wall and the central panel also exhibit damage in the mortar joints: in the first case, this is due to the extension and bending of the piers, while in the second case to the shear and bending effect. Increasing the lateral displacement, the damage propagates in the central panel in the form of a "Y", as shown in figure 1b for 0.3%. As shown in figure 1c, corresponding to 0.3% drift, the vertical stress is not very high, so the effects of brick failure are limited. Finally, it is also worth noting that a considerable rotation appears at the top of the wall, due to the non uniform compression in the piers.

Applications

The proposed FE model sets up an effective support both in the design of experiments and in the following simulation and interpretation of the test results. Moreover, once the validity of the model is ascertained, it may be used as a generator of virtual experiments, from which it is possible to understand the role of the wall geometry and the sensitivity to the material data.

In the previous section it is shown that the model is able to catch fairly both the wall strength and, even after the limit point, the dissipative capacity. All of this makes the dynamic analysis meaningful in order to obtain information on the wall safety to earthquake. The results should give widespread meaning to quasi-static parameters such as strength, ductility and hysteretic damping.

From the practical point of view, this tool may be useful for forecasting the effectiveness of some retrofitting techniques aimed to rehabilitate a damaged building or to improve the safety level to the seismic actions.

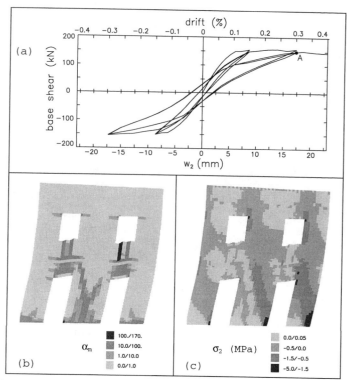

Figure 1. (a) top floor displacement w_2 vs. base shear;
          (b) mortar joint damage α_m distribution;
          (c) vertical stress σ_2 distribution.

Dynamic response to seismic excitation

        The masonry wall previously described has been
subjected to the artificial accelerogram shown in figure
2a applied to the base; its response spectrum is typical
of stiff soils, with a significant frequency content in
the range 3÷10 Hz, and the intensity corresponds to a
peak acceleration $a_{max}=0.5$ g.
        The displacement history of the second floor is
shown in figure 2b, from which it comes out that after
the phase of maximum amplitude ($w_2=32$ mm, drift 0.55%)
the vibration takes place around a sidewise deformed
shape. The diagram of the second floor displacement
versus the base shear is shown in figure 2c. These cycles
look like the ones obtained from the quasi-static
analysis; however, it emerges that, apart from the high

intensity phase, the damaged wall behaves as elastic. Finally, the mortar joint damage distribution at the end of the earthquake looks similar to the one obtained from the quasi-static excitation (see figure 1).

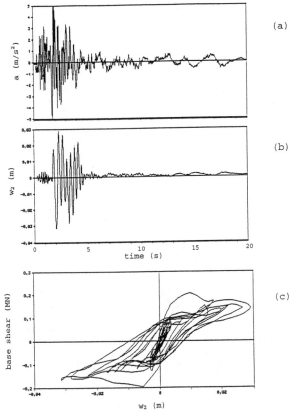

Figure 2. Dynamic analysis: (a) seismic input; (b) second floor displacement history; (c) cyclic response.

Effectiveness of some retrofitting techniques

Both the test results and the numerical simulation of the above mentioned masonry wall have shown zones of weakness which are located in the architrave over the doors, at the base of the wall and in the central pier. Thus, the retrofitting must operate on these structural elements.

The architrave may be reinforced by inserting strong lintels which effect, shown in the diagram of figure 3, consists in a concentration of the damage in the three piers at the first story, producing a collapse mechanism that concerns the first story. While the strength is not increased by this technique, it results in a large increase of the dissipation capacity, due to the frictional sliding in the piers.

As an alternative it is possible to limit the damage in the central pier by pre-stressing the masonry by steel tie rods; in this case two pairs of rods having 14 mm diameter have been provided for each pier, inducing an increase of the vertical stress in the masonry of 0.2 Mpa on average. The figure 3b shows that such technique partly reduces the damage in the central pier to the detriment of an increase of the damage over the lintels, producing a collapse mechanism that concerns the overturning of the three piers. However, in this case both the lateral strength and the hysteretic damping do not increase significantly.

Thus, the first retrofitting seems to be preferable for the greater dissipation capacity. To obtain an effective increase of the lateral strength both the above mentioned techniques should be considered together.

References

Abrams D.P. and Costley A.C., Dynamic response measurements for URM building systems, Proc. *U.S.-Italian Workshop on Guidelines for Seismic Evaluation and Rehabilitation of Unreinforced Masonry Buildings*, NCEER, 94-0021, 3-27, Buffalo, 1994.

ANSYS User's Manual, Revision 5.0, Swanson Analysis Systems, Inc., DN-R300, Houston, 1992.

Binda L., Mirabella G., Tiraboschi C. and Abbaneo S., Measuring masonry material properties, Proc. *U.S.-Italian Workshop on Guidelines for Seismic Evaluation and Rehabilitation of Unreinforced Masonry Buildings*, NCEER, 94-0021, 6-3, Buffalo, 1994.

Calvi G.M. and Magenes G., Experimental research on response of URM building systems, *Proc. U.S.-Italian Workshop on Guidelines for Seismic Evaluation and Rehabilitation of Unreinforced Masonry Buildings*, NCEER, 94-0021, 3-41, Buffalo, 1994.

Gambarotta L. and Lagomarsino S., Damage in brick masonry shear walls, *Proc. Europe-U.S. Workshop on Fracture and Damage in Quasibrittle Structures: Experiment, Modeling and Computer Analysis*, 463-462, E&FN Spoon, London, 1994.

Gambarotta L., Lagomarsino S. and Morbiducci R., Brittle-ductile response of in-plane loaded brick masonry walls, *Proc. 10th European Conference on Earthquake Engineering*, 3, 1663-1668, Balkema, Rotterdam, 1995.

Simo J.C., Kennedy J.G. and Govindjee S., Non-smooth multisurface plasticity and viscoplasticity. Loading and unloading conditions and numerical algorithms, *Int. J. Numer. Methods Engrg.*, **26**, 2161-2185, 1988.

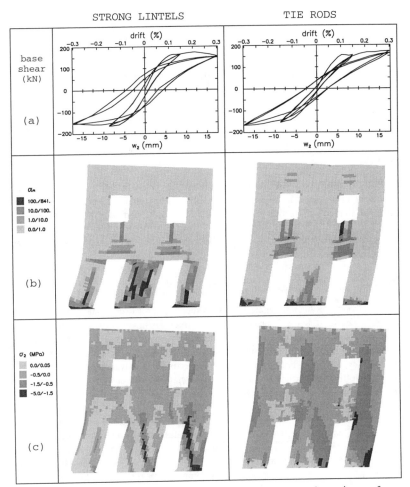

Figure 3. Retrofitting by strong lintels or by tie rods:
(a) top floor displacement vs. base shear;
(b) mortar joint damage α_m; (c) vertical stress σ_2.

Implications Derived from Recent Research in Mexico
on Confined Masonry Structures

Sergio M. Alcocer[1]

Abstract

Main results of a research program on the behavior of confined masonry
structures under seismic loads are outlined. The investigation, conducted at the
National Center for Disaster Prevention in Mexico, is part of a project aimed at
improving the design and construction of low-cost housing buildings. The research
has provided valuable information to improve the analysis, design and construction of
confined masonry structures in Mexico.

Introduction

Masonry has been the material most widely used for residential construction in
Mexico. Low-cost housing projects are constructed using traditional methods for
confined masonry. Confined masonry consists of load-bearing walls surrounded by
small cast-in-place reinforced concrete columns and beams, hereafter referred to as
tie-columns, TC's, and bond-beams, BB's, respectively. The system is such that walls
must resist both the vertical and lateral loads. Details of wall design and construction
in Mexico may be found elsewhere (Alcocer & Meli, 1995).

The seismic behavior of confined masonry buildings has been generally
satisfactory, particularly in Mexico City. Nevertheless, significant damages have been
observed in near-epicentral regions during strong ground shaking. New building code
requirements have become more stringent thus requiring designs to be revised, and in
most cases, be modified substantially to comply with the new ordinances. Since
construction programs for multi-family low-cost housing in Mexico involve prototype
designs which are repeated several times, the impact on construction cost is very
large.

----

[1]Head, Structural Engineering and Geotechnical Area, National Center for Disaster
Prevention, Av. Delfin Madrigal 665, Pedregal de Sto. Domingo, Mexico, DF 04360.

To improve the design, reinforcement detailing and construction of low-cost housing, an analytical and experimental research program is underway at the National Center for Disaster Prevention in Mexico. In this paper, main results obtained in the different phases of the project are outlined.

Structural Characteristics of Low-Cost Housing Buildings in Mexico

An evaluation of the structural characteristics of low-cost housing buildings and their evolution over time, particularly since the 1985 Mexico earthquake, was carried out (Meli et al., 1994). This study was intended to provide a framework and guide for future analytical and experimental investigations at CENAPRED.

Main conclusions derived from this study are: 1) Load-bearing wall construction (either as confined masonry or reinforced masonry) with cast-in-place reinforced concrete (RC) slabs is the structural system most commonly used. Hand-made solid clay bricks and concrete blocks are the units preferred. Prefabricated slabs (made of joists and concrete blocks) are also typical. 2) Although same construction systems are used since 1985, at present RC walls are more frequently used in combination with masonry walls to increase the seismic resistance of the building. 3) Particularly since 1985, there are large differences in the structural design criteria adopted. While there are some projects with high safety factors (ratios of ultimate strengths to factored loads), others do not comply with minimum code standards. In Fig. 1, wall densities for 21 structural projects studied are presented. Wall density is defined as the ratio of the effective transverse wall area in one direction and the floor plan area. The effective transverse wall area refers to a reduced transverse wall area considering the lower effectiveness of slender walls to resist lateral shear forces.

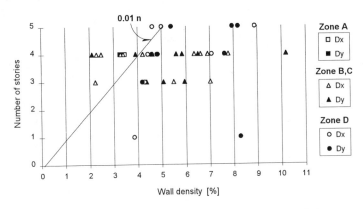

Figure 1. Wall Densities for 21 Structural Projects Studied

It has been demonstrated that wall density is a valid indicator of the seismic safety of load bearing walls (Meli, 1994). To calculate wall densities for buildings with concrete and masonry walls, concrete walls were transformed into equivalent masonry walls considering the average shear strengths of the materials and differences in their load-displacement behavior. From the graph it is clear that some buildings in seismic zones B,C (moderate hazard) and D (high hazard) do not satisfy the density demanded by code requirements (line d=0.01n). 4) It seems that architectural considerations mostly dictate the prototype projects; structural considerations are unimportant. Thus, there should be projects specifically tailored for high seismic hazard zones.

Effect of the Type of Flexural Coupling in Wall Behavior

Three models consisting of two walls were linked by slabs or by slabs and parapets, thus changing the type of flexural coupling between them and consequently, the flexural to shear capacity ratio M/Vd. Walls were built with hand-made solid clay bricks (25x12.5x6.2 cm). The mortar used to join the bricks was proportioned by volume and had a cement:sand ratio of 1:3 Average prism strength for specimen W-W was 4.6 MPa and for both WBW and WWW was 5.1 MPa. Grade 60 (f_y=412 MPa) steel was used for the longitudinal reinforcement of TC' s, BB's and slabs. Tie-columns with nominal sections of 12.5x15 cm were reinforced with 4-#3 longitudinal bars and with grade 30 #2 hoops spaced at 20 cm. At the ends of the TC's spacing was reduced to 7 cm over a 35-cm length. Bond-beams, with 12.5x25 cm section, were reinforced with 4-#4 bars and #2 hoops spaced at 20 cm. Slab was 10 cm thick and was reinforced with 6-#3 bars in the long direction and #4 bars in the short direction. Specimens were tested under alternated lateral loads to simulate earthquake-type loading; a constant compressive vertical stress equal to 0.49 MPa was applied to simulate the gravity loads on a ground story wall of a five-story housing building. Details of the test layout and loading history can be found elsewhere (Alcocer & Meli, 1995). Crack patterns and hysteretic responses are shown in Fig 2. Main conclusions obtained are: 1) The flexural to shear capacity ratio M/Vd clearly affected the final crack pattern. However, the failure mode for all specimens was governed by shear deformation of the masonry panels. 2) First diagonal cracking was caused by comparable lateral forces regardless of the type of coupling. The Mexican code equation showed excellent agreement with cracking loads. 3) Stable hysteretic loops were observed up to 0.6% drift ratios; for larger deformations severe degradation was noted. 4) Although the M/Vd ratio affected the initial stiffness of the specimens, stiffness decay was similar for all structures and followed a parabolic curve.

Effect of the Type of Horizontal Steel on Wall Behavior

Two structures with different types of horizontal reinforcement were tested to assess their effect on wall strength, energy dissipation and deformation capacity

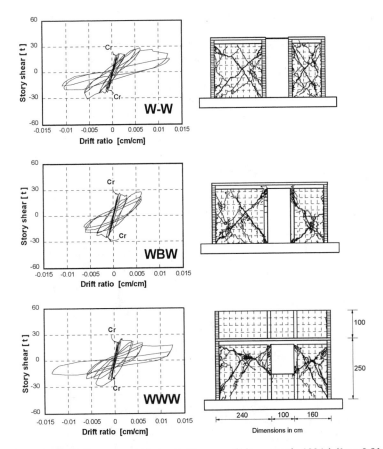

Figure 2. Crack Patterns and Hysteretic Response (Alcocer et al., 1994a) (1 t = 9.81 kN)

(Alcocer et al., 1994b; Díaz & Vázquez-del-Mercado, 1995). Specimens were similar to WBW, but horizontally reinforced within the running bond joints. Models were reinforced to maintain the predicted cracking load. Specimen WBW-E was reinforced with a ladder-type prefabricated reinforcement which consisted of a set of two longitudinal high-strength cold-drawn smooth wires separated 9 cm by welded cross-wires spaced at 40 cm. Nominal wire yield stress was 491 MPa; standard 10-gauge wire was used. The ladder type steel was placed at every other joint so that the reinforcement ratio was p_h=0.102%. For specimen WBW-B, two high-strength deformed wires (4-mm diameter) were spaced at one every third joint (p_h= 0.091%).

Nominal wire yield stress was 589 MPa. For both structures, reinforcement was continuous along the wall and was anchored to the TC'S longitudinal steel through 180-deg hooks bent in-site after installation of horizontal steel within the joint. A similar loading sequence to that used in the previous phase was applied. Final crack patterns and hysteretic curves are shown in Fig. 3. From the analysis of results, main conclusions were withdrawn: 1) Although specimens failed in shear, the model with deformed wires exhibited larger flexural deformations. 2) The prefabricated reinforcement did not modify the behavior observed in the unreinforced specimen. The fragile failure observed at the junction of longitudinal wires and crosswires indicate the inadequacy of this type of reinforcement to be used to resist earthquakes. 3) The behavior of the 180-deg hooks used to anchor the horizontal reinforcement was satisfactory. 4) The strength of an horizontally reinforced wall can be estimated by adding the masonry cracking resistance and the yield force of horizontal wires multiplied by an efficiency factor. This factor, recommended to be equal to 0.7, considers the nonuniform strain distribution of reinforcement over the wall height.

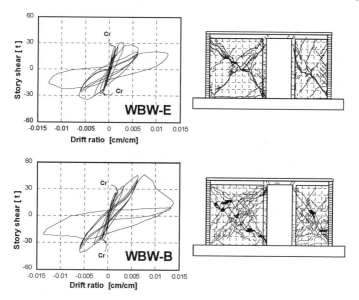

Figure 3.   Crack Patterns and Hysteretic Response (Díaz & Vázquez-del-Mercado, 1995) (1 t = 9.81 kN)

5) To avoid crushing of the masonry diagonal strut, the ratio of horizontal steel should be less than 0.3 $f'_m / f_{y,h}$, where f'_m is the masonry prism strength and $f_{y,h}$ is the horizontal steel yield stress.

Three-Dimensional Two-Story Confined Masonry Structure

Aimed at studying the effect of 3-D construction (orthogonal walls and rigid floor systems) and at comparing the behavior observed in 2D one-story specimens, a full-scale 3D two-story confined masonry structure was tested under alternated lateral loads (Alcocer et al., 1993; Sanchez, 1995). The structure is shown in Fig. 4. To reduce the torsional displacements, orthogonal masonry walls were built. The specimen was designed and detailed according to present code requirements for masonry construction in Mexico City; the structure was designed to fail in shear at the ground floor. Longitudinal and transverse steel in TC's and BB'S was the same as that used in specimen WBW. Analogously to WBW, five hoops were closely spaced (at every 7 cm) at TC'S ends to increase the TC shear strength, to reduce damage due to the penetration of masonry inclined cracking into the TC, and to achieve a more stable wall behavior. The structure was tested to failure by applying cycles of quasi-static alternated lateral loads, which were distributed over the height following an inverted triangular shape. Two cycles were applied at same displacement level. A vertical stress equal to 0.49 MPa at ground level was maintained constant throughout the test. The final crack pattern, and the base shear - drift ratio of level 1 and the second story shear - second story drift curves are shown in Fig. 5. Outstanding features of the behavior of 3D were: 1) Damage was concentrated in the first floor and was characterized by inclined (shear and diagonal tension) cracks. Second story walls were practically undamaged. 2) For drift ratio curves for level 1, hysteretic loops were stable and symmetrical. Good energy dissipation and good deformation capacity were exhibited by 3D. For the second story drift ratio, curves were almost linear-elastic. 3) Measured strength was 1.95 times that calculated using design code equations. 4) Similarly to WBW, stiffness decayed with displacement and followed a parabolic cove. 5) To reduce the likelihood of crack penetration into TC's, which triggers the strength deterioration, it ie recommended that interstory drift ratio be limited to 0.35%. 6) Behavior of wall systems was comparable to that observed in 2D one-story masonry wall arrangements.

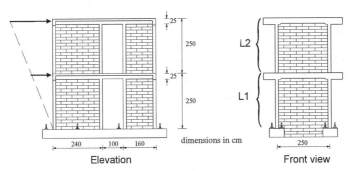

Figure 4. Geometry of Specimen 3D (Alcocer et al., 1993)

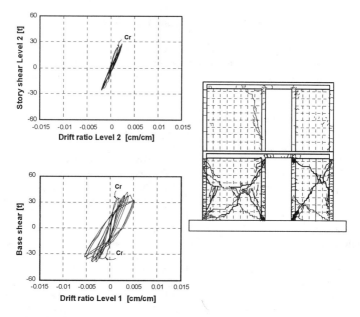

Figure 5.   Crack Pattern (North Side) and Hysteretic Response of Specimen 3D
(Alcocer et al., 1993) (1 t = 9.81 kN)

Repair and Strengthening of Specimen 3D

Specimen 3D was repaired and strengthened, and was retested to failure (3D-R).
Rehabilitation works were only performed in walls in the loading direction. Nothing
was done to slabs, bond-beams, and orthogonal walls. Cracked and crushed concrete
at the interior TC's ends was replaced with new concrete. Inclined masonry cracks
were cleaned with water jet to remove the dust and crushed particles. Afterwards,
cracks were filled with a cement mortar and brick pieces. A welded wire mesh (15
cm x 15 cm, 10-gauge wire) was placed on one wall face and was covered with a 2.5-
cm thick cement mortar. The mesh was anchored to the wall by means of 5 cm long
wood nails; approximately, 4 cm were driven into the wall by hand and the nail heads
were bent to fix the mesh wires. Old-fashioned metal bottle caps were used between
the wall surface and the mesh to ease placement of mortar behind the mesh and to
improve the mortar-masonry bond. Nail density (number of nails per area) was
varied: 9 nails/m^2 was used in the North side, while 6 nails/m^2 were placed in the
South face. The mesh was fixed to TC's and BB's with concrete nails. The mesh was
not anchored to the foundation. The mesh terminated at the TC's edges (i.e. the mesh
did not surround the TC's). Prior to placement of mortar, wall surfaces were

saturated. Mortar was proportioned by volume with a cement:sand ratio of 1:4. Mortar was placed manually using masonry trowels. Final crack patterns in the North side, and the base shear - drift ratio of level 1 and the second story shear - second story drift ratio curves are shown in Fig. 6. From test observations and data analysis, the following conclusions and recommendations were developed: 1) Inclined cracking was more uniformly distributed over the wall surface in the first floor. Walls with 9 nails/m^2 exhibited a more uniform distribution of damage compared with those with 6 nails/m^2. 2) Although the failure mode of 3D-R was also controlled by shear (as in 3D), the mesh acted as a tie to control crack opening. 3) Hysteresis loops were stable with no pinching and minor strength decay up to roof drifts to 0.75% (first floor drift ratios of 0.94%). 4) Strength of 3D-R was 1.64 times, on the average, that measured in 3D, and was reached at a drift 2.5 times that associated to the strength of 3D. 5) The efficiency of horizontal wires of the mesh varied with the displacement and wall aspect ratio. Average values were 0.64 and 0.44 for roof drift ratios of 0.75% and 0.60%, respectively. 7) Initial stiffness of 3D-R was 2/3 that measured initially for 3D and 6.7 times the final rigidity of 3D. 8) Displacement ductility ratios were 2.5 and 4.4 for 3D and 3D-R, respectively. 9) Maximum allowable drift ratios for structures rehabilitated with mortar reinforced with welded wire meshes is 0.6%. 10) For design purposes, the strength of a wall rehabilitated with a welded wire mesh and mortar can be estimated as the sum of the masonry cracking load, the dowel strength of TC longitudinal steel, and the yield force of the mesh multiplied by an efficiency factor equal to 0.5. In this factor, the mortar participation to strength is included.

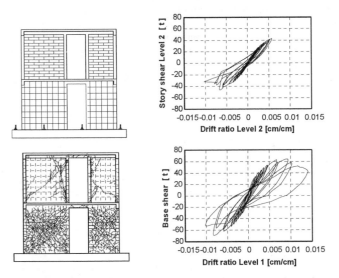

Figure 6.   Crack Pattern (North Side) and Hysteretic Response of Specimen 3D-R (Ruiz, 1995) (1 t = 9.81 kN)

<u>Analytical Study of Confined Masonry Structures</u>

Aimed at extending the results obtained in the experiments, an analytical assessment of the response of confined masonry structures subjected to different seismic environments was undertaken (Flores, 1995). Simplified macroscopic hysteretic models were developed and calibrated with data from tests performed at CENAPRED. Stiffnesses for the loading and descending branches follow a rule which depends on the maximum drift ratio reached in previous cycles. This rule is of the form: $R = R_{initial} [a(\gamma_{max})^4 + b\gamma_{max} + 1]$ where $R_{initial}$ is the initial stiffness and γ_{max} is the maximum drift ratio reached in previous cycles. The hysteretic model is limited by the response envelope.

In Fig. 7, calculated and measured hysteresis loops for two wall specimens are shown. Step-by-step nonlinear analyses were performed on one-story and three-story typical masonry low-cost housing buildings. Earthquake records with distinctly different acceleration maxima, duration and frequency content were used as input. The long-period accelerogram used was the SCT record of the 1985 Mexico earthquake (M8.1). High-frequency records were those obtained at Diana station in Acapulco (M 6.9 from April 1991), at Llo-lleo (1985 Chile earthquake M7.8) and at the Kobe Meteorological Agency (1995 Hyogo earthquake M7.2). A synthetic accelerogram was calculated for a M8.2 earthquake taking the April 1991 Acapulco record as a Green function. Other variables included in the analyses were the masonry strength and stiffness, both in compression and in shear, and the existence of horizontal reinforcement in walls. Analyzed structures were designed according to current design spectra in Mexico. Main results are: 1) No damage was observed in the models subjected to the SCT record. This coincides with the satisfactory behavior of masonry buildings during the 1985 earthquakes. 2) Lateral force distribution over the height of the three- story building subjected to the SCT record was nearly uniform.

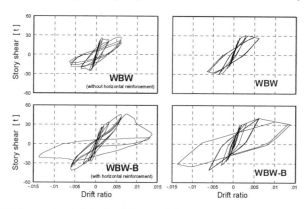

Figure 7. Measured and Calculated Hysteresis Curves (Flores, 1995) (1 t = 9.81 kN)

This is consistent with the measured response of an instrumented masonry building in Mexico City. This distribution contrasts with the inverted triangular force distribution specified in the code. However, when subjected to the high-frequency records, force distribution was inverted triangular. 3) Damage in the three story buildings was concentrated in the ground floor, as it has been observed in post-earthquake field trips. 4) For high-frequency earthquakes, horizontal reinforcement in ground floor walls yielded higher strengths and deformation capacities thus allowing the buildings to survive. 5) It is recommended that maximum drift ratio for confined masonry walls with no horizontal reinforcement and with horizontal reinforcement be 0.3% and 0.6%, respectively.

Final Comments

The ongoing research project on low-cost housing in Mexico, has provided valuable information to improve the analysis, design and construction of confined masonry structures. An assessment of the contribution of horizontal steel bars and welded wire meshes to wall strength and deformation capacity is being carried out (Aguilar, 1995; Pineda, 1995). Experimental studies using other materials are planned. The analytical evaluation of building behavior will be continue.

References

Aguilar, G., (1995). "Efecto del refuerzo horizontal en el comportamiento de muros de mampostería ante cargas laterales" (in Spanish), BSc thesis (in process), School of Engineering, UNAM.

Alcocer, S.M., Sánchez, T.A., and Meli, R. (1993). "Comportamiento ante cargas laterales de una estructura tridimensional de dos niveles a escala natural construida con mampostería confinada" (in Spanish), *Proceedings X Mexican National Congress on Earthquake Engineering*, Puerto Vallarta, October, 416-423.

Alcocer, S.M., et al. (1994a). "Comportamiento ante cargas laterales de sistemas de muros de mampostería confinada con diferentes grados de acoplamiento a flexión" (in Spanish),*Cuadernos de Investigación, No. 17*, CENAPRED, July, 53-76.

Alcocer, S.M., et al. (1994b). "Comportamiento ante cargas laterales de sistemas de muros de mampostería confinada con distintos tipos de refuerzo horizontal" (in Spanish), *Cuadernos de Investigación, No. 17*, CENAPRED, July, 77-94.

Alcocer, S.M., and Meli, R. (1995). "Test program on the seismic behavior of confined masonry structures," *The Masonry Society Journal*, 13 (2), February, 68-76.

Díaz, R.R., and Vázquez-del-Mercado, R. (1995). "Comportamiento de muros de mampostería confinada reforzados horizontalmente" (in Spanish), BSc thesis, School of Engineering, UNAM.

Flores, L. E. (1995). "Estudio analítico de estructuras de mampostería confinada" (in Spanish), BSc thesis, School of Engineering, UNAM.

Meli, R. (1994). "Structural design of masonry buildings: The Mexican practice," *Masonry in the Americas*, American Concrete Institute, SP-147, 239-262.

Meli R., et al. (1994). "Características estructurales de la vivienda de interés social en México" (in Spanish), *Cuadernos de Investigación, No. 17*, CENAPRED, July, 25-52.

Pineda, J. (1995). "Comportamiento ante cargas laterales de muros de mampostería confinada reforzados con mallas electrosoldadas" (in Spanish), MSc thesis (in process), School of Engineering, UNAM.

Ruiz, J. (1995). "Reparación y refuerzo de una estructura tridimensional de mampostería confinada de dos niveles a escala natural" (in Spanish), MSc thesis, School of Engineering, UNAM.

Sanchez, T.A. (1995). "Comportamiento de estructuras de mampostería confinada sujetas a cargas laterales" (in Spanish), MSc thesis, School of Engineering, UNAM.

Ductile Masonry Construction In California

Hanns U. Baumann, S.E., ASCE MB (1)

Abstract

The recent design and construction of an eight-story building of ductile masonry construction has given the author a "Real World" lesson in ductility, constructability and productivity of masonry construction on a San Francisco, California site.

History

In the Detroit area, over the last twenty years Architect-Engineer-Developer James Braden has been involved in the design and construction of over twenty mid-rise apartment buildings[3].

A very efficient system has evolved which uses hollow precast concrete plank (P/C plank) resting on load bearing masonry shearwalls (LBMS). Key to the efficiency of this system is the elimination of an expensive on-site tower crane.

Most masonry related materials are lifted daily via a low cost temporary exterior mounted elevator. This does not include structural steel lintel beams and pallets of block which are lifted once a week with the P/C plank by a mobile truck crane.

Thus, typically on the jobsite during the first four days, masons and hod carriers are constructing and grouting the LBMS. On Friday the precast hollow plank is delivered and set by truck crane. The structural concrete topping is poured immediately after the scaffolding is moved up to the next level onto the P/C plank at the floor above. Thus, on Monday morning the laying of the Concrete Masonry Units (CMU) to support the next floor above can be started.

Fillmore Marketplace Project

In 1992, Housing Associates, Inc. (HAI) of San Jose, California, who were planning a 120 unit subsidized housing project, asked Braden/Drosihn, Inc. (BDI) of LaSalle , Michigan to investigate the feasibility of using Braden's Detroit System for HAI's project at the site in Fillmore Center, San Francisco. The project had some unusual features in that the units per acre density was extremely high, (80 units per acre) and also that 55% of the units have their own entrances (very important to today's homebuyers). In addition,

---

(1) President, Baumann Engineering,
567 San Nicolas Dr., Ste. 104, Newport Beach CA 92660

unlike so many of the traditional masonry designs which have a minimum of exterior wall offset planes, this aesthetically pleasing architectural design incorporated a number of offset planes, including the traditional San Francisco bay windows.

Structural System Competition

Contemporary Californian structural systems were competitively bid against the Detroit System. Not too Braden's surprise, his Detroit "hand-built" System withstood the competition of poured concrete "tower crane systems" such as flying forms and tunnel forms.

Despite the media hype of "high-tech" systems, Braden, by building his own buildings in Michigan and Illinois, had discovered what the vast majority of design professionals have not learned. Braden had discovered long ago that the simplicity of LBMS supporting low cost P/C plank was very hard to beat in a very competitive market.[2]

In essence, present day CMU masonry is a variation of precast concrete construction which uses CMU as "stay-in-place forms." Modern masonry industry techniques such as pre-mixed mortar and new admixtures to aid rapid grout placement by pumping, have greatly improved the productivity of modern day masons.

Conversion of LBMS to DLBMS

However, to exploit the competitive advantages of the "Detroit System" in California, the Detroit System first had to be modified. Baumann Engineering had to convert the LBMS to DLBMS (Ductile Load Bearing Masonry Shearwalls) in order to resist the very large anticipated ground motion at the San Francisco site.

Site specific response spectra, provided by GeoSpectra, Incorporated of Richmond, California, predicted a ground acceleration of .55g. If the CMU shearwalls could be designed and constructed to give a reliable ductile response, the design base shear coefficient could be reduced to $1/3$ that of a (non-ductile) elastic response (Pauly, et al, 1992).[5]

The Detroit System

The Detroit System in plan (Fig. 1) has an exterior "LBMS tube" surrounding an interior "LBMS tube" which encloses the elevators and stairways. The P/C plank spans 33' from the "interior tube" to the "exterior tube". The design by BDI exploited the strength and durability of CMU walls as well as its beauty (Fig. 2).

Collector Dragstruts and Diaphragm Chords

It was recognized during design that, if the exterior and interior walls could be designed as DLBMS, they would offer a very efficient lateral force resisting system only if the seismic forces could be economically collected and distributed by the P/C plank floor diaphragm to the interior and exterior DLBMS.

However, as was demonstrated during the Northridge earthquake,[4] designing dragstrut collectors into structures with P/C plank floor diaphragms requires very special attention to structural detailing.

Since this was the first DLBMS building to be built in the City of San Francisco several valuable lessons were learned and innovations

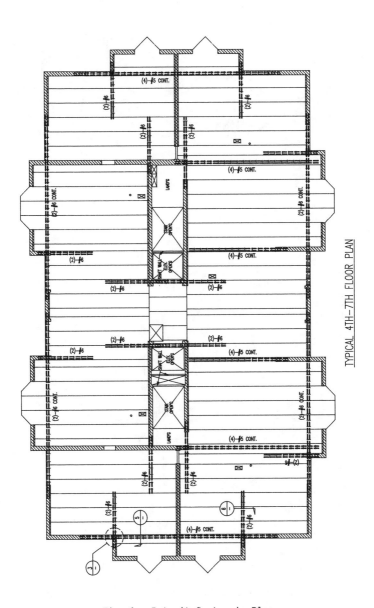

Fig. 1 - Detroit System in Plan

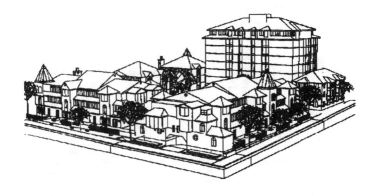

Fig. 2 - Fillmore Marketplace, Archt. Rendering

conceived during the design and construction of the structural
system while it was being converted from the lower (Detroit) seismic
zone into the higher (San Francisco) seismic zones.

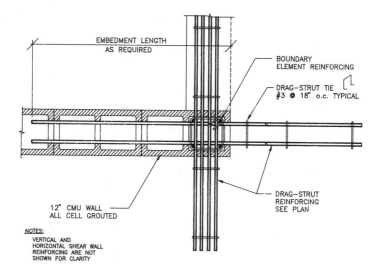

Fig. 3 - Dragstrut Collectors

Structural Detailing of Ductile Masonry

Since this was also the first DMS building to be built by the masonry contractor, he was also having his first experience at trying to place much larger amounts of reinforcement steel into 12" wide, (f'c = 3,750psi) CMU.

Besides confinement reinforcement requirements, new to the mason, there was the ever present need to pass reinforcement on all three axes at many locations where north/south dragstruts intersected east/west dragstruts and passed through boundary elements (Fig. 3).

Designing reliable dragstrut collectors into the 8" P/C plank, with just 2 1/2" thick structural concrete (f'c = 4,000psi) topping, was made more difficult when the engineers learned that the preliminary budget had not included the cost of forming cast-in-place dragstruts between separated P/C planks (Fig. 4).

The solution was to devise an innovative dragstrut design as shown in Fig. 5. Drag struts and diaphragm chords in the transverse direction to the span of the P/C plank were constructed in a similar way (Fig. 6).

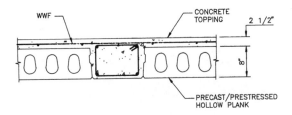

Fig. 4 - Original Dragstrut Design

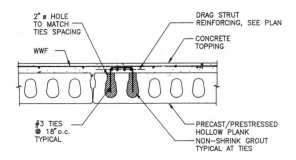

Fig. 5 - Dragstrut Redesign, Longitudinal

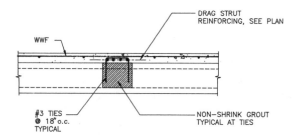

Fig. 6 - Drag Strut Redesign, Transverse

Ladder-Type Reinforcement Innovation

Also during the design of this mid-rise building, it became apparent to the Baumann Engineering staff that if a new type of proprietary ladder reinforcement[1] was laid in the grout core, as opposed to mortar joints, several important improvements could be achieved (Fig. 7)

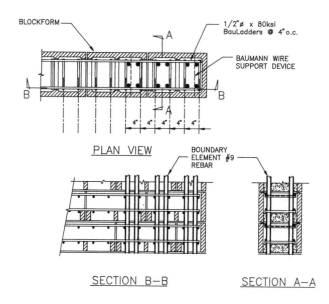

PLAN VIEW

SECTION B-B   SECTION A-A

Fig. 7 - BlockForm

First, because ladder-type joint reinforcement is limited by code to a maximum diameter of 1/4" the ladder-type reinforcement in the grout core could use much larger (1/2"ϕ) and much higher strength (fy = 80,000psi) wire than joint reinforcement.

Thus, because larger diameter and stronger wire can be used, the shear capacity of the DMS can be increased. By using f'c = 4,000psi grout, the wall can now be designed as a Ductile Load Bearing Concrete Shearwall (DLBCS) instead of the lower strength DLBMS.

A ductile response of a concrete shearwall would allow the use of higher seismic reduction factors in the design (Pauley, et al, 1992)[5].

Also, the proprietary Ladder Type Horizontal Shear Reinforcement (LTHSR) fabricated from high strength steel can also act as transverse confinement reinforcement at the same time it provides shear reinforcement.

Further, the closely spaced LTHSR serving as confinement reinforcement in the boundary elements will allow a very significant reduction in lap splice length of the boundary reinforcement. This will allow the masons to lay the CMU much faster over the much shorter dowels.

And finally, the LTHSR will significantly reduce congestion where reinforcement steel must pass on all three axes.

CMU as Ductile Precast Concrete Shearwall Form

As a result of these newly learned lessons a new type of proprietary high strength (f'c = 4,000) modified bond beam concrete block is proposed as a "precast concrete stay-in-place form" in Ductile Precast Concrete Shearwalls (DP/CS).

Testing of this DP/CS is underway at the University of Southern California. The horizontal shear/confinement reinforcement is closely spaced (4" o/c) proprietary LTHSR manufactured of 1/2"ϕ x 80ksi wire. The results of these tests will be available by mid-1996.

Conclusions

Assuming successful test results, the use in seismic zones of the Detroit System with this newly proven reinforcement and CMU form innovations gives promise of a new low cost way to construct much taller DLBP/CS buildings.

As demonstrated by the lessons learned at the Fillmore Marketplace project, design engineers should consider the ductility, constructability and productivity advantages of DLBP/CS systems and their beauty when selecting structural systems for their mid-rise and high-rise buildings.

Appendix

[1] Baumann, Hanns U. (1993) "Optimum Inelastic Deformability of Tall Precast Concrete Buildings in Seismic Regions," *Proceedings, 62nd Annual Convention, Structural Engineers Association of California* Scottsdale AZ, pp.55-63

[2] Braden, J. Office Records of Building Costs, (1961 to Present)

[3] Braden, J. Presentations to NCMA MIA 1972-1980

[4] Earthquake Spectra, (April, 1995) Earthquake Engineering Research Institute, University of California, Berkeley

[5] Pauly, T., Priestly, M.J.N., (1992) "Seismic Design of Reinforced Concrete and Masonry Buildings"

Estimating Transverse Strength of Masonry Infills

Richard Angel[1] (ASCE Associate Member) and
Joseph Uzarski[2] (ASCE Member)

Abstract

A simple calculation method is given for estimating out-of-plane strength of unreinforced masonry infill panels. Development of the method is reviewed starting with a description of a series of large-scale experiments, a brief presentation of sample test results and formulation of an analytical model.

A series of eight masonry infill panels were made of either concrete or clay masonry units. Test panels were first subjected to in-plane horizontal forces to crack them, and then loaded with transverse pressures. The purpose of the research was to investigate how transverse strength is reduced when a panel is cracked with in-plane shears. Test results showed a substantial out-of-plane strength even for cracked panels because of arching effects. Axial thrusts developed in a panel were reacted against the surrounding reinforced concrete frame. An analytical model for arching behavior in plates was developed based on these observations. The model then served as the foundation for an evaluation procedure that is highlighted with this paper.

The paper emphasizes how the suggested evaluation procedure was developed and verified. An illustrative example on its use is provided.

Experimental Program

The experimental program consisted of testing unreinforced clay and concrete masonry infills that were placed within a concrete frame as shown in Figure 1. The concrete frame was designed according to the 1989 ACI-318 requirements so that it was both ductile and tough when subjected to load reversals.

Static, in-plane lateral forces were applied until cracking in the masonry infill occurred. To assure that a fully cracked condition was reached, cycles of reversed shears were continued until lateral deflections were twice that observed at first

---

[1] Research Analyst, The Center for Naval Analyses, 4401 Ford Avenue, Alexandria VA 22302.
[2] Professional Engineer, SOHA Engineers, San Francisco, CA.

cracking. Following the in-plane loading, panels were subjected to pressures applied across their plane using an air-bag. Pressures were increased monotonically until ultimate loads were reached.

A total of eight infill specimens were tested. Parameters of the study (described by Abrams et al., 1993) were the type of unit, the *h/t* ratio for the infill and the mortar type. Both clay brick and concrete block infills were tested. The clay units were a low strength reclaimed brick (Chicago common) laid in a single wythe running bond. Concrete units were standard 10 cm (4-inch) or 15 cm (6-inch) blocks laid in a single wythe running bond. A typical Type N mortar (a 1:1:6 mix of Portland Cement, lime and sand) was used as the control mortar. Another mortar comprised only of lime and sand (1:3) represented mortars used at the earlier part of the century.

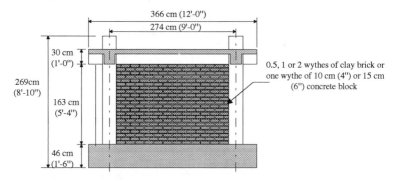

Figure 1 Dimensions of Test Specimen

Physical and mechanical properties of the masonry infills are presented along with the corresponding out-of-plane test results in Table 1. Test results for five specimens are presented in Figure 2. Values recorded for the out-of-plane tests indicate the strength of the infill panel; or the maximum measured pressure that was applied to the infill. The maximum pressure applied to the panels was recorded for cases where the capacity of the infill exceeded the capacity of the loading rig. Maximum lateral deformations in the infill were limited to 3% drift which was considered to be an upper bound for any loading condition. Drift (Δ/h) was defined as the lateral deformation at the center of the panel (Δ) over the height of the panel (h).

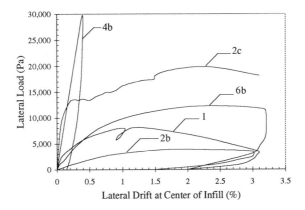

Figure 2 Specimen Test Results

Table 1 Specimen Out-of-Plane Test Results

Test Number	Infill Type	Infill h/t	Mortar Type	f'_m (kPa)	In-Plane Test Max. Values		Out-of-Plane Test Lateral Pressure		
					Δ_{max}/h	Shear Stress (kPa)	Unrepaired (Pa)	Repaired (Pa)	I.P.** (Pa)
1	brick	34	S	11,510			8,190		
2a	brick	34	N	10,860	0.344	1,870			
2b							4,020		
2c								19,970	
3a	brick	34	lime	10,140	0.218	1,300			
3b							5,990		
3c								20,930	
4a	block	18	N	22,900	0.094	930			
4b							29,800*		
5a	block	11	lime	21,460	0.062	1,350			
5b							32,240*		
5d									32,330*
6a	brick	17	lime	4,590	0.250	660			
6b							12,410		
6c								30,850*	
6d									32,510*
7a	brick	17	N	11,010	0.250	1,170			
7b							30,990*		
8a	brick	9	lime	3,500	0.390	490			
8b							32,090*		

* - Maximum pressure values recorded for the infill, not the strength of the infill.

** - Max. pressure values recorded for the infill, with in-plane forces corresponding to Δcr.

Analytical Model

An analytical model was developed to predict out-of-plane behavior and strength of existing cracked panels. This model is developed in detail by Angel et al.

(1994), and here is briefly introduced. The model consisted of a strip of unit width of panel that spans between two supports fully restrained against translation and rotation and is uniformly loaded. As the load (w) increases, thrust forces (T) develop due to confinement by the frame. The free body diagram for the lateral load resisting mechanism is presented in Figure 3.

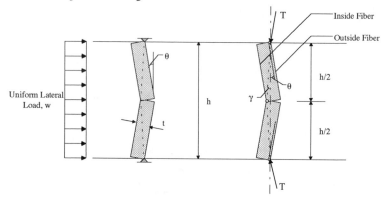

Figure 3 Idealized Loading and Behavior of Unit Strip of Infill Panel

The uniform lateral load is related to the thrust force by summing horizontal forces that act at mid-span (Figure 3). By equating the thrust force to the internal compressive force (for small angles only), Equation 1 is obtained relating the load, w, to the maximum compressive stress at the support (f_b).

$$w = \frac{4k_1\left(\dfrac{b}{t}\right)f_b\dfrac{\cos\gamma}{\cos\theta}\sin\gamma}{\left(\dfrac{h}{t}\right)}$$ Equation 1

where: all angles are defined in Figure 3

b is the bearing width of the panel on the frame

f_b is the maximum compressive stress at the support

k_1 represents the ratio of peak stress to average stress in the masonry.

Based on experimental results (Angel et al., 1994), the model was modified to account for existing in-plane panel damage. The out-of-plane strength varied with the amount of existing in-plane damage and with the slenderness ratio of the panel. The strength reduction caused by the in-plane damage was not linearly related to the slenderness ratio. Slender infills were greatly affected by in-plane damage. The strength for these slender panels can be reduced by a factor of two. Experimental results support this observation. According to this model, the out-of-plane strength of infills with a lower slenderness ratio are not affected as much by in-plane damage.

Evaluation Procedure

Based on the analytical model, a simple procedure has been developed for the out-of-plane strength evaluation of panels in existing buildings. The procedure considers the amount of damage, the flexibility of the confining frame, and the properties of the panel. Assumptions used to simplify the model as well as the required steps for evaluation of the panels are outlined in this paper. The procedure and the derivation of its primary parameters are presented in detail by Angel et al. (1994).

Obtaining the out-of-plane strength of a panel using Equation 1 requires the evaluation of cumbersome expressions for the parameters in the equation. This is an iterative process that becomes tedious and time consuming; thus, it must be simplified to be easily used by the practicing engineer. Developing a user friendly evaluation procedure was achieved by assuming a value of 0.5 for k_1, and a value of 1 for the ratio $\cos\gamma/\cos\theta$ (true for small angles only). Further, Equation 1 is simplified by grouping b/t, f_b', and $sin(\gamma)$, and creating a new dimensionless parameter λ (Equation 2). Substituting λ into Equation 1 produces the expression in Equation 3. Parameters b/t, f_b', and $sin(\gamma)$, depend on the crushing strain of the masonry, and on the slenderness ratio of the panel. Evaluating λ for a constant crushing strain for the masonry of 0.004 and for a slenderness ratio ranging between 5 to 40, produces the values presented in Table 2.

$$\lambda = \left[\left(\frac{f_b}{f_m'} \right) \left(\frac{b}{t} \right) \sin\gamma \right] \qquad \text{Equation 2}$$

$$w = \frac{2 f_m'}{\left(\dfrac{h}{t} \right)} \lambda \qquad \text{Equation 3}$$

Table 2 Parameter Approximation

h/t	λ	R_1 for ratio of Δ/Δ_{cr}	
		$\Delta/\Delta cr = 1$	$\Delta/\Delta cr = 2$
5	0.129	0.997	0.994
10	0.060	0.946	0.894
15	0.034	0.888	0.789
20	0.021	0.829	0.688
25	0.013	0.776	0.602
30	0.008	0.735	0.540
35	0.005	0.716	0.512
40	0.003	0.727	0.528

According to the NEHRP Handbook for the Seismic Evaluation of Existing Buildings (FEMA, 1992), a typical ten story building was analyzed (Angel et al., 1994) under the worst possible conditions, and the maximum lateral acceleration that an infill panel would experience is approximately equal to 2.0 g's (equivalent to a range of X to XII in the Modified Mercalli Intensity Scale). Based on the developed analytical model, and considering the allowable masonry stresses and related factors of safety (for masonry with compressive strength larger than 3,450 kPa), panels with slenderness ratios smaller than 9 are considered safe (capacity exceeds 2.0 g's) and no further analysis is required. Otherwise the steps outlined below should be followed for the evaluation of out-of-plane strength of masonry panels.

1.) In-plane damage assessment

There are two methods for quantifying the amount of damage for cracked panels: 1) visual inspection which is described in detail in this paper, and 2) analysis of the maximum deflection experienced by the structure in terms of the displacement observed at cracking of the infill panel which is explained in detail by Angel et al. (1994).

One method used to evaluate the damage of a panel is visual inspection. Based on experimental results, visual inspection of the panel can classify the amount of existing panel damage into three different ranges as illustrated in Figure 4. The three different cracking stages were obtained from experimental results and normalized in terms of the lateral deflection required for cracking of the infill.

Out-of-plane strength reduction factors for a known amount of in-plane damage (R_j) and panel slenderness ratios have been tabulated in Table 2. R_j was derived by modifying the analytical model to consider the existing in-plane damage in the panels, and estimating a best fit curve to duplicate the known experimental results.

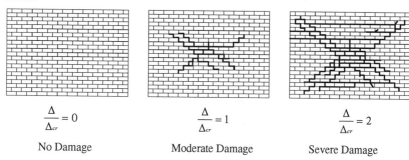

$$\frac{\Delta}{\Delta_{cr}} = 0 \qquad\qquad \frac{\Delta}{\Delta_{cr}} = 1 \qquad\qquad \frac{\Delta}{\Delta_{cr}} = 2$$

No Damage   Moderate Damage   Severe Damage

Figure 4 Damage Classifications

2.) Flexibility of confining frame

Infill panels with all sides continuous (neighboring panel in every direction) may assume to have fully restrained boundary conditions $(R_2 = 1)$. For infill panels with at least one side not continuous (neighboring panel missing on any panel direction) a reduction factor for the out-of-plane strength is applied (R_2). R_2 accounts for the decrease in out-of-plane strength of the panel that would be observed due to the flexibility of the boundaries of the panel. To fully obtain the maximum out-of-plane strength of a panel, the confining frame must be relatively rigid (EI>25,800 kN-m for all columns and beams); otherwise, a reduction factor less than 1 is applied. Evaluation of the stiffness of the smallest frame member (column or beam) on the non-continuous panel side should be performed, and results are to be used in conjunction with Equation 4 and Equation 5.

$R_2 = 0.175+3.2X10^5 EI$ for 5,700 kN-m < EI < 25,800 kN-m   Equation 4

$R_2 = 1$ for EI > 25,800 kN-m   Equation 5

3.) Out-of-plane strength of the panel

The out-of-plane strength of previously cracked, or uncracked infill panels within confining frames at any location of a structure may be evaluated using Equation 6. Values of λ for a range of slenderness ratios are given in Table 2.

$$w = \frac{2f_m}{\left(\dfrac{h}{t}\right)} R_1 R_2 \lambda$$ Equation 6

Correlation with Experimental Results

Results obtained using the analytical model gave out-of-plane strength values similar to the strengths obtained experimentally. Comparison between out-of-plane strengths as calculated by the evaluation procedure and the experimental results are illustrated in Figure 5. Measured strengths have been normalized to a compressive strength of 6,890 kPa for comparison with strength curves. Notice that some panels were not tested to failure since their strength exceeded the capacity of the loading rig. For these panels, the maximum applied pressure was recorded rather than their strength. Predicted strengths for panels with no existing in-plane damage are illustrated with $\Delta/\Delta_{cr} = 0$, while strengths for panels damaged in the in-plane direction corresponding to a maximum in-plane drift of twice the required for cracking of the panel are illustrated with $\Delta/\Delta_{cr} = 2$. The out-of-plane strength reduction obtained from the applied in-plane damage along with all the experimental results are presented in Figure 5. As shown in the figure, the strength of the panels varies with the slenderness ratio of the panel and with the amount of in-plane damage.

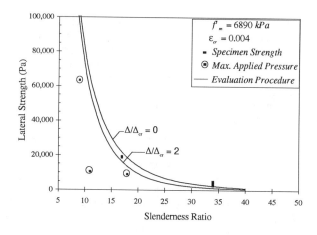

Figure 5 Predicted Behavior

Example

A reinforced concrete building was subjected to lateral accelerations produced by a nearby earthquake, and damage to some of its elements occurred. A view of the building under observation is presented in Figure 6. Although the lateral accelerations did not cause serious damage to the frame elements, damage to a number of the masonry infills occurred. Because of in-plane damage to some of the infills, questions regarding out-of-plane strength of the panels have been raised.

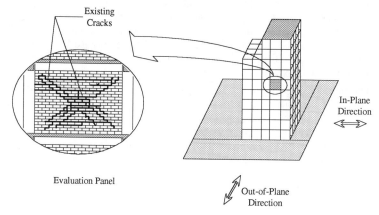

Figure 6 Example Problem

The typical dimensions of the panels in the building were 366 cm x 610 cm x 19 cm (height, width, thickness). The panels consisted of older brick masonry built in double-wythe with a medium strength Type N mortar. In addition to the physical properties of the panels, evaluation of the corresponding mechanical properties are required. For this purpose, a series of masonry prism tests, and in-place shear tests were carried out. Results for these masonry properties are presented in Table 3.

The building had not been instrumented during the earthquake, therefore the damaged panels must be evaluated by visual inspection. Existing panel damage was considered to be severe as illustrated in Figure 6. The building experienced a maximum lateral drift of approximately twice that required for cracking of the panel ($\Delta/\Delta_{cr}=2$). This factor of two becomes important for the estimation of the reduction in the out-of-plane strength of the panel caused by in-plane damage (R_1). The value of R_1 from Table 2 ($\Delta/\Delta_{cr}=2$) becomes 0.688 for a slenderness ratio of 20.

Table 3 Frame-Infill Properties

Frame		Infill	
Physical Properties	Mechanical Properties	Physical Properties	Mechanical Properties
I_c=575.4E3(cm^4)	E_c=24.8E6 (kPa)	t=18.7(cm)	f'_m=9,650(kPa)
I_b=650.4E3(cm^4)		h=365.8(cm)	E_m=5.17E6(kPa)
h'=426.7(cm)		L=609.6(cm)	f_a=276(kPa)
L'=670.6(cm)		(h/t)=20	f_v=1,379(kPa)

The frame being evaluated is surrounded by continuous panels in all directions; thus factor R_2 becomes 1.

The process used to evaluate the out-of-plane strength of a cracked or an uncracked panel is similar with the only difference being that the reduction factor for in-plane damage on the panel (R_1) becomes unity for uncracked panels. The procedure consists of substituting the required information that has been obtained for the panel into Equation 7. Notice that Equation 7 includes a 1/4 factor in the numerator to consider an allowable compressive strength in the masonry equaling $1/4\,f'_m$. Given the physical and mechanical properties of the panel (Table 3), the expression has been evaluated and the results are presented. The estimated out-of-plane strength of 5.1 kPa (1.37 g's) for an uncracked panel and 3.5 kPa (0.94 g's) for the cracked panel show that the panels should be repaired to reach a higher strength (larger than 2.0 g's).

$$w = \frac{2\frac{1}{4}f'_m}{\left(\frac{h}{t}\right)} R_1 R_2 \lambda \qquad\qquad \text{Equation 7}$$

$$w = \frac{2\frac{1}{4}f'_m}{\left(\frac{h}{t}\right)}R_2\lambda = \frac{2\frac{1}{4}(9{,}650kPa)}{(20)}(1)(0.021) = 5.1kPa(uncr.) \qquad \text{Equation 8}$$

$$w = \frac{2\frac{1}{4}f'_m}{\left(\frac{h}{t}\right)}R_1R_2\lambda = \frac{2\frac{1}{4}(9{,}650kPa)}{(20)}(0.688)(1)(0.021) = 3.5kPa(cracked) \quad \text{Equation 9}$$

A repair method developed by Prawel (1990) and tested by Angel et al. (1994) is briefly described in this example to show the effectiveness of its use. The panel rehabilitation technique to increase the out-of-plane strength consisted of a one-inch thick ferrocement coating parged to both faces of the infill panel. This operation was conducted after totally cleaning the wall obtaining a smooth surface. A single sheet wire mesh was placed on each face. The new thickness of the panel became approximately 24 cm (slenderness ratio equal to 15). Plaster compressive strength obtained from cylinder testing reached 19,300 kPa. Thus for the strength evaluation of the panel, the lesser compressive strength of the material was used (9,650 kPa for existing masonry). Strength of the repaired panel was estimated to be 10.9 kPa (2.34 g's) as given in Equation 10. The lateral strength of the repaired infill was satisfactory exceeding the lower limit of 2.0 g's, and the evaluation of the panels was successfully completed.

$$w = \frac{2\frac{1}{4}f'_m}{\left(\frac{h}{t}\right)}R_2\lambda = \frac{2\frac{1}{4}(9{,}650kPa)}{(15)}(1)(0.034) = 10.9kPa(repaired) \qquad \text{Equation 10}$$

Summary

An evaluation procedure developed to estimate the out-of-plane strength of uncracked and cracked panels is presented. According to the procedure, the strength of the panels vary with the compressive strength of the masonry, with the corresponding slenderness ratio, and with the amount of existing damage in the panel. Reduction factors are calculated to account for the amount of existing in-plane damage in the panel, and the flexibility of the confining frame. A rehabilitation or retrofit technique consisting of parging a ferrocement coating to one or both faces of the infill panel largely increases its out-of-plane strength.

Acknowledgments

The research presented in this paper was part of a study at the University of Illinois on seismic evaluation and repair of masonry infills. The project was part of the national coordinated program on *Repair and Rehabilitation Research for Seismic Resistance of Structures* funded by the National Science Foundation (Grant #BCS 90-156509).

References

[1] Abrams, D.P., R. Angel, and J. Uzarski, "Transverse Strength of Damaged URM Infills," *Proceedings of Sixth North American Masonry Conference,* Drexel University, Philadelphia, June 6-9, 1993, also printed in *The Masonry Society Journal,* Volume 12, Number 1, pp. 45-52, August 1993.

[2] Angel, R., D. Abrams, D. Shapiro, J. Uzarski, and M. Webster, "Behavior of Reinforced Concrete Frames with Masonry Infill Walls," *Civil Engineering Studies, Structural Research Report No. 589,* University of Illinois, Urbana-Champaign, March 1994.

[3] "NEHRP Handbook for the Seismic Evaluation of Existing Buildings," *Federal Emergency Management Agency FEMA-178,* Earthquake Hazards Reduction Series 47, June 1992.

[4] Prawel, S.P., "Summary of Research Tasks and Methods to be Used: Renovation/Retrofit/Repair of Brick Masonry," *Project No. 891004,* University at Buffalo, April 1990.

Lateral Strength of Brick Cladded Frames

by
Stephen P. Schneider[1], Associate Member ASCE, and Stephen J. Favieri[2]

Abstract

This paper describes research-in-progress on an experimental program to investigate the in-plane seismic behavior of steel frames with unreinforced masonry infills having large window openings. This experimental investigation is a necessary first step to fully understand stiffness, strut behavior, and strength of these composite systems with extensive cracking in the masonry infill. Preliminary results suggest that the contribution of the masonry to the strength and stiffness of the frame is significant up to drifts of approximately 1.3%. At drifts beyond this level, the strength and stiffness of the composite system are reduced to that of the bare steel frame.

Introduction

Steel frames with unreinforced clay-brick masonry infills having large window and door openings were a common construction type earlier this century. Several of these composite systems however, were damaged near Oakland California during the 1989 Loma Prieta earthquake (Freeman 1994). Although masonry panels were used primarily as an architectural component, many buildings in Oakland are considered structurally unsound due to perceived excessive cracking of the masonry infill. Further, there is a lack of simple analytical methods to reliably estimate the lateral-load resistance of steel frames with brick masonry infills. Consequently, the engineering community needs more information on the behavior of these composite systems, and guidelines to evaluate structural integrity.

This paper describes an experimental study to determine the lateral strength and stiffness of steel frames composite with unreinforced masonry infills having large window openings. Test specimens were designed to isolate and quantify the fundamental seismic behavior of frames with masonry infills. Strut behavior of the masonry infill, flexural stiffening due to the masonry piers, and stiffening of the beam-to-column joint have been identified as the most important influences to the seismic resistance of these structural

---

[1] Assistant Professor of Civil Engineering, University of Illinois @ Urbana-Champaign, 205 North Mathews Avenue, Urbana, IL   61801-2352
[2] Research Assistant, University of Illinois @ Urbana-Champaign

systems. This paper focuses on the strut action and flexural stiffening portion of the experimental program.

Little analytical or experimental information is available to help structural designers evaluate the seismic behavior of steel-masonry composite systems. Mander *et. al.* (1993, 1994) suggest that steel frames with brick masonry infills exhibit moderately ductile behavior when loaded in the plane of the wall. However, when subjected to large drifts the masonry infill became granulated and may be susceptible to out-of-plane forces. Strut models have been used (Freeman 1991, Hamburger 1993, Mander *et. al.*) to estimate the strength of steel frames with masonry infills, however some of these models have not been verified experimentally. Further, previous experimental research studied the performance of solid infill panels, while infills in most buildings have large window or door openings.

Experimental Program

An experimental program was necessary to investigate the in-plane lateral-deformation behavior of steel frames with masonry infills having large window openings. Of particular interest was the contribution of each component to the strength and stiffness of the composite system over a wide range of drift levels. A photograph of the test set-up is shown in Figure 1. The test specimen and the reaction system for this test program are shown in Figure 2, and a schematic of the test specimen only is illustrated in Figure 3. All tests were performed in the Newmark structural engineering lab at the University of Illinois.

Figure 1. Photograph of the Test Specimen

To investigate desired behavior, a specimen was chosen to represent an interior steel column with masonry piers on either side. A W8x31 was selected for the steel column of the test specimen. The W8x31 was attached, using full-penetration groove welds, to the

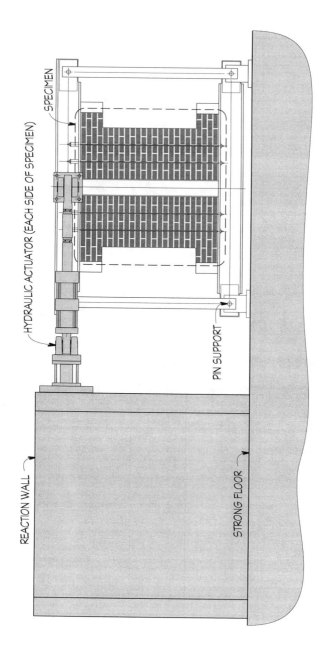

Figure 2. Elevation of Test Set-Up

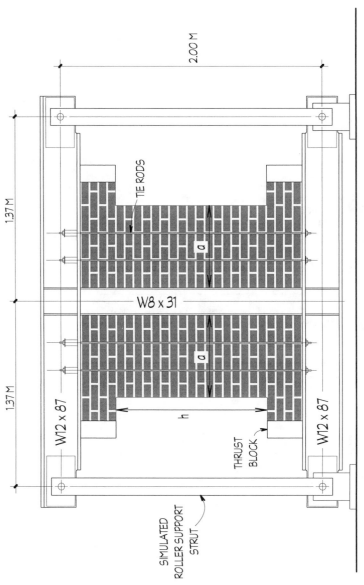

Figure 3. Schematic of a Typical Test Specimen

center of two W12x87 beam elements. The W12x87s were sufficiently stiff and strong, relative to the W8x31, to approximate a fixed-fixed end condition for the steel column. The upper W12x87 was supported at its ends by struts to simulate roller supports, while the bottom W12 was supported at its ends by pin supports that were anchored to the strong floor.

Clay-unit masonry was placed on each side of the W8 steel column. Symmetry was maintained to simulate an interior column. To provide full bearing, a mortar bed was placed between the bricks and the steel elements at each interface. Structural steel tubes were welded to the W12s at four locations. The masonry was placed with full bearing against these steel tube "blocks" to help develop the thrust that can occur in the brick infill due to the adjacent masonry pier. Window openings were simulated by the absence of thrust that can develop in the masonry over the window opening height [h]. Tie rods were tensioned between the W12s to produce a compressive axial stress in the masonry infill. Compressive stress was applied to account for some live load that may be distributed to the infill from the floor gravity loads, and from the self-weight of the masonry infills above the floor. The axial stress applied by the tension rods was adjusted to compensate for the load shared by the W8 steel column and the struts.

The primary parameters for this test program were the width [a] for each side of the masonry infill and the number of wythes. The pier height was fixed with a window opening height of *1.15M*. Type N mortar was used for all specimens, and all bricks were from the same shipment. Compressive stress for each test was maintained at *420Pa*, which represents gravity loads from about four floors. For comparison, one test had no applied axial stress. Test parameters for each specimen are listed in Table 1.

Test Specimen	Axial Stress (Pa)	a (cm)	h/a	Wall Thickness (No. of Wythes)	Wall Area (cm^2)
1	420	40	2.88	1	365
2	420	60	1.92	1	550
3	420	60	1.92	2	1,100
4	420	80	1.44	1	750
5	420	40	2.88	2	730
6	0	40	2.88	2	730

Table 1. Variation of Height-to-Width Ratios for Test Specimens

Load cells, displacement transducers, and strain gages were placed on the test frame and the specimen to determine the distribution of the lateral forces in each element of the composite system. Of particular importance was the amount of lateral shear distributed to the steel column compared to the shear in the masonry pier, and the increase in stiffness of the steel frame due to the masonry infill.

Earthquake loads were simulated by imposing lateral deformations to the center of the upper W12, using two 500 KN hydraulic actuators. The hydraulic actuators maintained equal forces on each side of the specimen to minimize the torsion imposed on the specimen. Predetermined cyclic deformations were imposed on each test specimen, and a typical deformation history is shown in Figure 4. To identify crack growth during the test, the deformation corresponding to each quarter cycle was numbered sequentially. Various levels

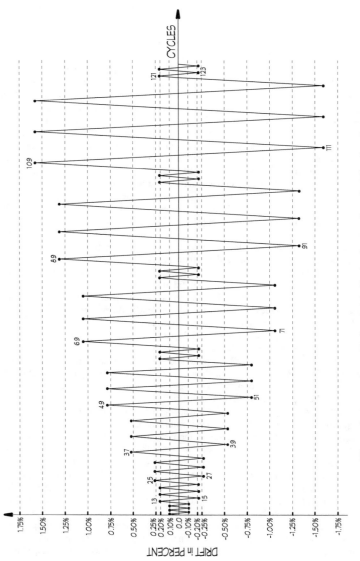

Figure 4. Approximate Deformation Cycles Imposed on the Test Specimens

of deformation were needed to investigate seismic behavior from initial to extensive cracking, and to eventually study behavior at failure of the masonry infill. In order to determine the deterioration of the elastic strength and stiffness characteristics, two "elastic" deformation cycles were imposed after three cycles of same-amplitude drift. These "elastic" cycles were imposed when the inelastic drifts exceeded 0.75%.

Sample Results
    A sample of the test results for Specimen 2 are shown in Figures 5 through 7. To determine behavior of the steel frame only, a force-displacement test was conducted without the masonry pier in place. Further, the bare steel frame test was performed to proof-load the full-penetration groove welds and to calibrate some of the gages on the test frame. Lateral load-deflection relations for the bare steel frame are shown in Figure 5. Displacements were measured using an LVDT attached to the center of the upper W12x87, and the lateral load was computed by summing the forces in both hydraulic actuators. Hysteretic behavior was quite stable, and no local buckling occurred during these tests. Also, the elastic stiffness deteriorated less than 15% after all inelastic cycles were imposed. Permanent lateral-displacement in the steel frame resulting from this test was corrected to obtain the original undeformed configuration, and the masonry was placed inside the steel frame.
    Figure 6 illustrates the load-deformation behavior of Test Specimen 2 that includes the unreinforced masonry pier. Comparison between Figures 5 and 6 illustrate the contribution of the masonry infill to the strength and stiffness of the steel frame. At low-level displacements, the masonry pier increased the strength of the steel frame by about 5.1 times, and the elastic stiffness was increased approximately 4.5 times. After an imposed drift of about 1% the composite frame strength was only about 55% larger, and the elastic stiffness was about 30% larger than the bare steel frame alone. After the maximum drift of approximately 1.65% was imposed, the strength of the composite frame was only 10% of the bare steel frame, and the elastic stiffness was the same as the steel frame.
    The cyclic deformation history in Figure 4 illustrates two cycles of elastic-level drift, of approximately 0.2%, between each three-cycle set of large displacements. These elastic-level drift cycles exhibited little influence from the masonry pier. The stiffness and strength during these deformations were very close to that of the bare steel frame. This was due to the amount of cracking exhibited by the large-deformation cycles prior to the initial elastic-level drift. Enough cracking occurred in the masonry to minimize its influence on the steel frame at small-deflection levels. These results suggest that after significant damage occurs, the strength and stiffness of the bare steel frame should be used to determine the characteristics of a composite system at small deformations.
    Figure 7 illustrates the crack pattern of the masonry piers after the test. The left infill developed a shear crack, the large vertical crack, which reduced the effective width of the masonry pier. This may be due to a strut developing in the masonry infill on each side of the steel column. At large deformations, the portion of the masonry between the vertical crack and the window opening crumbled and was on the verge of collapse. The crumbled portion of masonry is indicated by the darker shaded brick. The right pier had a different failure mode. Extensive crushing of the masonry occurred in the vicinity of the initial flexural cracks, just above the lower corner of the window opening. The crushed portion of masonry is shown by the darker shaded brick on the right side of the column. This extensive masonry crushing clearly limited the amount of diagonal strut action that developed in the

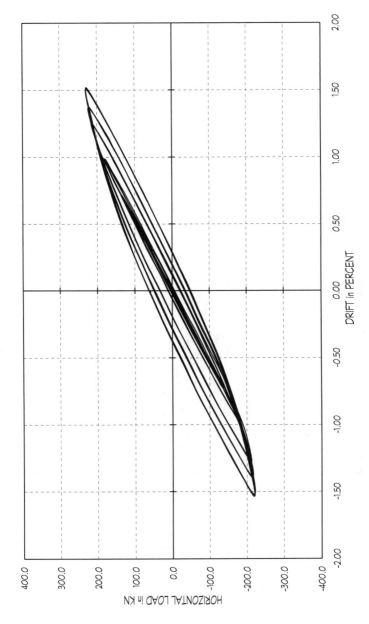

Figure 5. Force-Displacement Behavior of the Bare Steel Frame

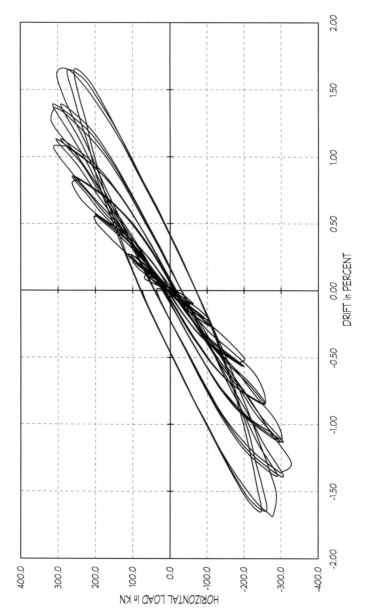

Figure 6. Force-Displacement Behavior of the 60 cm Single Wythe Unreinforced Masonry Wall

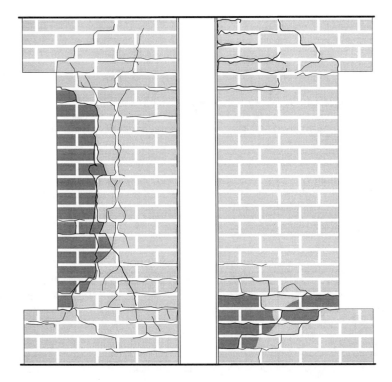

Figure 7. Crack Pattern for Test Specimen 2

piers, and explains the deterioration of the strength and stiffness of the composite system to that of the bare steel frame alone. However, it should be noted that this extensive cracking occurred at very large-displacement levels.

Concluding Remarks

Although results from only one specimen of a more comprehensive test program were presented, several preliminary conclusions can be inferred:

1. Steel frames with unreinforced masonry infills exhibited reasonably ductile behavior. Most of the hysteretic energy was dissipated on the first cycle of each set of three same-amplitude displacement cycles. Subsequent cycles at the same drift level exhibited less energy dissipation due to the opening and closing of cracks. However, after the frame experienced large drifts only the ductility of the bare steel frame was apparent. The masonry infill at large drifts was extensively cracked, and ineffective with respect to the ductility of the composite system.

2. Load sharing of each component of the composite system depended on the drift level. At small drift levels the masonry pier shared as much as 90% of the total lateral load. After drifts of approximately 1.3% were imposed however, the total horizontal load shared by the masonry piers was reduced to about 10%.
3. The elastic stiffness of the initial composite system was significantly larger than that of the bare steel frame. This contribution from the masonry pier however, became minimal after the composite frame experienced approximately 0.75% drift.
4. Diagonal struts formed vertical cracks which reduced the effective pier width. These vertical cracks reduced the participation of the masonry infill to the strength and stiffness of the composite system.

Acknowledgments
    This research is sponsored by the National Center for Earthquake Engineering Research Center [NCEER] under grant SBC SUNY 93-3113 & 94-3111 NSF. This funding is gratefully acknowledged. Opinions expressed in this paper are those of the authors, and do not necessarily reflect the opinion of NCEER or NSF.

Appendix A: References
1. Freeman, S.A. (1991). "Behavior of Steel Frame Buildings with Infill Brick." *Proceedings of the 6th Canadian Conference on Earthquake Engineering*, Toronto, Canada, June 12-14.

2. Freeman, S.A. (1994). "The Oakland Experience During Loma Prieta -- Case Histories." *Proceedings from the NCEER Workshop on Seismic Response of Masonry Infills, Technical Report NCEER 94-0004*, National Center for Earthquake Engineering Research, State University of New York at Buffalo, Buffalo, N.Y.

3. Hamburger, R.O., and Chakradeo, A.S. (1993). "Methodology For Seismic Capacity Evaluation of Steel-Frame Buildings with Infill Unreinforced Masonry." *Proceedings of the 1993 National Earthquake Conference; Earthquake Hazard Reduction in the Central and Eastern United States*, Memphis, TN, May 1993.

4. Mander, J.B., Nair, B., Wojtkowski, K., and Ma, J. (1993). "An Experimental Study on the Seismic Performance of Brick-Infilled Steel Frames With and Without Retrofit." *Technical Report NCEER 93-0001I*, National Center for Earthquake Engineering Research, State University of New York at Buffalo, Buffalo, N.Y.

5. Mander, J.B., Aycardi, L.E., and Kim, D.K. (1994). "Physical and Analytical Modeling of Brick Infilled Steel Frames." *Proceedings from the NCEER Workshop on Seismic Response of Masonry Infills, Technical Report NCEER 94-0004*, National Center for Earthquake Engineering Research, State University of New York at Buffalo, Buffalo, N.Y.

Measured Seismic Behavior of a Two-Story Masonry Building

Gregory R. Kingsley, Member ASCE[1]
Guido Magenes [2]
G. Michele Calvi, Member ASCE[3]

Abstract

Preliminary results are presented from an experimental test on a full-scale two-story unreinforced masonry building subjected to cyclic loading. Deformation and failure modes are described, with emphasis on the interaction of individual components in the structural system. The applicability of simple structural models in the prediction of structural response is discussed.

Introduction

The significant life-safety hazard that unreinforced masonry (URM) buildings can present in seismic zones is well known. Unfortunately, in many seismically active areas of the world, URM buildings constitute the major part of the existing building stock, and wholesale replacement, or even strengthening, of URM buildings is not feasible. The need to address this problem in an efficient and economic manner has generated significant interest in the development of rational assessment, analysis, and retrofit methods which are appropriate for these structures. The formidable inventory and inherent diversity of these buildings demands a rational yet simple approach to assessment, well supported and validated by experimental research. This paper outlines some recent results from an ongoing coordinated research effort designed to contribute to this need, including experimental and analytical work at several institutions, (among them the University of Pavia, Italy, the University of Illinois, Urbana, the Joint Research Center of the European Communities, Ispra, and the National Laboratory of Civil Engineering, Lisbon).

[1] Project Manager, TSDC of Colorado, 805 14th St., Golden, Colorado, 80401

[2] Researcher, Department of Structural Mechanics, University of Pavia, Via Abbiategrasso 211, 27100 Pavia, Italy

[3] Professor, Department of Structural Mechanics, University of Pavia, Via Abbiategrasso 211, 27100 Pavia, Italy

The research program is designed to incorporate the complete building assessment process, including the following key phases:

1. Survey and observation
2. Nondestructive testing
3. Analytical modeling of components and building systems
4. Design of strengthening techniques
5. Analytical modeling of strengthened structures

Towards this end, the program includes not only static, dynamic, and pseudodynamic experimental tests on URM materials, components and structural systems, but also a full complement of nondestructive tests and assessment methods. Both qualitative and quantitative assessment techniques are directed towards support of analytical models for quantitative assessment of the nonlinear response of URM buildings subjected to seismic ground motions. Emphasis is placed on the need for reliable structural models which are sufficiently simple to allow a clear interpretation of the results in a rational way: analysis must relate directly to an engineer's understanding of the fundamental possible response and failure mechanisms.

In this context, this paper presents selected results from experimental tests, primarily from a full-scale, 2-story URM building prototype tested at the University of Pavia. Test results are related to current methods of seismic evaluation, and to results of simplified predictive models.

Scope And Objectives of the Project

The full-scale prototype building test was conceived as a complement to and verification of building component and material tests, with the specific objective of observing the damage and failure of structural components within a structural system and their interaction in the global response.

The principle objectives of the test included the following

1. study of the interaction of intersecting walls and of wall sub-elements;
2. study of the effects of openings and of damage localization;
3. verification of the effectiveness of preliminary tests on materials and sub-assemblages;
4. verification of the effectiveness of non-destructive tests and structural identification methods;
5. validation and improvement of numerical models;
6. verification of the effectiveness of repair and strengthening intervention.

A simple representation of the geometry of the prototype building is shown in Figure 1. Loading was in the plane of the walls with openings. The experimental research program is comprised of several distinct phases:

1. tests on materials and small assemblies
2. static and dynamic tests on structural elements
3. non destructive tests on the prototype
4. quasistatic tests on the prototype
5. strengthening and re-testing the prototype

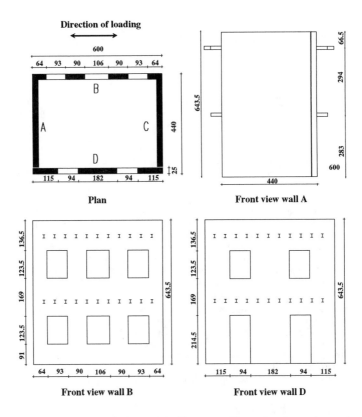

Figure 1. Plan, Front, and Lateral Views of the Prototype, (dimensions in centimeters).

Material Properties

Materials for the structure were chosen to represent typical old, urban construction in Italy, but they also represent well the materials used throughout the U.S., prior to 1930. Solid fired-clay bricks were used with a mean compressive strength on cubes of 16 MPa. The mortar was a mix of hydraulic lime and sand (1:3 volume), giving a compressive strength ranging from 2 to 3 MPa. The measured mean compression strength of masonry prisms was 6.2 MPa. All the tests on materials were performed at the Politecnico of Milano, and a detailed report of these is given in (Binda et al., 1995). Among these tests, particularly meaningful are the tests on brick-mortar joints, in the form of shear tests and tensile tests. The shear strength of the joint as evaluated from a regression on triplet tests could be expressed as: $\tau = 0.23 + 0.57\sigma$ (MPa) where σ is the normal stress on the mortar joint.

Full-Scale Test Structure and Testing Procedure

The building consisted of four two-wythe solid brick walls with a total wall thickness of 250 mm. The plan dimensions were 6 x 4.4 m, and the height was 6.4 m, with non-symmetric openings as shown in Figure 1. One of the longitudinal walls (wall D or "door wall", parallel to the direction of loading) was disconnected from the adjacent transverse walls (walls A and C), while the other (wall B or "window wall") was connected to the adjacent walls with an interlocking brick pattern around the corner. This allowed investigation of the limiting cases of wall connection in a single test.

The floors consisted of a series of isolated steel beams (I section, depth = 140 mm), designed to simulate a very flexible diaphragm. Both vertical and horizontal loads were applied through the floor beams. Concrete blocks were used to simulate gravity loads, for a total added vertical load of 10 kN/m per floor. The state of stress resulting from dead weight and added load resulted in vertical stresses ranging up to 0.4 - 0.5 MPa at the reduced section between to the openings level at the ground floor. Such stresses corresponded to those in a structure larger than the test building, with floors extending on both sides of a wall or to a higher number of stories.

The prototype building was tested under quasistatic applied displacements programmed to simulate dynamic load/displacement patterns, with reference to a 3/8 scale exact replica of the prototype building tested dynamically on the shaking table at the University of Illinois (Abrams and Costley, 1994, Calvi and Kingsley, 1994]. Based on the results of the scale model tests, it was determined that, for the stiffness and mass distribution of the walls and floors of the test structure, equal forces applied at each story level provided a close approximation of typical inertia forces recorded in the shaking table tests.

Simplified Numerical Analysis

To estimate the ultimate horizontal load carrying capacity of the walls a simplified "story mechanism" approach (Tomazevic and Weiss, 1990) was followed, where the failure mechanism of each story is modeled as a "pier- failure" mechanism. The behavior of each pier is expressed by means of a horizontal shear-horizontal displacement envelope, in the form of an idealized bi-linear (elastic-perfectly plastic) behavior, which is determined by the maximum or ultimate strength, the elastic stiffness, and the ultimate displacement, beyond which the reacting force is assumed to drop to zero. The initial elastic stiffness of a pier can be calculated assuming an effective height of the pier and imposing suitable rotational restraints (e.g. fixed-fixed or fixed at the bottom base and free at the top). The ultimate strength of each pier is defined as the lowest shear associated with $a)$ diagonal cracking (V_f), $b)$ flexural failure (toe crushing) (V_f), $c)$ shear sliding (V_s) (Calvi, Kingsley, and Magenes, 1995). Such values are calculated given the vertical stresses due to gravity and dead load. The ultimate displacement is defined according to available experimental information on the ductility associated with each failure mode.

With such an approach, the interstory shear-drift envelope for each wall of each story is calculated by imposing horizontal displacement compatibility of the piers. In this case the initial elastic stiffness, as well as the strength and ductility parameters, were derived from the interpretation of static cyclic tests on masonry piers (Calvi and Magenes, 1994).

Figure 2 shows the interstory shear-drift envelopes for the longitudinal walls D (door wall) and B (window wall), where the collaboration of the transverse walls in wall B is neglected. From the interstory envelopes, the total shear carrying capacity of the whole wall can be estimated, assuming a ratio between the forces (or the interstory drifts) at the first and second story. In fig. 3 the total base shear vs. the absolute displacement at the top of the building is reported, in the case where the seismic forces at the first and second story are assumed to have a 1/1 ratio, as in the experiment.

Such an approach, though simplified, can also give an estimate of the sequence of failures in the piers. According to the analysis shown, in wall B failure is reached by diagonal cracking of the central piers and subsequent flexural failure of the exterior piers. Similarly, in wall D failure is given by diagonal cracking of the central pier and subsequent flexural failure of the lateral piers. In both walls the ultimate displacement is limited by the piers failing in shear.

The results of such methods, in terms of strength and deformability, depend on the degree of fixity assumed for the piers. A common assumption is the fixed-fixed condition (as used in this case), which tends to underestimate the displacements, since the deformation due to spandrel cracking and flexibility is neglected. Nevertheless the calculated strength tends to be less sensitive to this assumption, since only the flexural strength is affected. Also, the method assumes that the mean vertical load in the piers remains constant during the application of the seismic load.

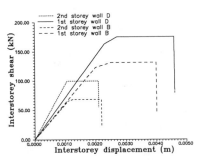

Figure 2. Interstory Shear-Drift Envelopes for Walls B (Window Wall) and D (Door Wall) from a Simplified Nonlinear Analysis

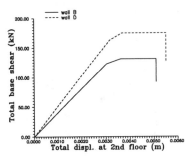

Figure 3. Calculated Shear-Displacement Envelopes for Walls B and D, Assuming a 1/1 ratio for the Seismic Forces at 1st and 2nd Story.

Test Results

The overall response of the structure is summarized in plots of base shear versus top displacement shown in Figures 4 and 5 for the door and window walls respectively. The maximum base shear in the door wall was approximately 150 kN, while the total in the window wall was slightly less at approximately 140 kN. The maximum horizontal force was initially achieved at a drift (top displacement / building height) of approximately 0.2%. The test was terminated when significant damage had developed in the piers and masonry lintels of the door wall at a maximum drift of approximately 0.4%; only minor degradation of the lateral load carrying capacity had occurred at this point. Since URM structures do not "yield" per se, it is not strictly correct to consider their response in terms of "ductility", however it is worth noting that the test structure achieved an ultimate displacement approximately twice that of the displacement when the maximum load capacity was first attained. The nearly constant load carrying capacity can be attributed to the observed joint-sliding failure mechanism in the critical piers. Even though diagonal cracks developed through both brick and mortar, large local joint sliding displacements were observed during final stages of the test.

The observed progression of damage in the walls was quite complex, with the nature and location of damage changing significantly with increasing drift. Initially, cracking was limited to the spandrels between the openings in both in-plane walls (see Figure 6 showing damage at a drift level of 0.075%), in part because these regions were not subjected to any vertical stress due to dead loads, and the joint shear strength was therefore less than elsewhere in the structure. As cracks developed in the spandrels, the coupling between the masonry piers decreased; eventually, cracks in the spandrels ceased to propagate further, and the failure mechanism became one dominated by shear cracking in the central piers. At the maximum drift level, exterior piers on the door wall failed in shear, while in the window wall the exterior piers remained essentially undamaged (Figure 7, showing damage at drift = 0.4%).

The door and window walls responded in a significantly different fashion to the imposed displacements. Measured vertical displacements and an evaluation of the deformation modes of the individual piers indicated that the door wall behaved as a coupled shear wall, with significant vertical displacements due to flexure at the top of the wall. The window wall, on the other hand, exhibited a response characterized by shear deformations localized in the piers, with only small uplift due to flexure. The difference is illustrated in Figures 8, showing horizontal displacement profiles for the two walls, and in Figure 9 showing representative measured vertical displacements over several cycles of response.

The response of the exterior piers in the two walls was also very different: the exterior piers in window wall exhibited a rocking mode, showing no diagonal cracks throughout the test, while the exterior piers in the door wall eventually failed in shear. The difference in the two walls was due both to the arrangement and aspect ratios of the piers, and to the fact that the window wall was connected to the transverse walls while the door wall was not. Note that the rocking piers in the window wall did not rock between horizontal cracks defined by the window height, but over a larger distance.

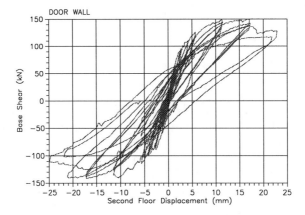

Figure 4. Total Base Shear vs. Second Floor Displacement in the Door Wall.

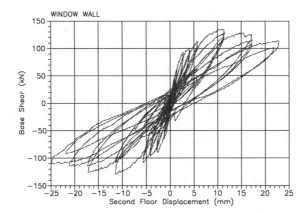

Figure 5. Total Base Shear vs. Second Floor Displacement in the Window Wall.

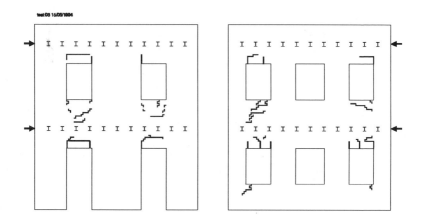

Figure 6. Crack Pattern at the End of Run 3 (max. drift 0.075%).

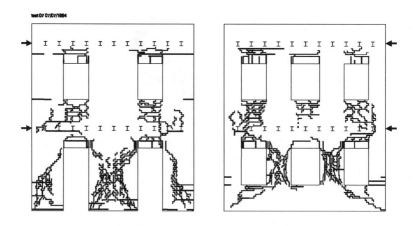

Figure 7. Final Crack Pattern at the End of Run 7 (max. applied drift 0.43 %)

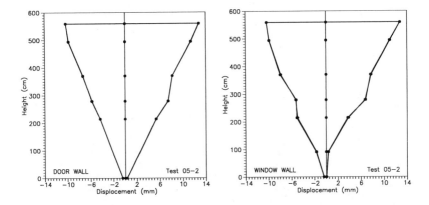

Figure 8. Horizontal Displacements Measured Over the Height of the Door and Window Walls

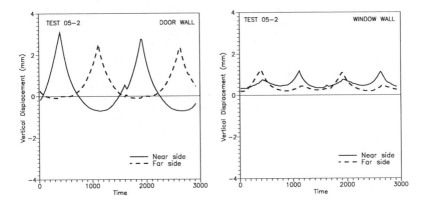

Figure 9. Vertical Displacements at the Top of the Walls on the Near and Far Sides, Showing Greater Flexural Action in the Door Wall.

While the achievement of relatively large drift levels is encouraging, this information should be tempered by the observation in dynamic tests on piers (Magenes and Calvi, 1994) that shear failure can be explosively brittle, and accompanied by a complete loss of integrity. On the other hand, dynamic tests of the 3/8 scale prototype (Abrams and Costley, 1994) showed rather stable rocking response for a complete building system. Further testing is needed to identify the mechanisms of response, but it is clear that the response of URM buildings can not be predicted sufficiently well by considering only the superposition of individual component response.

Estimated Strength Vs. Measured Response

The simple numerical models cited above were capable of providing a reasonable prediction of the failure load in the window wall, but overestimated the strength of the door wall. Behavior of the window wall conformed to the assumed shear dominated pier response of the model, but the coupled wall response of the door wall did not. In the test, significant degradation of the central pier before exterior piers reached their ultimate strength resulted in an ultimate strength somewhat less than predicted by summing the capacity of all three piers. Predicted ultimate displacements were approximately three times less than the observed displacements. The inability of the simple model to capture ultimate displacements is understandable, since it cannot capture the deformability of the masonry above and below the openings, and it is assumes restrained rotations at the ends of each pier. A sensible improvement could be expected if different constraints were assumed. Furthermore, the test indicated that the variation of axial loads due to overturning in the exterior piers had a strong influence on their failure mode. Including such effects is therefore desirable, even in very simple models.

Despite their limitations and the need for improvements, the strength predictions of the simplified models can be considered rather satisfactory, since at present the results that can be obtained even by very sophisticated nonlinear finite element models do not surpass this level of accuracy.

Current U.S. Practice, as represented in the Uniform Code for Building Conservation (Uniform Building Code, 1994), relies on even simpler models for predicting in-plane response. Strength is predicted by evaluating simple expressions for rocking (flexure) and sliding (shear), with no explicit consideration of diagonal shear failure or ultimate displacements. While the simplified models referenced herein fall short of characterizing the complex failure mechanisms of URM structures, they offer some improvement over accepted U.S. standards without an undue increase in complexity.

Conclusions

Work on the evaluation of experimental and analytical data is ongoing, but several broad conclusions can be drawn from the data presented herein:

1.   The failure mechanisms of URM structures may be quite complex, depending on the interaction of horizontal and vertical components and on the influence of both constant and varying axial loads. Components without axial loads (such as spandrels) must be evaluated in addition to axially loaded piers. Rocking mechanisms should include a consideration of pier height other than the clear window height.

2.   Simple models which idealize URM walls as a series of independent piers must be used with discretion. While simplified models (as opposed to finite element based models) may be used, and in fact must be recommended for the insight they provide into member failure mechanisms and interactions, attention should be paid to the implications and range of validity of the simplified assumptions which are introduced. For example, it is common to assume that the axial load on masonry piers is constant and equal to the dead load on the pier, neglecting variations caused by flexural response of the system of piers and spandrels; however, the influence of the variation of axial load in each single pier on the overall building behavior remains to be quantified, and preliminary results suggest that the influence can be significant.

3.   The apparently ductile response of the test structure suggests the possibility of a capacity design approach for unreinforced masonry in which a deformation mechanism is selected, and the components selectively strengthened to allow the "ductile" mechanism to occur while inhibiting undesirable shear failure modes. Such an approach is very attractive, but requires additional experimental and numerical verification before being implemented in practice.

Acknowledgments

Funding for the research program was provided by the Italian National Research Council (C.N.R., Progetto Finalizzato Gruppo Nazionale per la Difesa dai Terremoti). A formal cooperation agreement has been established with the NCEER, U.S.A. The National Science Foundation provided travel funds for the first author.

Appendix. References

1.   Abrams , D.P. and A.C. Costley , "Dynamic Response Measurements for URM Building Systems", US-Italian Workshop on Guidelines for Seismic Evaluation and Rehabilitation of Unreinforced Masonry Buildings, University of Pavia, Italy, June 22-24, 1994.

2.   Binda, L., C.Tiraboschi, G.Mirabella Roberti, G.Baronio, and G.Cardani . "Measuring Masonry Materials Properties: Detailed Results from an Extensive Experimental Research", Gruppo Nazionale per la Difesa dei Terremoti-Experimental and Numerical Investigation on a Brick Masonry Building Prototype, Report 5.0, Dip. di Ingegneria Strutturale, Politecnico di Milano, June 1995.

3.   Calvi, G.M., and G.R. Kingsley, "Problems and Certainties in the Experimental Simulation of the Seismic Response of MDOF Structures," Engineering Structures, accepted for publication October 1994.

4.   Calvi, G.M., G.R. Kingsley, and G. Magenes, "Testing of Unreinforced Masonry Structures for Seismic Assessment", Earthquake Spectra, Journal of the Earthquake Engineering Research Institute, submitted for publication, 1995.

5.   Calvi, G.M., and Magenes, G., "Experimental Research on Response of URM Building Systems," US-Italian Workshop on Guidelines for Seismic Evaluation and Rehabilitation of Unreinforced Masonry Buildings, University of Pavia, Italy, June 22-24, 1994.

6.   Magenes, G., and G.M. Calvi, "Shaking Table Tests on Brick Masonry Walls", 10th European Conference on Earthquake Engineering, Vienna, Austria, August 28- September 2 ,1994, Vol.3, 2419-2424.

7. Tomazevic, M., and P. Weiss. "A Rational, Experimentally Based Method for the Verification of Earthquake Resistance of Masonry Buildings." *Proceedings of the 4th U.S. National Conference on Earthquake Engineering,* Palm Springs, CA, May 1990, Vol.2, 349-359.

8. Uniform Code for Building Conservation, International Conference of Building Officials, 1994.

Response of Building Systems with Rocking Piers and Flexible Diaphragms

Andrew C. Costley[1], Student Member, ASCE
Daniel P. Abrams[2], Member, ASCE

Abstract

An examination of the appropriateness of two recent evaluation procedures (FEMA 178 and UCBC) is presented by comparing calculated response estimates with measured accelerations of two, reduced-scale unreinforced masonry buildings systems that were subjected to simulated earthquake motions on a shaking table. Predicted failure modes are compared with observed behavior, and recommendations for the usage of FEMA 178 Appendix C and UCBC Appendix Chapter 1 are made. The paper discusses the interaction of rocking piers and flexible diaphragms. Reductions in diaphragm motions due to pier rocking are detailed. Results from a nonlinear dynamic analysis model representing these effects are used to help explain the test observations.

Introduction

Seismic evaluation of existing building systems has traditionally been done with those analytical tools that are common to all structural materials. Elements of a structural system are assumed to behave in a linear elastic manner and horizontal forces are assumed to act statically and be applied across rigid diaphragms. More recent tools for evaluation and rehabilitation of unreinforced masonry (URM) buildings have emerged, such as FEMA 178 Appendix C or UCBC Appendix Chapter 1, that provide better estimates of seismic response by considering masonry piers to rock and floor or roof diaphragms to flex relative to walls.

---

[1]Ficcadenti & Waggoner Consulting Structural Engineers, 1601 Dove St., Suite 285, Newport Beach, CA 92660.

[2]Professor of Civil Engineering, University of Illinois at Urbana-Champaign, 1245 Newmark Laboratory, 205 N. Mathews Ave., Urbana, IL 61801.

In an effort to study the evaluation and rehabilitation of URM buildings with flexible diaphragms, two reduced-scale structures (S1 and S2) were constructed and tested on the University of Illinois shaking table (Costley and Abrams, 1995) (see Figure 1). Observed response of the test structures revealed two dominant effects that cannot be characterized with traditional linear, static analysis models. One, the flexible floor diaphragms in combination with the stiffer masonry walls resulted in inertial loads that acted at two separate frequencies. Two, rocking of the piers at the base story provided a nonlinear shear resistance that dissipated seismic energy with little damage.

The first part of the paper will compare the predicted response from the FEMA 178 and UCBC methodologies with the response measured in the laboratory. The second part of this paper will describe the behaviors of flexible diaphragms and rocking piers and discuss how they affect one another.

FEMA 178 and UCBC

Rocking strengths of the two test structures (S1 and S2) were determined using the FEMA 178 procedure. The basis of the FEMA 178 methodology is a comparison of a shear capacity with a rocking capacity for each shear wall being analyzed. Shear capacities are based on in-situ tests while rocking capacities are computed using vertical stresses and pier aspect ratios. With some restrictions, the lesser of the two capacities is assigned as the ultimate strength of the wall. The UCBC approach is similar to that of FEMA 178, except that a working stress approach is used instead of one based on ultimate strength.

The calculated ultimate strengths (FEMA 178) were within 20% of those measured during the final test runs of each building (see Table 1). Results from the UCBC analysis were comparable to measured strengths, with, in this case, a factor of safety exceeding three. Calculated rocking strengths based simply on the sum of PD/H, where P is the vertical load on a pier and D/H is the pier aspect ratio, length over height, included no reduction or enhancement coefficients as the codes did and were nearly as accurate as those determined using FEMA 178.

Table 1. Comparison of Calculated and Measured Base Shear Capacities

	S1 Capacity (kN)	S2 Capacity (kN)
FEMA 178	67.6	37.8
UCBC	16.0	12.5
PD/H	46.1	54.6
Measured	55.2	41.8

In both buildings, the FEMA 178 methodology indicated a rocking-controlled behavior rather than a shear-controlled behavior. Experimental observations confirmed this. The UCBC analysis, however, indicated that pier rocking would control over pier shear for S2, but that shear would control for S1. This was not the case as both buildings were controlled by rocking. Both documents promoted rocking as the preferred response over a shear-controlled behavior. The FEMA 178 document seemed to overpromote rocking by reducing the forces applied to the rocking-controlled walls (Equation C-21 of FEMA 178).

For simple structures like the ones tested here, both documents were fairly straightforward and easy to implement. However, calculated capacities of walls that are shear controlled can be much lower than either the rocking or shear capacities. This was the case for one of the S1 walls. In this instance, lowering the assumed axial stresses would have increased the calculated capacity which is contrary to the fact that both rocking and shear strengths decrease with decreased axial stress. No comments can be made regarding shear strengths per FEMA 178 and UCBC since both buildings tested were rocking controlled.

Unlike for linear elastic models, determining equivalent seismic forces was not an issue for these two evaluation methods. Wall strengths were based on pier capacities and were independent of lateral force distributions.

The main drawback to FEMA 178 and UCBC was that each prescribed a method to estimate pier (and wall) rocking capacity without specifying conditions necessary for piers to withstand the potentially large rocking displacements. Although demand was determined to be unable to exceed capacity while the test buildings were rocking, one can not conclude that every rocking-controlled masonry building will withstand any earthquake. Other requirements such as minimum tensile strengths and minimum aspect ratios (length over height) are needed to ensure that rocking piers will not degrade or topple.

Rocking Piers and Flexible Diaphragms

Pier rocking was observed during (the largest) four of the nine earthquake simulations conducted on the two test structures. Since neither test structure collapsed, and only minor damage was present after repeated testing, pier rocking was determined to be a reliable nonlinear behavior for URM structures. "Ductilities" on the order of 6-10 were measured for the two test structures. Note that rocking did not occur until horizontal cracks had formed across the bases and tops of the in-plane piers. Prior to cracking, both buildings behaved in an elastic manner.

During the initial earthquake simulations, the flexible diaphragms deflected substantially relative to the stiffer, in-plane masonry walls (Figure 2). On the average, diaphragm deflections relative to the average wall deflections were approximately three times that of the walls. After cracking, and during pier rocking, diaphragm deflections relative to the walls were greatly reduced while wall displacements became quite large (Figure 3). In general, relative diaphragm deflections did not exceed wall displacements while piers were rocking. In addition, during pier rocking, inertial forces were generally lower than the pre-rocking values as accelerations of the flexible diaphragms also decreased. The rocking (first-story) piers created a stable soft-story mechanism which isolated the flexible diaphragms from the dynamic base motion. Another effect of this mechanism was the relative reduction of second-story drifts.

Based on results from the shaking-table tests, a simple dynamic model was developed to estimate the large-amplitude displacements of the masonry walls while the piers were rocking. The model had three degrees of freedom (DOF), one for each in-plane wall and one for the flexible diaphragms (Figure 4). The two in-plane walls were assumed to follow a nonlinear elastic force-displacement curve while the diaphragms were assumed to be linear elastic. A time-step integration program was used with the 3-DOF model to calculate displacements resulting from base acceleration histories measured during the dynamic tests. Calculated displacement histories indicated that prior to rocking, diaphragm deflections relative to the walls were quite large (Figure 5), while during rocking, wall displacements were greater than relative diaphragm deflections (Figure 6). This matched the behaviors measured during the earthquake simulations of the test structures.

Conclusions

FEMA 178 and UCBC can be used to estimate the rocking capacities of URM buildings, but with some caution. The force reduction procedure for rocking-controlled walls in FEMA 178 can produce unconservative results while the method for distributing force among piers for the shear-controlled behavior can produce overly-conservative results.

Flexible diaphragms amplified wall displacements prior to cracking in the walls. While first-story piers were rocking, diaphragm deflections relative to the walls were greatly reduced. A simple nonlinear dynamic model was used to represent these behaviors.

References

1.   Costley, A. C. and D. P. Abrams, "Dynamic Response of Unreinforced Masonry Buildings with Flexible Diaphragms", SRS No. 605, UILU-ENG-95-2009, October 1995, pp 281.

2.   *NEHRP Handbook for the Seismic Evaluation of Existing Buildings* (FEMA 178), Building Seismic Safety Council, Washington, D.C., 1992, Appendix C.

3.   *Uniform Code for Building Conservation*, 1994 Edition, International Conference of Building Officials, Whittier, CA, Appendix Chapter 1.

Figure 1. Overhead View of S2 on Shaking Table

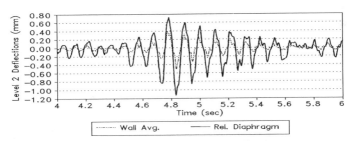

Figure 2. Measured Deflections Prior to Cracking (S2, Run 21)

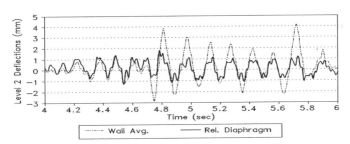

Figure 3. Measured Deflections After Cracking (S2, Run 23)

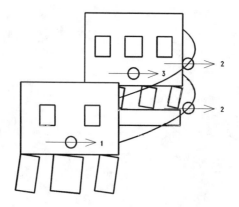

Figure 4. Three Degree-of-Freedom Model

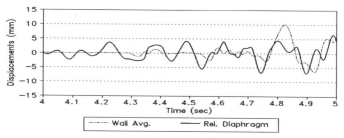

Figure 5. Calculated Displacements at Level 1 Prior to Rocking (S1, Run 15)

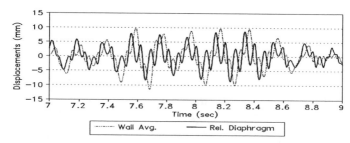

Figure 6. Calculated Displacements at Level 1 During Rocking (S1, Run 15)

Behavior of Two Long-Span High Strength Concrete Prestressed Bridge Girders

Theresa M. Ahlborn[1], ASCE student member, Carol K. Shield[2], ASCE associate member, and Catherine W. French[3], ASCE member

Abstract

Two full size long-span composite prestressed bridge girders were constructed using high-strength concrete (28-day compressive strengths exceeded 83 MPa [12,000 psi]) to investigate transfer length, camber, prestress losses and fatigue life. The AASHTO relationship for transfer length was found to be conservative for large diameter prestressing strands (15.3 mm [0.6 in.]) spaced at 50.8 mm (2 in.) on center in high strength concrete. Load - deflection response to simulated truck load and overload testing was predictable. Loads required to initiate flexural cracking were much lower than expected in one of the girders for which initial shrinkage cracks were observed prior to strand release. Upon release, the shrinkage cracks were observed to close completely. The initial shrinkage crack closure and its effect on camber, prestress losses, and flexural cracking loads was quantified using geometric compatibility.

Introduction

High-strength concrete (HSC) offers many advantages in prestressed bridge girder construction including use of smaller sections, increased span lengths and/or wider girder spacings. Shallower HSC sections can be used in place of normal strength concrete members of the same length enabling greater vertical underclearances or lower bridge embankments. Alternatively HSC can be used to increase span lengths for a given girder cross section, giving way to wider underpasses and in some cases fewer piers. Using HSC to increase girder spacing enables fewer girder lines per bridge which leads to lower fabrication, transportation and erection costs.

Current design provisions of the American Association of State Highway Transportation Officials (AASHTO 1993) are based on empirical relationships developed from isolated tests of concrete specimens with compressive strengths limited

---

[1] Graduate Research Assistant, Department of Civil Engineering, University of Minnesota, 500 Pillsbury Dr. S.E., Minneapolis, MN 55455-0220

[2] Assistant Professor, Department of Civil Engineering, University of Minnesota

[3] Associate Professor, Department of Civil Engineering, University of Minnesota

to 35-45 MPa (5000-6500 psi). Little research has been conducted on full-size girders to determine the effect of using higher strength materials.

Description of Girders and Test Program

At the University of Minnesota, two full size long-span composite girders were constructed with concrete compressive strengths exceeding 83 MPa (12,000 psi) and lengths of 40.5 m (132 ft. 9 in.), nearly 50% longer than typically fabricated. The girders, MnDOT 45M sections, were 1140 mm (45 in.) deep, reinforced with forty-six 15.3 mm (0.6 in.) diameter 1860 MPa (270 ksi) low-relaxation prestressing strands spaced at 50.8 mm (2 in.) on center. For this standard Minnesota girder, the maximum length constructed to date using normal strength concrete has been 27 m (89 ft.). The high strength girders were designed assuming 1.2 m (4 ft.) center to center spacing achieving noncomposite span-to-depth ratios of 35. Composite concrete decks were added to each individual girder, simulating actual bridge construction and giving a composite span-to-depth ratio of 29. Fig. 1 depicts the cross section of the composite members.

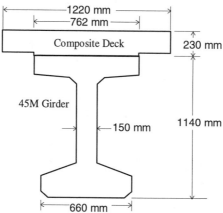

Figure 1. Composite Cross Section

The girders were designed identically with the exception of three variations: the mix design (Girder I - limestone aggregate mix; Girder II - glacial gravel with microsilica mix); end strand patterns (Girder I - 4 draped / 8 debonded on each end; Girder II - 12 draped strands on one end and 4 draped / 8 debonded on the other end); and stirrup anchorage details (modified U stirrups with leg extensions on all ends except one end of Girder I which used standard U stirrups).

The concrete mixes were based on research from a companion HSC materials study (Mokhtarzadeh 1993). The concrete mix used in Girder I consisted of Type III Portland cement, sand, crushed limestone aggregate, and a superplasticizer. The mix had an average water/cement ratio of 0.32. The mix used for Girder II also used Type III Portland cement, sand and superplasticizer. However, rounded glacial gravel aggregate was used instead of limestone. In addition, microsilica was used at a rate of 7.5% replacement by weight of cement for Girder II. The mix for Girder II had an average water/cementitious material ratio of 0.36. Required concrete compressive strengths for both girders were 61.5 and 72.4 MPa (8920 and 10,500 psi) at release and 28 days, respectively to meet design requirements. The strength at release was 64 MPa (9300 psi) for Girder I and exceeded 71.9 MPa (10,400 psi) for Girder II. At 28 days, the strength of Girder I continued to increase well above the design requirement to 83.4 MPa (12,100 psi) and was 86.3 MPa (12,500 psi) at 200 days. The strength of Girder II also continued to increase with time. The strength was 78 MPa (11,100 psi) and 83.4 MPa (12,100 psi) at 28 days and 200 days, respectively.

The Grade 270 (1860 MPa) 15.3 mm (0.6 in.) prestressing strands were tested for ultimate strength and modulus of elasticity. The ultimate strength was found to be 1850 MPa (268.8 ksi), and the modulus was approximated to be 200,700 MPa (29,120 ksi) using strand strain gages oriented along the helix angle. The actual measured cross-sectional strand area was 147 mm^2 (0.228 in.^{2}) compared with the nominal area of 139 mm^2 (0.215 in.^{2}). Strands were specified to be tensioned to 75% f_{pu} after seating (where f_{pu} is the ultimate strand strength). The initial tensioning level of the prestressing strands after seating was 71.3% f_{pu} (1320 MPa [191.6 ksi]) according to strand strain data.

The two prestressed bridge girders were cast on the same bed in an outdoor precasting yard in August of 1993. No modifications were made to standard construction techniques. Both girder mixes showed good workability and consolidation during placement. The girders were heat-cured under tarps using their own heat of hydration. Although both girders were cast on the same bed, Girder I was cast approximately 1.5 hours prior to the casting of Girder II. Girder II achieved its nominal release strength at an age of 14 hours and form removal began at 17 hours. The forms on Girder I were removed at an age of 22 hours, although Girder I did not achieve its release strength until a half hour later. A total of 5.5 hours elapsed between the time the forms were removed from Girder II and release (2 hours for Girder I). During this time, shrinkage cracks were observed to develop in Girder II. A total of fifteen vertical cracks were observed along the length of the girder, concentrated within the middle 40% of the span length. The cracks extended from the top flange towards the bottom flange. Eleven of the cracks extended approximately 864 mm (34 in.) deep (nearly to the bottom flange). Four of the cracks were less than 152 mm (6 in.) deep. Prestressing strands were flame cut to release them from the prestressing bed. Upon strand release the vertical cracks in the girder closed completely. If not for the lines drawn on Girder II to identify the initial shrinkage crack locations, there was no indication the girder had been initially damaged.

The girders were transported to a testing site. No stability problems were encountered during handling. At an age of 200 days, 1.2 m (4 ft.) wide composite decks were cast on each girder using unshored construction techniques. A standard MnDOT bridge deck mix (3Y33) was specified for the composite deck with a 27.6 MPa (4000 psi) required compressive strength. Test results indicated the strength to be 35.3 MPa (5100 psi) at 7 days and 40.4 MPa (5850 psi) at 28-days.

The girders and decks were instrumented with several gage types. Bondable electrical resistance foil type strain gages were installed on stirrups, prestressing strands, and deck reinforcing steel. Foil and vibrating wire strain gages were embedded in the concrete. External instrumentation included tiltmeters, LVDTs, DEMEC (DEtachable MEChanical) gages, and acoustic emission (AE) transducers.

A comprehensive testing program was developed to monitor girder transfer lengths, cambers, and prestress losses. The girders were each subjected to a series of static and cyclic load tests to investigate serviceability. Ultimate flexure and shear tests are yet to be conducted.

Results

Transfer Length

The strain measured at release in the transfer region of Girder I (draped/debonded end A) is shown in Fig. 2. Superimposed on the figure is the predicted strain distribution assuming the AASHTO transfer length of 50 d_b (strand diameter). The shallow dips in the calculated strains indicate the effect of gravity load causing a decrease in the concrete compression strain at the level of the strands. The slight increases following the dips are caused by the initiation of bonding pairs of debonded strands along the length of the girder. The increases appear minor due to the small percentage of debonded strands in the cross section.

Transfer lengths were determined graphically from the measured surface strains using methods discussed in Ahlborn (1995). In the "95% Average Maximum Strain Method," strain readings were first smoothed by averaging the data over three gage lengths to reduce anomalies in the data. The average maximum strain was determined by computing the numerical average of the smoothed strains contained within the strain plateau. The intersection of a line corresponding with 95% of the average smoothed strain data and the smoothed strain profile represents the transfer length. Using this procedure, transfer lengths in the range of 565 to 725 mm (22.2 to 28.5 inches) were obtained for the four girder ends.

For the "Final Average Method" employed, data points outside of the range of one standard deviation from the averaged strain plateau are discarded. The average strain of the remaining data points is then determined, and the transfer length is defined as the intersection of the final average strain and the data. The results obtained with this method were very similar to those obtained using the "95% Average Maximum Strain Method" and ranged from 569 to 696 mm (22.4 to 27.4 inches).

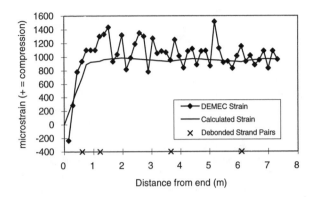

Figure 2. Transfer Length

The measured transfer lengths were all less than those predicted using the AASHTO relationship of 50 d_b (780 mm [30.7 in.]). This indicates that the AASHTO estimated transfer length relationship was a conservative predictor for these high strength concrete girders with the arrangement of 15.3 mm (0.6 in.) diameter strands on 50.8 mm (2 in.) centers.

Prestress Losses

Prestress losses occur instantaneously due to elastic shortening at release and over time due to steel relaxation, and creep and shrinkage of concrete. The components generating the prestress losses are interdependent, leading to the complex nature of predicting prestress losses and the state of stress in a member at any given time. The force in the prestressing strands continuously decreases until such time when the losses stabilize; the majority of the losses occur within the first 6 to 12 months of the member life.

Data were recorded from vibrating wire gages installed at the center of gravity of the strands (cgs) to monitor the change in concrete strain with time. Gages were installed at each of the following locations: 0.45L, 0.50L, and 0.55L. The change in concrete strain at the cgs was assumed equal to the change in strand strain. Multiplying this change in strain by the elastic modulus of the strand gave the change in strand stress. This method of experimentally determining losses does not fully account for strand relaxation losses, but this component is typically small for members with low-relaxation strands.

The predicted losses were calculated for each girder using the Naaman time-step method (Naaman 1982). Calculations were based on transformed geometric section properties and measured material properties including elastic moduli, ultimate creep coefficients of 1.043 and 1.236 for Girders I and II respectively, measured prestressing steel yield and ultimate strengths (for relaxation curves), strand elastic modulus, and initial strand stress after seating.

The measured and predicted prestress losses are given in Table 1 at release, deck casting (200 days), and at the time of the crack tests (Girder I - 598 days; Girder II - 727 days). At release, the measured losses were greater than predicted especially in the case of Girder I (13.4 vs. 12.3%). At the time of deck casting, the difference between the predicted and measured losses was maintained for Girder I; however, the measured losses were significantly lower than predicted for Girder II (16.3 vs. 19.9%). At cracking, the difference in strand stress losses between the measured and predicted values was on the order of 34 and 41 MPa (5 and 6 ksi) for Girders I and II, respectively; however in the case of Girder I the losses were underpredicted; whereas, for Girder II the losses were overpredicted.

Table 1. Predicted and Measured Prestress Losses (%)

	Girder I		Girder II	
	Measured	Predicted	Measured	Predicted
At release	13.4	12.3	11.8	11.6
At deck casting	21.3	20.0	16.3	19.9
At cracking	23.9	21.5	18.6	21.8

Camber

Camber has been monitored for both girders since the time of strand release. Fig. 3 illustrates centerline cambers observed during the first 11 days after girder

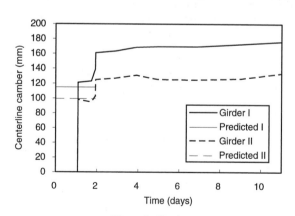

Figure 3. Camber

casting. A predicted initial camber of 115 mm (4.52 in.) was computed for Girder I using measured material properties and self weight. Girder I had a measured (on-bed) camber of 121 mm (4.76 in.) which increased to 139 mm (5.47 in.) after lifting the girder from the bed and immediately setting it back down. The "actual" initial camber lies between these values. Friction between the precasting bed and the girder tends to reduce the initial on-bed camber and increase the lift/set camber measurement. A predicted initial camber of 99 mm (3.91 in.) was computed for Girder II using measured material properties and self weight. Girder II had an initial (on-bed) camber of 98 mm (3.86 in.) and a lift/set camber of 103 mm (4.06 in.).

Loading Response

Static and cyclic loads were applied to each composite girder using servo-controlled hydraulic actuators located 0.4L from the supports to simulate the moment induced by an AASHTO HS25 truck, where HS25 is the standard design truck loading used for bridge serviceability design in the State of Minnesota. The load was proportioned assuming a girder spacing of 1.2 m (4 ft.) center to center. Overload testing at 125% HS25 was also performed.

The actual stiffness during the service and overload testing (up to 125% HS25 truck) was slightly higher than design checks had predicted due to the increased measured material and transformed section properties over design values. No stiffness degradation was observed after undergoing 1.0 million cycles in the uncracked state. An additional 2 million cycles (1 million at HS25, 1 million at overload) were applied to the cracked girders. During these cycles, no appreciable stiffness degradation was observed for either Girder I or Girder II.

The static load - deflection responses of Girders I and II are shown in Figs. 4

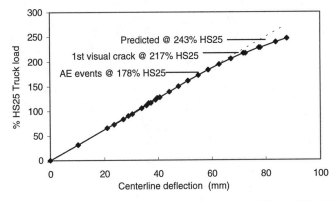

Figure 4. Girder I Load vs. Deflection Measured During Initial Flexural Crack Tests

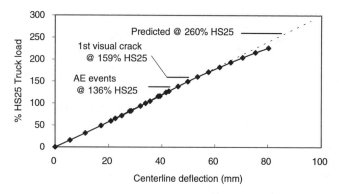

Figure 5. Girder II Load vs. Deflection Measured During Initial Flexural Crack Tests

and 5, respectively. The y-axis represents the percent of HS25 truck loading applied. Using the experimentally determined concrete tensile strength and measured losses, Girder I was predicted to crack at 243% HS25 truck load. The AE monitoring equipment indicated that crack initiation began to occur at 178% HS25, and the first visual crack appeared at 217% HS25, below the predicted cracking loads but well above the intended service capacity of the bridge girder. Girder II had a higher predicted cracking load of 260% HS25 due to the lower measured losses relative to Girder I, but the AE equipment indicated crack initiation at 136% HS25 and visual cracks were observed at 159% HS25. The predicted cracking loads were higher for Girder II, however cracks were observed at much lower loads than those observed in Girder I, although still above the intended service capacity.

<u>Geometric Compatibility</u>

In the previous sections predicted properties of Girders I and II (i.e. prestress losses, camber, cracking loads) were discussed relative to the measured properties. Accounting for the differences in the girder material properties and strand patterns at the end of the girders, it was anticipated that the relationship between observed behavior and predicted behavior would be similar for Girders I and II. This was not the case as shown in Table 2. Relative to Girder I, Girder II had a higher predicted cracking load; however Girder II was observed to crack at a much lower load than Girder I. The only difference between the girders not taken into account in the predicted response was the effect of the initial shrinkage cracks observed in Girder II before strand release. Assuming that the relationship between measured and predicted behavior of Girder I was indicative of initially undamaged girder behavior, an expected response for Girder II was determined:

$$Expected_{II} = \frac{Measured_I}{Predicted_{II}} Predicted_I$$

This relationship was used to determine the *expected* response for Girder II had there been no initial shrinkage cracks. [Authors' Note: The term "expected" herein refers to the calibrated predicted response for Girder II.] The *expected* values are shown in Table 2 for camber, losses and cracking loads. As evident in the tabulation, the actual measured response of Girder II was less than *expected*. The effect that the shrinkage cracks may have had on the observed behavior of Girder II was further investigated.

Table 2. Predicted, Measured and Expected Girder Responses

	Girder I	Girder II	Girder II *expected*
Initial Camber (mm)			
Measured Camber (on-bed/after lift-set)	121/139	98/103	103/120
Predicted Camber	115	99	
Initial Prestress Losses (%)			
Measured Losses	13.4	11.8	12.6
Predicted Losses	12.3	11.6	
Flexural Cracking Load (%HS25)			
Measured	178 (AE) 217 (Visual)	136 (AE) 159 (visual)	190(AE) 232 (visual)
Predicted	243	260	

For simplification, the following were assumed: 1) the cambered shape of the girder can be modeled as an arc of a circle, 2) the shrinkage cracks all extend to the same depth, and 3) the shrinkage cracks are evenly spaced along the beam. After strand release, the deformation of the beam can be thought of as the superposition of two deformations: cambering of the beam due to prestressing less the self weight deflection (*expected* deformation), and bending of the beam to close the shrinkage cracks. The first deformation is shown in Fig. 6a. Here the girder cambers by bending about its neutral axis to yield the *expected* camber. A second deformation is now superimposed on the camber deformation. Because there is compression in the top fiber of the beam, the shrinkage cracks close by pivoting about their tips; hence the arc length at the crack tips remains unchanged. This motion shortens the top fiber length by the sum of the lengths of the shrinkage crack top openings (Δ_{top}) and underlines elongates the beam bottom fiber an amount Δ_{bot}. The net effect of the shrinkage crack closure is a reduction in the girder camber. The *expected* girder shape, and girder shape after crack closure are shown in Fig. 6b. Figure 6c is a blown-up detail of the girder left end rotations due to the initial camber deformation and crack closure deformation. The vertical line represents the girder left end prior to release. At release, the *expected* end rotation is about the girder neutral axis. Due to shrinkage crack closure, the end rotates in the opposite direction about the shrinkage crack tip depth to the actual deformed position. As can be seen from the figure, every fiber below the crack tip is longer than it would have been if there were no shrinkage cracks, and hence there is less compressive strain in fibers below the crack tip depth than what would have been expected. A more detailed description of the geometric effect can be found in Ahlborn (1996).

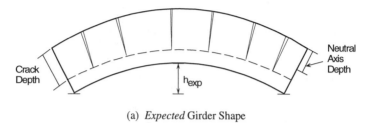

(a) *Expected* Girder Shape

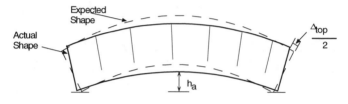

(b) Girder Shape after Crack Closure

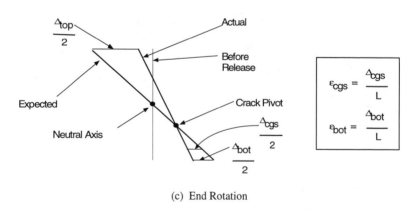

(c) End Rotation

Figure 6. Geometric Effect of Shrinkage Crack Closure

To determine the feasibility of this hypothesis, average crack widths to account for the difference between expected camber and measured camber were calculated. The average depth of the 11 largest shrinkage cracks was 864 mm (34 in.). Assuming this depth to the crack tip, an averaged *expected* camber of 113 mm (4.45 in.) and an averaged actual camber of 101 mm (3.98 in.) corresponded with average crack openings of 0.20 mm (.008 in.). The actual crack openings were estimated to be on the order of 0.25 mm (0.01 in.).

Finding the geometric effect of shrinkage crack closure to reasonably explain the differences in *expected* versus measured cambers, its effects with respect to prestress losses and cracking loads were examined. As mentioned above, geometric compatibility requires the fibers or arc lengths below the pivot plane (crack depth) on the girder to elongate. The strains at the center of gravity of strands (ε_{cgs}) and at the bottom fiber of the girder (ε_{bot}) can be found using similar triangles as shown in Fig. 6c. These results indicate that strains generated at the center of gravity of the strands cause an elongation in the strands due to geometric compatibility. The effect creates greater tension in the prestressing strands (Δf_{strand} = 1.8 MPa [0.26 ksi]), which corresponds with reduced losses at release. Although small, the reduction in losses correlates with the trend of reduced measured losses in Girder II relative to *expected* losses. The initial offset in bottom fiber strain (ε_{bot} = 17.6 $\mu\varepsilon$ less compressive) indicates that it would take less load (less additional strain) to crack the girder in flexure than would be *expected*. This agrees with the measured cracking loads being less than the *expected* values. Based on the measured prestress losses, it was possible to determine the load which would cause initial flexural cracking in the girders, 250% HS25, which is smaller than the predicted value of 260% HS25.

These results indicate that the initial shrinkage cracks may have had an effect on the performance of Girder II. In modeling the geometric effect, the changes in strain caused by crack closure were averaged over the entire length of the girder. In reality, shrinkage crack closure would cause primarily local changes in strain. Consequently, strain changes due to geometric effects would be much greater in the regions directly below the shrinkage cracks than the averages calculated previously. *Expected* cracking loads would be further reduced considering this effect. The local effect also implies initial flexure cracks should occur below the shrinkage cracks. This correlates with the actual initial flexural crack locations observed in the cracking load tests of Girder II. The initial flexural cracks were either directly below or in the vicinity of the shrinkage cracks.

Conclusions

Two full size prestressed concrete bridge girders have been monitored over time and load tested to investigate behavior. Results indicated that transfer lengths in both girders were approximately 80 percent of those predicted by AASHTO. AASHTO relationships were conservative for these larger diameter strands when used in high strength concrete girders and placed on 50.8 mm (2 in.) centers. The static response of the girders to design truck loading was favorable. Serviceability testing under HS25 truck and overload (125% HS25) testing showed no stiffness degradation in either girder after being subjected to 1 million cycles in the uncracked state. Flexural cracking also showed no effect on stiffness at the HS25 and overload levels.

It was anticipated that the relationship between the observed and predicted behavior would be similar for both girders. This was not the case. Girder II had a higher predicted cracking load than Girder I yet it was observed to crack at a much lower load. It was suspected that initial shrinkage cracks, which had only been observed in Girder II prior to strand release, could account for part of this discrepancy.

Geometric compatibility was used to investigate the effects of these top flange shrinkage cracks on initial camber, initial prestress losses and flexural cracking loads. Comparing measured camber to expected camber using geometric compatibility, crack widths in the top flange were found to be within reason of those observed. Shrinkage crack closure was shown to increase strand stresses, thereby reducing the expected prestress losses, as was observed. Crack closure also caused the concrete strains below the crack tips to become less compressive, thereby causing flexural cracking to initiate at a lower than expected load for Girder II.

Acknowledgments

This project has been collectively sponsored by the Minnesota Department of Transportation, Minnesota Prestress Association, University of Minnesota Center for Transportation Studies, Precast/Prestressed Concrete Institute, and the National Science Foundation Grant No. NSF/GER-9023596-02. The authors also wish to acknowledge the generous donations of materials, equipment and technical support by Elk River Concrete Products, Truck/Crane Services, Union Wire & Rope, W. R. Grace & Co., Simcote, Inc., Lefebvre & Sons Trucking, Golden Valley Rigging, Atlas Foundation, Borg Adjustable Joist Hanger Co., and United Technologies. Appreciation is also expressed for the assistance of graduate students Alireza Mokhtarzadeh, Jeffrey Kielb, Jeffrey Kannel, and Douglass Woolf. The views expressed herein are those of the authors and do not necessarily reflect the views of the sponsors.

References

Ahlborn, T.M., French, C.W., and R.T. Leon, "Applications of High-Strength Concrete to Long-Span Prestressed Bridge Girders," *Transportation Research Record No. 1476*, Washington, D.C., 1995, 22-30.

Ahlborn, T.M., Shield, C.K., and French, C.W., "Tests of Two Long-Span High Strength Prestressed Bridge Girders," submitted to *PCI Journal*, 1996.

American Association of State Highway Transportation Officials (AASHTO), *Standard Specifications for Highway Bridges, 15th Edition*, Washington, D. C., 1993.

Mokhtarzadeh, A. and French, C.W., "High-Strength Concrete Effects of Materials, Curing and Test Procedures," *PCI Journal*, May-June 1993.

Naaman, A.E., *Prestressed Concrete Analysis and Design*, McGraw-Hill, 1982.

High Performance Concrete Applications
In Bridge Structures In Virginia

Celik Ozyildirim[1], Jose Gomez[1], M. Elnahal[2]

Abstract

This paper summarizes current work in high-performance concrete (HPC) in bridge structures by the Virginia Department of Transportation (VDOT). VDOT has planned 5 bridge structures using HPC for construction in 1995 and 1996. In four of these structures high strength concrete will be used in AASHTO prestressed beams. Three of the structures will have concrete with a minimum 28 day compressive strength of 55 MPa (8,000 psi) and a release strength of 41 MPa (6,000 psi), and the fourth will have beams with 69 MPa (10,000 psi) at 28 days and 46 MPa (6,600 psi) at release. The beams on the fourth structure will contain 15 mm (0.6 in) diameter strands. Beams on the other three structures will have 13 mm (0.5 in) strands. All strands will be at the 51 mm (2 in) spacing. The concrete in three of the bridges will meet the VDOT's new proposed low permeability concrete requirement of 1,500 coulombs or less for the prestressed members, 2,500 coulombs or less for deck slabs, and 3,500 coulombs or less for concrete substructures. One bridge utilizing HPC will be constructed with steel beams but with its deck and substructure meeting the low permeability requirements.

Construction began in early 1995 on two of the bridges with 55 MPa (8,000 psi) strength at 28 days. Before the bidding of the first of the structures with 55 MPa (8000 psi) concrete, an experimental project was conducted to ensure that high release and ultimate strengths could be achieved. Four 9.4 m (31 ft) long AASHTO Type II beams were prepared and tested to failure. Similarly, before the advertisement of the bridge with beams containing 69 MPa (10,000 psi) at 28 days and 15 mm (0.6 in) strands, two 9.4 m (31 ft) beams with fully composite slabs were prepared and tested to failure.

[1] Principal Research Scientist and Research Scientist, Virginia Transportation Research Council, Virginia Department of Transportation, 530 Edgemont Road, Charlottesville, Virginia 22903
[2] Assistant State Structures and Bridge Engineer, Virginia Department of Transportation, 1402 East Broad Street, Richmond, Virginia 23219

Introduction

High-performance concrete (HPC) have enhanced specific properties, such as workability, durability, strength and dimensional stability,[1] resulting in long-lasting, economical structures. Many conventional concrete bridge structures deteriorate rapidly, and require costly repairs before their expected service lives are reached. Four major types of environmental distress affect bridge structures and cause early deterioration: corrosion of the reinforcement, alkali-aggregate reactivity, freeze-thaw deterioration, and attack by sulfates.[2] In each case, water or solutions penetrate into the concrete and initiate or accelerate the damage. HPCs designed for low permeability resist the infiltration of aggressive liquids, and thus are more durable.

Until 1992, the Virginia Department of Transportation (VDOT) specifications for prestressed concrete required a minimum 28-day design strength of 35 MPa (5,000 psi) and a maximum water-cementitious material ratio (W/CM) of 0.49. The required air content was 4.5±1.5% which is increased by 1% when a high-range water-reducing admixture (HRWRA) is added. In 1992, due to concerns with durability, the maximum W/CM was reduced to 0.40. Also, at that time a pozzolan (Class F fly ash, silica fume) or slag was required for alkali-silica resistivity in concrete containing siliceous aggregates and cements with alkali contents exceeding 0.40%. In 1994, VDOT adopted a special provision, experimentally in some bridges, for low permeability concrete, based on the rapid chloride permeability test, AASHTO T 277 or ASTM C 1202. After satisfactory performance during the test period, VDOT plans to adopt this specification for use in all bridges.

Low permeability concrete generally contain pozzolans or slag and have a low W/CM.[3] These modifications result in high compressive strengths which together with low permeability result in more economical structures, initially by reducing construction costs through increased span lengths and the use of fewer beams and, in the long term, by reducing maintenance costs through increased durability.[4,5]

With the use of HPC mixes in prestressed beams, additional prestressing force can be obtained in the beams. To increase the prestressing force and avoid steel congestion caused by additional strands, it is essential to use larger diameter strands (greater than 13 mm [0.5 in]). However, at present there is a moratorium on the use of 15 mm (0.6 in) strands, declared by the Federal Highway Administration (FHWA) in a 1988 memorandum. Additionally, the FHWA memorandum required a minimum strand spacing of four times the nominal strand diameter on prestressing strands for pretensioned applications and a multiplier of 1.6 times AASHTO equation 9-32, the equation used to calculate development length of fully-bonded prestressing strands. Thus, the use of 15 mm (0.6 in) strands at 51 mm (2 in) spacing needs further evaluation, but it is crucial to fully utilizing high performance concrete in long-span prestressed bridges.[6]

This paper summarizes VDOT's current efforts with HPC in prestressed concrete, two test programs to support the HPC in prestressed concrete, and preliminary test data from the first 2 bridges.

VDOT Efforts

VDOT has planned 5 HPC bridge structures with high strength concrete beams in 1995 and 1996 (Table 1). Bridges numbered 1, 2, and 4 will have concrete with a minimum 28 day compressive strength of 55 MPa (8,000 psi) and a release strength of 41 MPa (6,000 psi). Bridge #5 will have beams with 69 MPa (10,000 psi) at 28 days and 45 MPa (6,600 psi) at release. In this bridge, the beams will contain 15 mm (0.6 in) diameter strands while bridges 1, 2, and 4 will have 13 mm (0.5 in) strands; all at the 51 mm (2 in) spacing. All the concrete in bridges 1, 3, and 5 will also be required to meet VDOT's new proposed low permeability concrete special provisions. The special provision requires 1,500 coulombs or less for the prestressed concrete, 2,500 coulombs or less for the deck, and 3,500 coulombs or less for the substructure.

Table 1. Bridges with high-performance concrete

#	Location	Length m (ft)	No. of Spans	Span Length m (ft)	Beam Type	Beams per Span	Beam Str. MPa (psi)	Low Perm.
1	Rte. 40 over Falling River, Brookneal	97.5 (320)	4	24.4 (80)	IV	5	55 (8,000)	Yes
2	Rte. 629 over Mattaponi River, Walkerton	365.8 (1200)	12	30.5 (100)	IV	5	55 (8,000)	No
3	Telegraph Rd. over Fairfax Co. Parkway	55.5 (182)	2	91	Steel	11	N/A	Yes
4	Rte. 10 over Appomattox River, Richmond	654.4 (2147)	22	27.1/29.6 (89/97)	IV	5	55 (8,000)	No
5	Virginia Avenue over Clinch River, Richlands	45.1 (148)	2	22.6 (74)	III	5	69 (10,000)	Yes

Contracts for bridges 1 and 2 have already been awarded and construction began in early 1995. The Brookneal bridge was designed to use 5 beams in each span versus 7 that would have been required with the 41 MPa (6,000 psi) concrete. The spacing between the girders was 3 m (10 ft) which necessitated a deck thickness of 215 mm (8.5 in); an increase of 13 mm (0.5 in) over the original design using 7 girders. Even with the requirement for low permeability concrete, the bridge construction unit cost per square foot (per 0.09 square meter) was $49, which was less than the 1994 average cost of $58 for 34 bridges in the Federal-aid Highway system in Virginia. The initial savings was estimated to be $30,000, approximately 4% of the total bridge cost. Additional savings are expected over the life of the structure due to anticipated longevity and low maintenance of HPC.

The second structure, a 12-span bridge with each span 30.5 m (100 ft) long, is in Walkerton. The contractors had two options: to use AASHTO Type V beams containing 41 MPa (6,000 psi) concrete, or AASHTO Type IV beams with 55 MPa (8,000 psi) concrete. The contractor chose the second option and a cost savings of $160,000 is estimated, approximately 7% of the cost of the bridge. The spacing between the beams is 2.2 m (7.5 ft). The bridge construction unit cost per square foot (per 0.09 square meter) was $47.5 again below the average of $58.

Test Beams
Test Program 1

Before the initiation of the HPC projects, an experimental project was conducted to support the design of high strength, low permeability beams with 15 mm (0.6 in) strands. Four prestressed concrete AASHTO Type II beams each containing 10 strands 15 mm (0.6 in) in diameter at 51 mm (2 in) center-to-center spacing, two on top and eight across the bottom, were fabricated at a prestressing plant and tested to failure at the FHWA Structures Laboratory in McLean, Virginia. Two of the beams were prepared with concrete to develop a compressive strength of 69 MPa (10,000 psi) and the other two with 83 MPa (12,000 psi) compressive strength. The beams were steam cured to obtain 70% of the compressive strength within 24 hours.

Trial batches were prepared at the plant and in the laboratory before the preparation of actual field concrete for the test beams. Concrete tests showed that high-strength air-entrained concrete with 28-day strengths exceeding 69 MPa (10,000 psi) and a minimum release strength of 70% of the 28-day strengths can be produced with W/CM of about 0.30 or below. To achieve such low W/CM requires high amounts of cementitious material, proper selection of aggregates and high dosages of an HRWRA. Thorough mixing and good construction practices must be followed during placement, consolidation, and curing. Excessive retardation should be eliminated by the proper use of admixtures and the control of concrete temperature. To achieve high early strengths, proper temperature management is needed. With low curing temperatures it is difficult to achieve high early strengths, but higher ultimate strengths can be achieved. Optimum temperature for both the early and ultimate strengths can be determined by trial batching and testing.

The mixture proportions for the test beams are given in Table 2. A fixed amount of coarse aggregate, 1,068 kg/m³ (1,800 lb/yd³), was used in these concrete. Thermocouples were imbedded at several points within Beam 1 and Beam 3 prior to casting to continuously monitor temperature. During steaming, a temperature of 70 C (160 F) was planned within the enclosure. However, the temperature inadvertently approached 85 C (185 F) which resulted in temperatures exceeding the boiling point in Beam 1 and nearly so in Beam 3. Test specimens were stored in the recesses of the forms nearest the steam duct vents and some exhibited visual cracks attributed to high heat. These specimens exhibited high variability in strength and some did not meet the strength requirements due to heat-related damage. The variability in strength and the extent of cracking were attributed to the location of the specimens in relation to the steam duct vents. The 28-day compressive strengths for Beams 1 through 4 were 48.6 MPa (7050 psi), 70.8 MPa (10270 psi), 70.1 MPa (10160 psi), and 55.1 MPa (7990 psi), respectively. The corresponding permeability data at 28 days were 2464 coulombs, 625 coulombs, 723 coulombs, and 1448 coulombs.

Table 2. Mixture proportions for the test beams

Ingredients	Test Program 1		Test Program 2	
	83 MPa	69 MPa	69 MPa	41 MPa (slab)
Portland Cement, kg/m³ (lb/yd³)	445 (750)	386 (650)	454 (765)	377 (635)
Type of cement	II M, finely ground		I/II	
Silica Fume, kg/m³ (lb/yd³)			44 (75)	27 (45)
Slag, kg/m³ (lb/yd³)	178 (300)	178 (300)		
W/CM	0.27,0.28	0.31, 0.32	0.28	0.36
CA, kg/m³ (lb/yd³)	1608 (1800)			953 (1607)
CA (max. size, mm)	#78, granite (12.5 mm)		#8, limestone (9.5 mm)	#67, limestone (19 mm)
CA - SG	2.98		2.76	
FA, kg/m³ (lb/yd³)	580 (977)	644 (1086)	714 (1204)	899 (1515)
FA	siliceous sand		crushed limestone	
FA - SG	2.60		2.75	
FA - FM	2.90		3.00	

In addition to temperature monitoring, the beams were instrumented with brass studs, spaced at intervals of 100 mm (4 in) on the outside surface of the beams along the path of both the bottom strands and the draped strands. Measurements with a Whittemore gage were taken before and just after detensioning, as well as at one day, 7 days, 14 days and 28 days from time of placement of concrete. The difference between the initial reading and the readings after detensioning, normalized with respect to the initial readings, yields a strain profile along the face of the girder at the location of the brass studs. Ideally, this strain profile reaches a plateau, indicating that the transfer of prestressing force from the strands to the concrete has occurred. The flexural bond length, or embedment length, can be determined experimentally, and by adding the transfer length, the total strand development length can be determined. Thus, AASHTO equation 9-32 could be evaluated for the 15 mm (0.6 in) strands. Unfortunately, operator error in the initial readings and damaged studs during demolding resulted in significant scatter in the data and a reasonable estimate of the transfer length could not be determined. Thus, the experimental determination of development length was canceled and testing was limited to the determination of cracking strength and ultimate strength. Also, plans for the use of 15 mm (0.6 in) strands for the girders in the first 2 bridges were discontinued (see Table 1, Bridges 1and 2).

The beams, which were simply supported with a clear span of 14 m (30 ft), were tested to determine the maximum load carrying capacity under a concentrated load at midspan. The calculated maximum load design concentrated load at midspan was determined to be 641 kN (144 kips) for the 69 MPa (10000 psi) beams and 659 kN (148 kips) for the 83 MPa (12000 psi) beams, based on the design moment capacity as predicted by AASHTO equation 9-13. The calculated maximum design concentrated load at midspan to cause first crack was 369 kN (83 kips) for the 69 MPa (10,000 psi) beam design and 378 kN (85 kips) for the 83 MPa (12,000 psi) design. The results of the load testing are shown in Table 3. The lowest load to cause first crack was observed to be 400 kN (90 kips) in Beam 1 designed for 83 MPa (12,000 psi) strength. This beam had the high internal temperatures exceeding boiling. All flexural failures were due to concrete crushing in the outermost fibers of the top flange. The prestressed beams with 15 mm (0.6 in.) diameter strands had satisfactory concrete strengths (exceeding 69 MPa [10,000 psi]), and performed as intended under the loading condition. They did not reflect the high variability and lower-than-desired strengths observed in the test specimens damaged by heat since the test beams failed at loads 15% to 20% higher than predicted.

Table 3. Results of load testing - Test Program 1

Test Beam	Predicted cracking load, kN (kip)	Observed cracking load, kN (kip)	Predicted ultimate load, kN (kip)	Observed ultimate load, kN (kip)
Beam 1	378 (85)	400 (90)	659 (148)	721 (162)
Beam 2	378 (85)	445 100)	659 (148)	778 (175)
Beam 3	369 (83)	467 (105)	641 (144)	787 (177)
Beam 4	369 (83)	445 (100)	641 (144)	734 (165)

Test Program 2

The fifth bridge (Table 1), to be built in Richlands in 1996, will have prestressed beams with a 28 day strength of 69 MPa (10,000 psi), a release strength of 46 MPa (6,600 psi), and 15 mm (0.6 in) strands at a spacing of 51 mm (2 in). To support the design of this project and to obtain additional data on the development length for 15 mm (0.6 in) strands, two 9.4 m (31-ft) long AASHTO Type II beams were prepared and tested. These beams were similar to the ones in Test Program 1 except that a composite slab 200 mm (8 in) thick and 1.2 m (4 ft) wide was cast when the beams reached their 28 day design strengths (Figure 1 shows a cross-section of the test beam). The minimum 28-day design strength of the concrete in the slab was 41 MPa (6,000 psi). A concrete block measuring 0.6 x 0.6 x 0.9 m (2 x 2 x 3 ft) containing eight untensioned 15.2 mm (0.6 in) strands was prepared with the same concrete used in the beams. The strands were subjected to pull-out testing to evaluate the bond strength.

Figure 1. Test Beam Dimensions and Test Setup

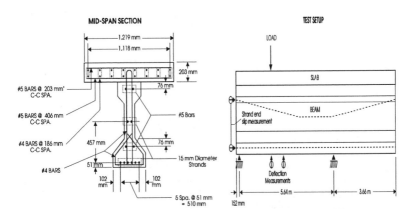

The mixture proportions are given in Table 2. Each beam was cast with one batch of concrete. The first batch had a compressive strength of 58.7 MPa (8,520 psi), and the second 55.0 MPa (7,970 psi) after steam curing, within 19 hours after batching. These specimens were 150x300 mm (4x8 in) cylinders cured within the enclosure. The temperature-matched cure specimens of the same size, which follow the temperature of the beam at the location of the bottom strands, had strengths of 68.1 MPa (9,880 psi) and 61.4 MPa (8,900 psi) respectively, after steam curing. This reflects the effects of higher heat development in the beams versus the specimens kept in the enclosure. Rapid chloride permeability taken as an average of two specimens at 28 days, was 159 coulombs for Batch 1 and 152 coulombs for Batch 2. The slab for both beams was cast using the same batch. It was moist cured for seven days. Moist cured cylinders had strengths of 27.8 MPa (4,030 psi) at 1 day and 81.2 MPa (11,770 psi) at 28 days and a permeability of 753 coulombs at 28 days. Even though the concrete in the slab had a higher W/CM it had higher strengths at 28 days than the concrete in the beams, showing the adverse effect of higher early curing temperatures on the ultimate strengths of concrete.

The transfer lengths were successfully determined, and are shown in Table 4 along with the results of the load tests. The test set-up is shown in Figure 1. This set-up was chosen to ensure two tests per beam. Linear variable differential transformers (LVDTs) were placed on the exposed strand ends to measure any slippage during the load test. LVDTs were also placed under the test beam, at the position of the load and at the span center-line to monitor displacements during loading. Two electrical resistance strain gages were placed on the surface of the slab, at either end of the actuator bearing, to monitor concrete strains during the loading.

The load for the first test was placed at 2.36 m (93 in) from the beam end for the first test. This distance corresponds to the AASHTO equation 9-32 development length for the 15 mm (0.6 in) strands used in the test beams. Hence, a flexural failure was

Table 4. Results from Load Testing - Test Program 2

Beam	Transfer Length cm (in)	Load Location (from beam end) m (in)	Observed Cracking Load kN (kip)	Observed Ultimate Load kN (kip)
Beam 1, end B	55.5 (21.9)	2.36 (93)	979 (220)	1601 (360)
Beam 1, end A	32.5 (12.8)	2.06 (81)	1334 (300)	1913 (430)
Beam 2, end B	38.9 (15.3)	1.75 (69)	1201 (270)	1913 (430)
Beam 2, end A	26.9 (10.6)	1.75 (69)	1245 (280)	2180 (490)

expected, with little or no strand slippage. However, several strands slipped at loads well below the predicted flexural capacity. Shear cracks appeared early, propagating from the end support. A plot of load versus midspan deflection is shown in Figure 2 (a). Note the lack of smoothness of the curve, relative to the plots from the other tests. This response is indicative of the cracking occurring in the beam.   The initial cracking and strand slippage was attributed to poor consolidation at the beam end during casting. One indication of poor consolidation was the exceptionally long transfer length, relative to the other transfer length measurements in Table 4. This assumption was further verified when the pull-out tests were conducted on the concrete block with eight untensioned strands cast with concrete from the batch used for the second beam. The results of the pull-out tests indicated excellent bond strength. The beams were designed with a maximum amount of shear reinforcing at the ends, making consolidation at the ends difficult. All test beams were consolidated with internal vibrators and it is recommended that more attention be given to consolidation and external vibration in addition to internal vibration. Even though slip occurred unexpectedly, the beam was able to carry loads in excess of the maximum design load, reaching an ultimate load of 1601 kN (360 k), 20% above the predicted value. Cracking extended into the deck slab.

The results of the tests are shown in Table 4. Significant strand slippage occurred in the third test (Beam 2, end B) just before reaching its ultimate strength, indicating that the minimum development length was 1.75 m (69 in) from the beam end.

First and Second Bridges

The prestressed beams of bridges 1 and 2 were fabricated with a release strength of 41 MPa (6,000 psi) and a 28-day strength of 55 MPa (8,000 psi). The beams of the first bridge also had to meet the permeability requirement. The beams for each bridge were cast at different plants. Table 5 gives the mixture proportions for the prestressed beams for each bridge. The W/CM ratios were low, close to 0.30. Silica fume was used in the first bridge and slag in the second one.

Twenty beams were cast for the first bridge. The contractor cured 9 of the beams using steam curing and the rest were moist cured. In two of the steam cured batches tested using cylinders placed in form recesses, the strengths were 56.3 MPa (8170 psi) and 54.1 MPa (7840 psi) within 18 hours of batching. The temperature matched cure specimens were 58.1 MPa (8430 psi) and 56.7 MPa (8230 psi)

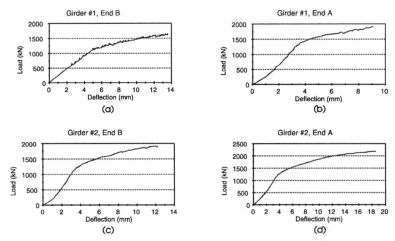

Figure 2. Load versus Midspan Deflection

respectively. The permeability values of the steam cured cylinders were 254 and 290 coulombs at 28 days. The moist cured beams were prepared on Fridays and released on Mondays. The 3 day strengths on two of the batches tested were the same at 53.9 MPa (7820 psi). The 28-day strengths were 83.8 MPa (12,120 psi) and 82.5 MPa (11,960 psi). The permeability values of the moist cured cylinders were 178 and 188 coulombs at 28 days.

Sixty beams were cast for the second bridge, all steam cured. Based on a single cylinder from each of the 60 batches, the average strength at release was 45.1 MPa (6540 psi) with a standard deviation of 4.0 MPa (586 psi) and 61.4 MPa (8900 psi) with a standard deviation of 3.2 MPa (469 psi), at 28 days. Extra cylinders were available from two batches and were tested for permeability. The values were 323 and 536 coulombs at six weeks.

Load tests were conducted on one beam for each of the two bridges in accordance with VDOT specifications. The purpose of the test was to determine the amount of residual deflection in the beam after removal of a maximum load of 95% of the calculated load to cause a flexural crack. Results are reported in terms of a percent difference between the centerline deflection at the maximum load and after the load is removed. The beam is considered to be adequate if a recovery of 90% is achieved.

The percent of recovery for the beam from the first bridge was determined to be 97.9%. The beam tested from the second bridge achieved a percent recovery of 96.5%. No visible cracks were detected on either beam.

Conclusions

Air-entrained HPC with high early (exceeding 41 MPa [6,000 psi] at release), and ultimate (exceeding 55 MPa [8,000 psi] at 28 days) strengths with low permeabilities

Table 5. Mixture proportions

Ingredients	Bridge 1	Bridge 2
Portland Cement, kg/m³ (lb/yd³)	446 (752)	303 (510)
Type of cement	I/II	IIM, finely ground
Silica Fume, kg/m³ (lb/yd ³)	33 (55)	
Slag, kg/m³ (lb/yd ³)		202 (340)
Water, kg/m³ (lb/yd³)	151 (255)	166 (280)
W/CM	0.32	0.33
CA, kg/m³ (lb/yd³)	994 (1,675)	1157 (1,950)
CA (max. size, mm)	#67, limestone (19 mm)	#68, granite (19 mm)
CA - SG	2.76	2.98
CA - Unit Wt, kg/m³ (lb/ft³)	1589 (99.2)	1736 (108.4)
FA, kg/m³ (lb/yd³)	845 (1,425)	586 (988)
FA	crushed limestone	siliceous sand
FA - SG	2.75	2.61
FA - FM	3.00	2.70

have been successfully made on a production basis. Even higher early (exceeding 55MPa[8,000 psi] release) and ultimate (exceeding 69MPa [10,000 psi] at 28-days) strengths can be produced using locally available materials. Achievement of low W/CM is essential and proper selection of materials and high quality control may be needed to maintain the low W/CM. Excessive retardation adversely affects the high early strengths. Proper temperature management is essential to achieve high early and high ultimate strengths. Proper consolidation is needed to produce quality concrete and to provide the bond needed between the strands and the concrete.

Recommendations and Future Work

Based on the results of Test Program 2, VDOT will be requesting a waiver from the FHWA for the use of 15 mm (0.6 in) strands at 51 mm (2 in) center-to-center spacing for the bridge at Richlands. This bridge will be instrumented and monitored for at least three years after construction. Three of the bridge girders will be evaluated for transfer length and end slip at release of pretensioning forces. Internal concrete strains and temperatures will be monitored during and after fabrication, as well as during and after erection. Measurements for loss in camber will be taken periodically.

Acknowledgments

Studies reported here were conducted by the Virginia Transportation Research Council of the Virginia Department of Transportation and sponsored by the Federal Highway Administration. The assistance provided by the FHWA through consultation, review, and use of testing facilities, including the mobile concrete laboratory and the structures laboratory, is very much appreciated.

The opinions, findings, and conclusions expressed in this paper are those of the authors and not necessarily those of the sponsoring agency.

References

1.   Zia, P., Leming, M. L., Ahmad, S. H., Schemmel, J. J., Elliott, R. P., and Naaman, A. E. 1993. *Mechanical behavior of high performance concrete. Volume 1. Summary report.* SHRP-C-361. Strategic Highway Research Program, Washington D. C.

2   Ozyildirim, C. 1993. *Durability of concrete bridges in Virginia.* ASCE Structures Congress '93: Structural Engineering in Natural Hazards Mitigation. Vol. 2, pp. 996- 1001.

3.   Ozyildirim, C. 1994. *Resistance to Penetration of Chlorides into Concrete Containing Latex, Fly Ash, Slag, and Silica Fume.* ACI SP-145, Durability of Concrete, American Concrete Institute, pp. 503-518.

4.   Lane, S. N., and Podolny, W. 1993. The federal outlook for high strength concrete bridges. *PCI Journal,* 38, No.3.

5.   Bruce, R. N., Russell, H. G., Roller, J. J., and Martin, B. T. 1994. *Feasibility evaluation of utilizing high-strength concrete in design and construction of highway bridge structures.* LA-FHWA-94-282. Baton Rouge: Louisiana Transportation Research Center.

6.   Buckner, C.D. 1994. *An Analysis of Transfer and Development Lengths for Pretensioned Concrete Structures.* FHWA-RD-94-049. Office of Engineering and Highway Operations R & D, Federal Highway Administration.

San Angelo High Performance Concrete Bridge in Texas

Mary Lou Ralls, P.E.[1]

Abstract

With its high strength and enhanced durability, high performance concrete (HPC) is anticipated to be cost-effective in bridges, both at the time of construction and throughout the life of the bridge. To assist the States in gaining the knowledge and experience needed to design and construct HPC bridges, the Federal Highway Administration, in conjunction with several state departments of transportation, is constructing HPC pilot bridges. Reported herein are details of one such effort in Texas.

The Eastbound Mainlanes of the North Concho River, U.S. 87 & S.O. Railroad Overpass in San Angelo, Texas, is a 290 m (950 ft) long, 8-span bridge that was designed to utilize HPC in the AASHTO Type IV beams, the composite cast-in-place concrete/precast panel deck, and the cast-in-place windowed substructure. With construction beginning the latter part of 1995, this HPC bridge is scheduled for completion in 1996.

Introduction

The optimum combination of concrete strength and enhanced durability of high performance concrete (HPC) is anticipated to result in lower life-cycle costs. The high concrete strength possible with HPC allows fewer beams and/or interior supports, which should result in savings during construction. The enhanced durability of HPC is anticipated to result in lower long-term maintenance costs, and possibly longer life.

----

[1]Bridge Design Engineer, Design Division, Texas Department of Transportation, 125 E. 11th Street, Austin, TX 78701

To push forward the implementation of HPC in bridges, the Federal Highway Administration (FHWA), in conjunction with several state departments of transportation, is constructing HPC pilot bridges (Lane, 1995). This cooperative effort is expected to accelerate the knowledge and experience gained by the States as efforts continue to alleviate the effects of an aging national highway system.

The Eastbound Mainlanes of the North Concho River, U.S. 87 & S.O. Railroad Overpass in San Angelo is the second HPC bridge in Texas.[2] Its design and construction are in conjunction with a research study sponsored by FHWA, the ten HPC pooled-fund States (California, Georgia, Iowa, Massachusetts, Minnesota, New York, Ohio, Pennsylvania, Texas, and Washington), and the Texas Department of Transportation (TxDOT), in cooperation with the Center for Transportation Research at The University of Texas at Austin. FHWA has also contributed funds for technology transfer through its Office of Technology Applications as part of the Strategic Highway Research Program (SHRP) Implementation Program. The bridge was let to contract in June, 1995, and is scheduled for completion in 1996.

The North Concho River, meandering through the west central Texas city of San Angelo in Tom Green County, is the focus of an attractive urban park developing along its banks. Expansion of U.S. 67 crossing the North Concho River has resulted in the need for two additional bridges to serve as Eastbound and Westbound Mainlanes. Figure 1 shows a rendering of these bridges. Because of the urban park setting, aesthetics was a primary consideration in their design.

The 290 m (950 ft) long, 8-span HPC Eastbound Mainlanes bridge is adjacent to the 292 m (958 ft) long, 9-span normal strength concrete Westbound Mainlanes bridge, as shown in Fig. 2. Spans 1 of these bridges are identical in length and width, and will be instrumented for comparison of behavior. Spans 2 cross the North Concho River. Span 5 of the HPC Eastbound Mainlanes and Span 6 of the Westbound Mainlanes cross U.S. 87. Spans 8 cross the South Orient Railroad. In Fig. 2, dashed lines at interior supports indicate construction joints, whereas solid lines at interior supports indicate expansion joints. The existing overpass had a 1993 Average Daily Traffic (ADT) count of 16,300 vehicles, with trucks accounting for 8.1

----

[2]The first HPC bridge, the Louetta Road Overpass in Houston, Texas, was let in February, 1994, and is currently under construction (Ralls and Carrasquillo, 1994; Ralls, 1995).

percent or approximately 1320 vehicles. The projection for the year 2013 is an ADT count of 26,800 vehicles, with trucks again accounting for 8.1 percent or approximately 2170 vehicles.

Figure 1.   Rendering of Overpass

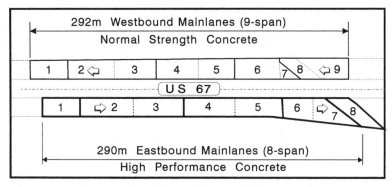

Figure 2.   Plan View of Overpass
(1 m = 3.28 ft)

Shown in Fig. 3 is a cross section of Span 2 of the HPC Eastbound Mainlanes, with its six 1372 mm (54 inch) AASHTO Type IV beams and composite 190 mm (7.5 inch) concrete deck.   Also shown is an elevation view at an interior support.   Each 12.2 m (40 ft) wide bridge includes two 3.7 m (12 ft) traffic lanes, a 3.0 m (10 ft) shoulder, and a 1.2 m (4 ft) shoulder. The width of portions of each bridge varies to accommodate ramps.

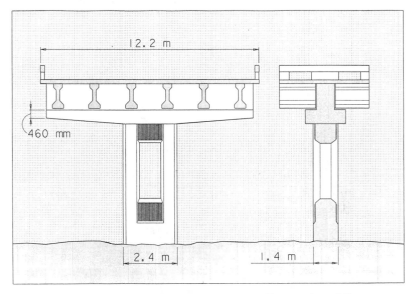

Figure 3.   Cross Section and Elevation View of Eastbound Mainlanes Span 2 (1 m = 3.28 ft)

Evaluating the Limits

In an effort to fully evaluate the potential of high strength concrete in bridges, several design parameters were considered. These include beam spacing, span length, and concrete compressive strength.

Beam Spacing. Span 1 of each bridge is 39.9 m (131 ft) in length, with an expansion joint at each end.   A comparison of these spans shows the most obvious advantage of high strength concrete -- four HPC beams compared to seven normal strength concrete beams.   Beam spacing increases from 1.7 m (5.7 ft) to 3.4 m (11.0 ft) with the use of high strength concrete. The deck overhang is 1.1 m (3.5 ft) for the HPC Eastbound Mainlanes, compared to 0.9 m (3.0 ft) for the Westbound Mainlanes. These spans will be instrumented to monitor behavior during construction and under live load.

Span Length. Span 1 of the Westbound Mainlanes was designed to approximately the maximum capacity of the normal strength concrete AASHTO Type IV beams. As previously discussed, the 40 m (131 ft) span length required four HPC beams at 3.4 m (11.0 ft).

The length of the HPC Eastbound Mainlanes Span 2, at 47.9 m (157 ft), allowed the interior supports to be placed on the banks of the North Concho River. This length is approximately the practical limit that can be designed with the 1372 mm (54 inch) AASHTO Type IV beams. At this length, stability concerns, discussed below, start controlling due to the slender I-shaped cross section. Six beams at 2.0 m (6.6 ft) were required for this span.

Spans 3 and 4 of the HPC Eastbound Mainlanes are 45.7 m (150 ft) and 45.4 m (149 ft) in length, respectively, each with five beams at 2.5 m (8.3 ft) spacing. Span 5 of the HPC Eastbound Mainlanes, which crosses U.S. 87, is 42.8 m (140 ft) in length, with five beams at 2.6 m (8.5 ft) average spacing.

Spans 6, 7, and 8 of the HPC Eastbound Mainlanes are 27.1 m, 21.3 m, and 19.4 m (88.9 ft, 69.8 ft, and 63.8 ft) in length, respectively. These shorter spans were required due to the railroad crossing diagonally underneath Span 8. The vertical clearance requirement for the railroad crossing resulted in 865 mm (34 inch) Texas Type B I-shaped beams in Span 8, rather than the 1372 mm (54 inch) AASHTO Type IV beams used in all other spans. Because of the geometric constraints, the beams in these spans did not require high strength concrete, and will not be discussed further in this paper.

Concrete Compressive Strength. Table 1 lists the concrete design strengths for the beams in Spans 1-5. The required concrete compressive strengths at transfer of prestress range from 61 MPa to 74 MPa (8900 psi to 10,800 psi) for the HPC beams in the Eastbound Mainlanes. The required concrete compressive strengths at 56 days ranged from 75 MPa to 101 MPa (10,900 psi to 14,700 psi) for these beams.

The four HPC beams in Span 1 of the Eastbound Mainlanes require 74 MPa (10,800 psi) at transfer and 94 MPa (13,600 psi) at 56 days. This is compared to the seven normal strength concrete beams in Span 1 of the Westbound

Mainlanes that require 40 MPa (5800 psi) at transfer and 54 MPa (7900 psi) at 28 days. The HPC beams are specified for 56-day strengths rather than the standard 28-day strengths to take advantage of the significant strength gain from 28 days to 56 days that occurs in high strength concrete.

Time	Beam Concrete Design Strength (MPa)					
	Span 1	1	2	3	4	5
	Westbound *	Eastbound (HPC)				
Initial	40	74	63	71	68	61
56 days	54 ‡	94	93	101	97	75

* 13mm dia. strands   (1 MPa = 145 psi)
‡ 28 days

Table 1.   Beam Concrete Design Strengths

AASHTO Type IV HPC Beams

Design.   The AASHTO Type IV HPC beams were designed for a maximum tensile stress of $10 (f'ci)^{1/2}$ at transfer of prestress compared to the $7.5 (f'ci)^{1/2}$ typically used, and $8 (f'c)^{1/2}$ at 56 days compared to the typical $6 (f'c)^{1/2}$ at 28 days. These limits were increased to take advantage of the higher tensile strength of the HPC used in these beams. A constant modulus of elasticity of 41 GPa (6 million psi) was used in the design. Ramon L. Carrasquillo, Ph.D., P.E., University of Texas at Austin researcher for this project, has experimentally verified that these values were obtained for the proposed concrete mix design, along with the initial and 56-day concrete strengths required (see Table 1).

Prestressing Strand.   In order to use the higher tensile and compressive capacities of high strength concrete to resist applied load in pretensioned concrete beams, a larger prestressing strand force at a lower eccentricity is required than can be obtained using 13 mm (0.5 inch) diameter strand at 50 mm (2 inches). The 40 percent greater force that is gained by using 15 mm (0.6 inch) diameter strand at 50 mm (2 inch) spacing is typically needed to fully utilize concrete compressive strengths greater than approximately 70 MPa (10,000 psi).

However, FHWA currently has a moratorium on the use of 15 mm (0.6 inch) diameter strand in pretensioned concrete applications (USDOT, 1988). In addition, the 50 mm (2 inch) spacing of 15 mm (0.6 inch) diameter strand violates the moratorium. Therefore, to obtain FHWA approval, experimental studies on the transfer and development lengths of this larger strand in I-shaped beams were initiated to document bond behavior at transfer of prestress and at ultimate load.

Ned H. Burns, Ph.D., P.E., University of Texas at Austin researcher on this project, is currently conducting these transfer and development length tests. Four 1016 mm (40 inch) Texas Type C beams with 15 mm (0.6 inch) diameter strands were fabricated at Texas Concrete Company in Victoria, Texas, two with HPC and two with normal strength concrete. The two HPC beams were transported to the Phil M. Ferguson Laboratory, where composite decks were cast. Tests are currently underway.

Stability and Deflections. A primary concern with the use of long, slender I-shaped beams is stability during transport to the job site and erection on the bridge supports. The maximum length of the 1372 mm (54 inch) HPC AASHTO Type IV beams is 46.7 m (153 ft), which gives a span-to-depth ratio of 34. This is compared to typical maximum length for normal strength concrete AASHTO Type IV beams of 36.6 m (120 ft), for a span-to-depth ratio of 27. The fabricator is giving special attention to the stability of these beams to avoid damage during shipment to the job site.

The use of higher concrete strength and larger prestress force at greater eccentricity causes uncertainty in the final deflection of the beam. Camber calculations using standard methods tend to show larger predicted cambers for long slender I-shaped beams than are seen in the actual beams. Thus, the design engineers are concerned that the initial predicted cambers may be excessive, while the fabricators are concerned that the final net cambers may be inadequate.

Concerns about stability and deflection of the long HPC beams resulted in a plan to fabricate one of the 46.7 m (153 ft) beams early in the construction schedule. The beam will then be transported to the job site, and erected on bearing pads identical to the support conditions in the bridge. The University of Texas at Austin researchers will monitor deflections, beginning at transfer of prestress, to determine behavior with time. The observed stability and deflection behavior will be used to modify, as needed, the design, fabrication, transportation, and erection of the beams.

HPC Mix Designs

The University of Texas at Austin researchers are working with the contractor and precast concrete fabricators to develop HPC mix designs that have the required strength and durability properties for the beams, the deck, and the substructure.

The fabricator of the HPC beams, Texas Concrete Company in Victoria, Texas, anticipates using one mix design for all beams in Spans 1-5 of the HPC Eastbound Mainlanes. Therefore, as indicated in Table 1, the concrete mix design will require a minimum of 74 MPa (10,800 psi) at transfer and 101 MPa (14,700 psi) at 56 days. Texas Concrete Company achieved an average compressive strength of 104.8 MPa (15,200 psi) for the mix design used for the Louetta Road Overpass HPC U-beams (Ralls, 1995). They are using a slightly modified version of this same mix design for the San Angelo HPC beams. The mix includes 32 percent by volume replacement of the cement with fly ash, which is the maximum replacement allowed by current TxDOT specifications. The mix does not include silica fume or air entrainment. Water reducing admixtures along with retarders are used to ensure a slump of approximately 230 mm (9 inches) with this HPC mix; this is an important consideration for placement.

Guidelines for several HPC mix designs for the Eastbound Mainlanes cast-in-place concrete deck and substructure were given to the contractor's local concrete batch plant by The University of Texas at Austin researchers, for use in the contractor's development of trial mix designs. These included mixes with and without fly ash, but with no silica fume. The batch plant performed trial mixes with local materials, and results indicate that required properties were obtained. Samples of the batch plant's local materials will be sent to the researchers for further testing.

Permeability studies to evaluate the durability of the HPC laboratory trial mix designs for this project are currently underway. Variables for these tests include type and volume of cement, type and size of aggregate, and the use of fly ash and chemical admixtures. Additional durability and strength tests will be done using the concrete from the actual construction.

Bridge Details

As shown in Fig. 3, the superstructure of the HPC Eastbound Mainlanes is similar in appearance to conventional prestressed concrete I-shaped beam bridges with composite decks, except that the spans are longer and the beams are more widely spaced. The deck is 190 mm (7.5 inches) in thickness, the minimum allowed in bridge construction in Texas. The composite deck has precast prestressed concrete panels and cast-in-place concrete. The design concrete compressive strength for both panels and cast-in-place concrete is 41 MPa (6000 psi). The Texas Type T202 traffic rails are slotted, with fractured fin formliners on the outside surfaces.

An unusual aspect of the decks is the use of special reinforcing bars developed by the University of Kansas as part of a large-scale research program to improve the development characteristics of reinforcing bars (CERF, 1994). The top mat of reinforcement in the decks of Spans 2 and 3 in both the Eastbound and Westbound Mainlanes will have bars with a modified deformation pattern that increases the rib bearing area. This special reinforcement will be epoxy-coated in the Westbound Mainlanes and uncoated in the Eastbound Mainlanes, as shown in Fig. 4. Although the improved development length characteristics of the bar will not be used in this application, the bars may help minimize deck cracking. Their effects on deck behavior will be monitored. Deck cracking over the interior supports between Spans 2 and 3, relative to other similar interior support locations on the bridges, will be compared.

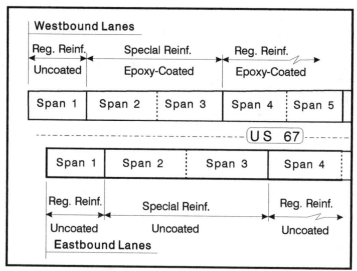

Figure 4. Plan View Showing Locations of Special Deck Reinforcement

The cast-in-place concrete substructure has been designed to complement both the urban park setting and the state-of-the-art materials used. The tapered cap sits atop a single column on a drilled shaft foundation. Windows in the columns visually lighten the substructure. Fractured fin formliners, similar to the railing formliners, enclose the upper and lower portions of the windows in the columns. A rub finish, rather than painting, is specified. Design concrete compressive strengths are 55 MPa (8000 psi) for the caps and 41 MPa (6000 psi) for the columns.

Instrumentation and Monitoring

An increase in the database for behavior of HPC bridges, both during construction and under live load, is critically needed. The University of Texas at Austin researchers in this project have, therefore, developed an extensive instrumentation plan. This plan includes thermocouples and match-cured cylinders to monitor the temperature gradients developed in the HPC beams during curing, and thus allow evaluation of the effect of these temperatures on the properties of the concrete. Electronic resistance strain gauges, vibrating wire strain gauges, and mechanical (DEMEC) strain gauges are being installed at the plant and will be used to measure the concrete strains. An external stretched-wire system is used to monitor deflections. Long-term monitoring of the bridges will document the degree of improvement in performance obtained from the use of high performance concrete.

Summary

During the implementation of high performance concrete into bridge construction, bridge costs for high performance concrete bridges may be somewhat higher than typical bridge construction, as indicated by preliminary bridge cost data analyzed to date on this project. However, once the use of high performance concrete is more standard practice, it is expected to be a cost-effective innovation for bridge design and construction. The Federal Highway Administration and the ten high performance concrete pooled-fund States are providing funding and support to the States to research, develop, and document the details required for implementation. The North Concho River, U.S. 87 & S.O. Railroad Overpass Eastbound Mainlanes is one of the bridges constructed under this program, and is scheduled for completion in 1996. Results to date include the following.

1. A significant reduction in number of beams per span is possible using the high compressive strengths that can be attained in high performance concrete. For the 40

m (131 ft) spans in these bridges, the number of beams was reduced from seven to four.

2. Simply-supported 1372 mm (54 inch) AASHTO Type IV high strength concrete beams can span 47.9 m (157 ft). For these long, slender I-shaped beams, stability and deflections are significant concerns, and may control the design as well as erection procedures.

3. High performance concrete compressive strengths up to 101 MPa (14,700 psi) can be attained in 56 days, with fly ash and without silica fume. The strength gain from 28 days to 56 days can be significant in high strength concrete.

4. The larger 15 mm (0.6 inch) diameter strands at 50 mm (2 inch) spacing are typically required to fully utilize concrete compressive strengths exceeding 70 MPa (10,000 psi). Transfer and development length tests for pretensioned concrete applications are required for FHWA approval at this time.

5. Long-term monitoring of high performance concrete bridges is needed to document behavior.

Acknowledgments

The researchers on this project are Ramon L. Carrasquillo, Ph.D., P.E., Ned H. Burns, Ph.D., P.E., and David W. Fowler,Ph.D., P.E., with The University of Texas at Austin. Graduate students at the Construction Materials Research Group and the Phil M. Ferguson Structural Engineering Laboratory at The University of Texas at Austin are assisting with the experimental testing. The contractors for this project are Jascon, Inc., of Uvalde, Texas, and Reece Albert, Inc., of San Angelo, Texas. Texas Concrete Company in Victoria, Texas, cast research beams and will supply the bridge beams. The FHWA Bridge Division, Structures Division, Office of Technology Applications, Region 6 Office of Engineering, and Texas Division, as well as the project's National Peer Advisory Group, continue to provide timely technical assistance. The bridges were designed by Jeff P. Cicerello, P.E., Norman K. Friedman, P.E., and Mark W. Jurica, P.E., in the Bridge Design Section, Design Division, TxDOT.

References

1. Civil Engineering Research Foundation Summary Report, "Improving Development Characteristics of Reinforcing Bars," CERF Report #94-6002, November, 1994.

2. Lane, S., Munley, E., and Wright, B., "High Performance Materials: A Step Toward Sustainable Transportation," Proceedings of the Innovative Bridge Projects Session of the 1995 Annual Conference of the Transportation Association of Canada, Victoria, British Columbia, October, 1995.

3. Ralls, M. L., and Carrasquillo, R., "Texas High-Strength Concrete Bridge Project," *Public Roads*, Federal Highway Administration, U.S. Department of Transportation, Vol. 57, No. 4, Spring, 1994, pp. 1-7.

4. Ralls, M. L., "High Performance Concrete U-Beam Bridge: From Research to Construction," Proceedings of the Transportation Research Board's Fourth International Bridge Engineering Conference, Vol 2, National Academy Press, 1995, pp. 207-212.

5. United States Department of Transportation, Federal Highway Administration Memorandum, "Prestressing Strand for Pretension Applications - Development Length Revisited," October 26, 1988.

HIGH PERFORMANCE HIGHWAY BRIDGE SUBSTRUCTURES

By Robert W. Barnes,[1] Student Member, ASCE, and John E. Breen,[2] Fellow, ASCE

Abstract

The precast, pretensioned concrete I-beam and concrete slab cast in place on stay-in-place pretensioned deck panel superstructure is the prevailing system for short and medium span highway bridge construction in much of the United States today. High strength materials, plant production methods, repetitive elements and standardized details all contribute to the efficiency of this system. Although this technology has been dominant for several decades, the overwhelming preponderance of substructures for these same bridges consists of reinforced concrete cast in place. A precast, post-tensioned substructure system for such bridges is presented. The system was developed to benefit from the advantages inherent in precast production including: high strength and high performance materials, economies of scale, efficient standardized production and faster on-site erection times. Precast techniques also provide much needed aesthetic improvements through flexibility in utilization of attractive forms and surface textures. Environmentally sensitive sites are spared many of the disturbances that accompany cast-in-place operations. A family of single and multi-column bent shapes were developed based on the general range of applications for precast, pretensioned I-beam bridges. Potential casting and joining techniques are discussed, and examples are presented.

Introduction

Highway bridge construction, particularly in the state of Texas, has benefited greatly from the use of standardized processes. Precast, pretensioned concrete I-beam construction with cast-in-place deck slabs has become the most cost-effective form of moderate span superstructure in the United States. This type of superstructure system benefits from the use of several types of components

---

[1]Grad. Res. Asst., Ferguson Struct. Engrg. Lab., Univ. of Texas at Austin, 10100 Burnet Rd., Bldg. 24, Austin, TX 78758.

[2]Nasser I. Al-Rashid Chair in Civ. Engrg., Univ. of Texas at Austin, 10100 Burnet Rd., Bldg. 24, Austin, TX 78758.

produced by industrialized processes. These include precast slab panels that also serve as work platforms and formwork for subsequent casting operations. Modular concrete railings are chosen from a few standard designs that can be readily precast or slipformed. The backbone of the system consists of the girders themselves. The wide use of a few standard cross-section shapes has resulted in great economies of scale. The repetitive nature of the design, fabrication and erection of this superstructure system, coupled with over thirty years of industry experience, makes this system very difficult to beat in terms of economics and performance.

In spite of the success of this type of superstructure, nearly all of the related substructure construction still takes place in the field. Although the cast-in-place processes utilized in substructure construction are relatively straightforward and very familiar to most contractors, the present method is time-consuming. The time and effort involved in the traffic management aspects of construction often add substantially to the total structure cost. The emotional burdens placed on motorists and pedestrians by the seemingly endless construction activity compound the economic losses experienced by the surrounding community due to both traffic delays and impediments to business access.

Cast-in-place substructures do not feature the enhanced durability which accompanies the use of high performance materials in precast, prestressed superstructures. Inspection reports indicate that the major deficiencies which occur with prestressed concrete bridges are in the substructures. Life-cycle costs can hardly be minimized if the least durable members are those which are often subject to the most aggressive attack. In addition to the agents that wash down from the superstructure, the substructure must withstand physical and chemical attack from below: earth-borne chemicals, salt spray, ice, flowing water and debris, air pollution, etc.

Unfortunately, the cast-in-place processes that are presently utilized often result in unattractive substructures. Efforts to reduce the construction costs of the columns and bents have produced shapes that are easy to form but appear ponderous and dull. The multi-column bents used for most grade separation structures result in a visual effect often described as a "forest" of columns -- a disorderly assembly of vertical elements that belies the smooth horizontal flow of the superstructure. Water runoff from the deck usually produces extensive and unsightly staining of substructure elements relatively early in the useful life span of the structure. The relative proximity of the substructure to human observers compounds the visual effect of damage due to aging.

The technology exists to produce aesthetically pleasing substructures that reap the construction benefits already realized in superstructure construction. The repetitive nature of substructure construction in large highway interchanges or a series of grade separations is such that the standardization of a few cross-sections for precasting could result in substantial cost savings. On-site construction time and related costs would be greatly reduced. Use of precast, high performance concretes and post-tensioning technology would increase the durability and life expectancy of substructures, especially in aggressive environments. The appearance of the substructure would be enhanced by the increased structural efficiency and through the use of high quality forms and surface textures. Such surface treatments would serve as visually and economically attractive alternatives to the painting of concrete bridges and the maintenance associated with this practice.

With the preceding considerations in mind, the Center for Transportation Research of the University of Texas at Austin began a research project titled "Aesthetic and Efficient New Substructure Design for Standard Bridge Systems" in the Fall of 1993. The objectives of the project, cosponsored by the Texas Department of Transportation and the Federal Highway Administration, include improving the aesthetics and efficiency of moderate-span bridge systems by developing improved methods of substructure design and construction.

The pretensioned I-beam and composite slab superstructure system has proved very efficient in its wide use throughout Texas and the rest of the country. The span lengths and slenderness ratios characteristic of this structural system can be readily integrated into a complete bridge system that makes aesthetic sense. Therefore, the specific goal of this research is to apply readily available materials and technology to develop an efficient and attractive precast substructure system for precast, pretensioned I-beam bridges in the state of Texas.

As in the case of precast, pretensioned superstructures, the ultimate success of a precast substructure system depends on the efficiency realized through its repeated use. Therefore, the system must be repetitive yet flexible, offering a variety of applications to the bridge designer. The goal of the researchers is not only to develop the *concept* of a precast substructure system, but to provide enough design *details* so that when the use of such a system is considered, its inherent efficiency is not overshadowed by questions and doubts regarding its constructability.

Aesthetic Considerations

The role of aesthetics in bridge design has become increasingly important in recent years. Designers are recognizing that taxpayers perceive bridges as more than simply supporting devices for their travels. Bridges are an inescapable part of the human environment. In a society that is constantly on the move, highway bridges represent the largest man-made structures encountered by most humans on a regular basis. The visual and emotional impact of a bridge project is of great importance, especially during an era of increased wariness of the role of government in our daily lives. A beautiful bridge becomes a civic asset; an ugly bridge is perceived as more government waste. Projects that imbue societal pride and acceptance of public works create long-term economic benefits that cannot be directly computed from cost estimates or bid prices.

Bridge engineers often confuse aesthetics with ornamentation. This leads to the common misperception that aesthetically appealing bridges necessarily cost more than bridges designed without regard to aesthetics. On the contrary, every design decision affects the aesthetics of a bridge. Roadway geometry, structural system, bent locations, member sizes, surface treatments and other choices all dictate the aesthetic value of a bridge. Possibly the most overlooked factor regarding aesthetics is aging. What may have once been a beautiful bridge loses all aesthetic value when plagued with drainage stains and peeling paint a few years later.

Use of a particular substructure system does not necessarily make for an attractive bridge. The designer must carefully integrate the individual elements of the structure into a coherent whole. Although this system has been developed for application to precast, pretensioned I-beam bridges, blind selection of the proposed columns and bent caps guarantees neither a beautiful nor an efficient bridge. The system should be used as a set of tools or ideas, not as a finished product.

Although numerous authors have written works pertaining to aesthetic bridge design, Leonhardt (1984) and Menn (1990) are among the few to devote serious effort to discussing the aesthetic design of substructures. A few principal themes are present in the writings of both.

The most visually appealing overall form of any structure is usually one which clearly expresses efficient structural function. The most evident manifestation of a bridge's structural efficiency is the slenderness of the superstructure. However, substructure efficiency is represented by the transparency of the space beneath the superstructure and the orderliness of the elements that subdivide this space. Transparency is a function of both the size and number of columns. As with any aesthetic concern, consideration of all possible viewing angles is important. Although a system of bents each supported by several small columns may appear quite transparent from a few angles and distances, expression of structural function may be obscure from other viewpoints. Both Leonhardt and Menn recommend using as few columns per bent as possible, with two as an ideal maximum.

Single column bents should be used for superstructure widths up to approximately 12 m (40 ft). The ratio of superstructure width to column breadth should preferably be between 3.5 and 4.5. Transparency alone does not guarantee aesthetic success. A structure that appears to lack stability fails to express structural efficiency. Single column bents are valuable because the series of single vertical members clearly delineates the flow of the supported traffic. The versatility of the single column bent is useful when the designer is faced with skew crossings.

Bentcaps should also be as transparent as possible, especially for low bridges. The cap should be integrated into the substructure as much as possible. The inverted-T style cap currently used widely in Texas is valuable in this regard (Figure 1). The depth of the inverted-T stem is generally dictated by the girder depth. The bent appears top-heavy if the stem is wider than the supporting column. The stem and ledges should be in good proportion to one another as well as to the girders and columns. Unless a bridge is very low, the flow of forces can be expressed by sloping the soffits of the cantilever cap overhangs.

Combining the considerations above with the common geometric parameters of precast, pretensioned superstructure systems used in Texas, a series of candidate column sizes were developed. Columns were sized so as to be able to support roadway widths ranging up to approximately 13 m (44 ft) on single column bents. To adequately cover this range four standard column breadths (corresponding to the transverse axis of the bridge) were selected: 2000 mm (78.7 in), 2400 mm (94.5 in),

2800 mm (110.2 in), and 3600 mm (141.7 in). Column widths (corresponding to the longitudinal axis of the bridge) were selected based on the appearance of typical I-beam bridges in elevation. Two standard widths were chosen. A 1200 mm (47.2 in) section is recommended for column heights up to approximately 9 m (30 ft). A 1600 mm (63 in) section can be used for taller bridges or for bridges that have an apparent superstructure depth greater than approximately 2400 mm (94 in).

Figure 1. Inverted-T Bent Caps

Two columns per bent should be used for roadways ranging from 12 to 21 m (40 to 70 ft) wide. These bridges generally carry three to four lanes of traffic. These bents may also be used for smaller roadways with large degrees of skew. A pair of 2000 mm (78.7 in) or 2400 mm (94.5 in) broad columns would support the cap in a typical bent. Very wide roadways or excessive span lengths might necessitate the use of three columns per bent.

In addition to the bridge's abstract structural form, the designer must also consider how the shapes and surfaces of the members integrate with the environment. Humans are frequently exposed to substructures at close range. The attractiveness of these "up-close" surfaces is vital to the acceptance of the bridge. Human scale should be incorporated into these elements. Large areas of smooth concrete surface should be avoided. The interest of the observer may be sparked by the use of texture. The proposed substructure system exhibits large-scale texture in the form of chamfered column edges and beveled cap soffits (Figure 2). Medium-scale texture can be produced with the use of form liners. The resulting texture may enhance the vertical nature of the column or simply express a motif particular to the bridge's environmental setting. Fine-scale surface treatments such as exposed aggregate may be used in special situations. The precast plant offers an ideal environment for the application of quality surface textures through the efficient utilization of more complex form systems.

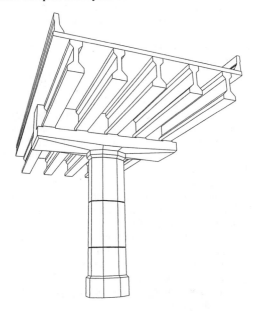

Figure 2. Precast Single Column Bent

The surface of a bridge should resist visual deterioration due to aging. The use of high performance concretes and precast plant processes produces more uniform and durable surfaces than those produced by conventional cast-in-place procedures. Hollow sections with interior drainage capability can be precast readily. Textured concrete surfaces also aid in the mitigation of damage due to aging and seem to discourage graffiti. If thoughtfully placed, grooves produced by form liners channel water away from the more exposed flat surfaces. The resulting stains are hidden in the shadowed recesses, minimizing visual disturbance. Both fine- and medium-scale texture serve as a deterrent to graffiti artists. A textured surface proves to be a less than ideal canvas for this type of expression.

Construction Considerations

The widespread use and economic success of a structural system are usually interdependent. A system must be economically feasible in order to be selected for construction. At the same time, economic benefits are not maximized until the system is widely used. The development of the precast, pretensioned I-beam superstructure is a example of this autocatalytic process. Ease of construction and standardization of girder cross-sections have led to widespread use of a few shapes. The resulting familiarity with the design and construction of this type of bridge in conjunction with the repeated use of forms and equipment has resulted in lower costs and increased use of the system. Constructability and standardization are crucial to the initiation of this process.

Precast concrete offers many construction advantages. Repeated use of standardized elements reaps the benefits of mass production. The plant environment offers more efficient utilization of non-skilled labor and mechanized processes not practical in the field. A recent example is the fabrication of 2440 mm (8 ft) long match-cast column segments for the US183 Elevated project in Austin, Texas. A foreman and one laborer easily produced a segment per day with the occasional assistance of a crane operator and a surveying crew. Reinforcement cages were produced by two ironworkers that divided their time between precasting beds. The foreman readily asserted that it was the easiest system of column construction in which he had ever participated.

On-site construction time is greatly accelerated through the use of precast, post-tensioned construction with match-cast joints. Because dimensional control is readily controlled in the plant and most dimensional tolerances are auto-correcting in the match casting process, proper alignment of surfaces, shear keys and post-tensioning ducts in the field is ensured. Forming and shoring operations are minimized, resulting in the rapid erection of elements with decreased environmental disturbance. Field operations are less prone to weather delay, and the construction season can be greatly lengthened in regions of extreme climate through the use of enclosures and/or plant steam curing. Costs and taxpayer irritation resulting from traffic interference and delays diminish.

Use of post-tensioned reinforcement can add a level of complexity to the design and construction of substructure elements. However, thorough design in which the construction process is fully considered should produce standardized details that allow rapid assembly of the substructure. A recent example of the successful application of this technology is the Chesapeake & Delaware Canal Bridge (DeHaven 1995). The bridge features 48 box piers consisting of precast segments 3 m (10 ft) in height that are post-tensioned with both bar and strand tendons. The contractor was able to erect 30 m (100 ft) of pier in a single day.

Unnecessary column weight is minimized by casting the segments with a central void. ACI-ASCE Committee 343 (1995) recommends a maximum column segment weight of approximately 355 kN (80 kips) so that each segment can be handled by a typical bridge crane. Under this condition, maximum segment lengths range from approximately 6 m (20 ft) for the 1600 mm x 3600 mm column to 9.5 m (30 ft) for the 1200 mm x 2000 mm column. Therefore, a great number of highway bridges would require only one segment per column.

The chamfer dimensions are the same for all eight column sizes in order to allow the use of some form parts for more than just one column size. The hollow core provides a potential location for a drainage pipe. Shear keys allow the easy alignment of match-cast segments. Possible tendon locations are standardized and typically spaced at 200 mm (8 in) intervals (Figure 3).

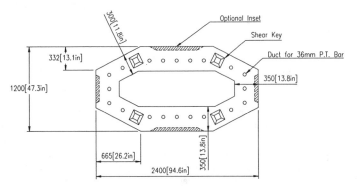

Figure 3. Plan View of 1200 mm x 2400 mm Column Segment

Post-tensioning bars were chosen as the primary reinforcement system for the columns. Column prestressing is concentric throughout, so there is no need for curved tendons. Seating losses are negligible. Easy coupling of bars allows the rapid connection and stressing of segments. Therefore, the full strength and integrity of newly erected portions of the structure may be used for staging subsequent erection operations. Stressing operations are simplified because jacks for bar tendons can be handled by one person.

The column post-tensioning bars must be carefully located and anchored in the cast-in-place foundation cap. The connection between the first column segment and the foundation is the most difficult in terms of geometric control because the two surfaces have not been match cast together. A slight alignment error at this level can result in a column that is significantly out-of-plumb and a difficult cap connection. One solution to this problem is the use of a cast-in-place concrete pedestal. A slight recess is cast into the foundation cap for the seating of the first column segment. The segment is placed in the recess and properly aligned with shims or other supports. Post-tensioning ducts are then spliced from the foundation to the segment. Cast-in-place concrete is then cast around the joint, ensuring accurate alignment. The cast-in-place portion of the joint may be hidden below grade or extended to a height of approximately 1000 mm (40 in) above grade to form an apparent pedestal. Similar methods have been used on the Chesapeake and Delaware Canal Bridge (DeHaven 1995) and the US183 Elevated project.

Once the cast-in-place joint has adequately hardened, subsequent column segments may be placed and epoxied together. Post-tensioning bars are coupled and stressed as required for structural integrity during construction. The column is exposed to small moments prior to placement of cap segments, so there is no need to couple and stress more than a few bars during column erection.

Another geometric challenge is encountered at the column-cap interface. Superelevation of the bridge deck is almost always present due to either horizontal curvature or drainage considerations. This cross-slope typically prevents the column and cap from intersecting at a right angle. There are several methods of addressing this problem. For example, the finished surface of the top column segment may be cast at an angle matching that of the cross-slope. The corresponding cap segment is then match cast against the finished surface of the top column segment. The column segment may be extended slightly into the cap segment form during casting in order to hide any crooked edges resulting from the angled finish. However, the ability to adequately break the bond between the two precast segments must be ensured.

A second option for handling cross-slope is the introduction of a collar segment to the column-cap joint. The column segment is cast with a horizontal finished surface. A small collar segment is then match cast against the column segment. The top surface of the collar segment is finished at an angle matching the cross-slope. Finally, the cap segment is match cast against the collar segment. This method is more complicated and involves one more segment than the first. However, there are advantages in its use. First, the casting difficulties associated with a sloping finished surface are concentrated in a small element. Developing a set of collar forms for a few standard cross-slopes could cost less than constantly altering the larger forms used to cast the column segments. Second, by match casting the cap segments against short collars rather than tall column segments, the cap casting may take place closer to ground level, reducing forming complications and costs. Finally, the collar may be used to express the flow of forces from the cap to the column, much like the column capital in classical architecture.

The system bent caps are inverted-T beams. One of two basic depths is chosen depending on the size of the I-girders used in the superstructure. In order to allow adequate anchorage of column post-tensioning in the cap, stem width depends upon the width of the supporting column(s).

Multiple strand tendons are used as the primary longitudinal reinforcement for the inverted-T beams because of the need to vary the prestress eccentricity in the positive and negative moment regions. Although stressing of these tendons is more complicated and involves heavier equipment than stressing of bars, these operations should only be necessary at the ends of each cap rather than at the ends of each segment as is the case with column segments.

The four basic cap segment configurations are shown in Figure 4. The cap length, number of columns and maximum segment weight determine which case applies. Cap segment weight should be limited to approximately 710 kN (160 kips). This allows the segments to be readily handled by a crane at each end. According to this limit, maximum segment lengths range from 7.5 m (25 ft) for the largest cap cross-section to 11 m (36 ft) for the smallest. Larger segment weights and lengths are possible if they can be economically transported and handled.

In all four cases, one primary cap segment is placed on each column. The joint is epoxied and vertical post-tensioning bars are installed and coupled. The bars are stressed and anchored at the top of the cap segment. The entire Case 1 cap consists of one primary segment. After connection to the column is complete, the horizontal multi-strand tendons are installed, stressed and anchored at the ends of the cap.

Case 2 results when cap weight requires the use of secondary column segments to achieve the desired cap length. After the primary segment is connected to the column, secondary segments are temporarily cantilevered from the primary segment. After both joints have been epoxied and the secondary segments are in place, tendons are installed and stressed as in Case 1.

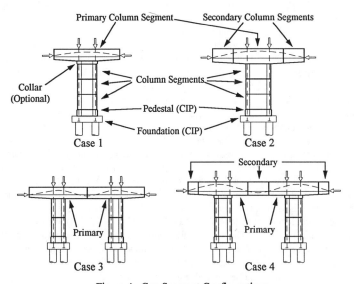

Figure 4. Cap Segment Configurations

Case 3 represents a dual column bent cap that consists of two primary segments. A small, cast-in-place, vertical closure joint between these segments precludes the difficult, simultaneous alignment of horizontal and vertical joints. First, each cap segment is connected to its supporting column as in Cases 1 and 2. After the post-tensioning ducts are spliced between these segments, the closure joint is cast. Cap tendons are then installed, stressed and anchored at the end of the cap to form a continuous beam for service loads.

The Case 4 dual column bent is analogous to Case 2. Cap length and segment weights are such that secondary segments are necessary. The secondary segments are temporarily supported from the primary segments until tendon installation and stressing operations are complete. Similar to Case 3, a cast-place closure joint is required between the interior secondary segment and one of the primary segments. Use of this method may result in dual column bents that support four to five lanes of traffic.

Structural Considerations

The use of precast, post-tensioned construction results in better structural performance. Plant conditions facilitate the quality placement of high performance concretes with low water-cement (W/C) ratios. Durability is increased by the use of these low permeability concretes and by the higher level of dimensional control possible for the placement of reinforcement and forms. Better vibration and curing techniques also result in higher quality elements. Shrinkage and creep effects are reduced through the use of steam cured, low W/C ratio concretes and the maturing of elements prior to erection and stressing.

Post-tensioning of the concrete results in stiffer elements that remain uncracked under service loads, decreasing service level deflections. Smaller stress ranges occur in prestressed reinforcement than in conventional reinforcement, therefore lessening fatigue effects. Decreased susceptibility to corrosion and fatigue results in a longer service life. Increased concrete shear capacity due to axial precompression decreases the amount of shear reinforcement required in the caps.

Unlike cast-in-place substructures, the design of a precast substructure system is often controlled by allowable stresses, particularly at the joints. Concrete stresses must be considered for all construction and service conditions. For match-cast, epoxied joints with no auxiliary bonded reinforcement, AASHTO specifications (1994) do not permit tension at the joints under combined service loads. Limited tensile stress is permitted for construction load combinations. Concrete precompression levels (after losses) of 7 to 9 MPa (1.0 to 1.3 ksi) are required to satisfy these conditions. The use of high performance concrete with f_c values ranging from 56 to 68 MPa (8 to 10 ksi) allows this precompression while preventing compressive overstress under service loads. Without high performance concrete, this system would be applicable to very few bridges.

Column design is typically controlled by biaxial bending stresses due to overturning load combinations. The critical construction design case usually occurs when the superstructure I-beams have been placed on only one "arm" of the cap. The critical service load combination (AASHTO 1992) is usually that of maximum overturning live load with reduced wind and dead loads (Group III), although the combination of full wind load and reduced dead load (Group II) may control the design of very tall columns. Nonetheless, all possible combinations should be checked, and the necessary ultimate capacity should also be verified.

Minimizing service tensile stresses is most efficiently accomplished by increasing the level of precompression while decreasing the magnitude of bending stresses. Thus, the column sections are hollow with wall thicknesses ranging from 300 to 350 mm (12 to 13.75 in) in order to minimize concrete area while maximizing moment of inertia. Nonprestressed longitudinal reinforcement amounting to approximately 0.5% to 1% of the gross area of the column section is distributed in two layers along the interior and exterior faces of the walls.

Post-tensioned bent cap flexural design is also usually controlled by allowable stress design. Critical sections include the face of the column for negative moment and shear, midspan for positive moment, and all segment joints. Judicious placement of live loads in conjunction with dead and wind loads determines the critical service load combination for each section. As with columns, possible transportation and erection conditions should be considered when determining critical construction load combinations. A particular case worth mentioning is the lifting of a primary cap segment that is to be connected to a column. Although this segment is designed for negative moments due to its central support under subsequent loading, it will likely be lifted to its final position by one crane at each end. Nonprestressed reinforcement should be designed for both the positive moment and shear demands of this operation. Adequate ultimate flexural and shear capacity should be ensured for factored loads.

Conclusions

A precast, concrete substructure system is economically feasible for short and medium span highway bridges. Use of high strength materials and precast plant processes can result in attractive bridges that feature improved performance, faster on-site construction, and decreased life-cycle costs. Repeated use of standardized elements and connection details generates economies of scale in the fabrication and erection of elements. Increased attention to the aesthetic value of bridge substructures produces a heightened sense of civic pride and acceptance of public works among taxpayers.

Acknowledgments

The authors gratefully acknowledge the support provided by the Texas Department of Transportation and the Federal Highway Administration. The contents of this paper reflect the views of the authors and do not necessarily represent the official views or policies of either supporting agency.

Appendix. References

American Association of State Highway and Transportation Officials (AASHTO). (1992). *Standard specifications for highway bridges*, 15th ed. AASHTO, Washington, D.C.

American Association of State Highway and Transportation Officials (AASHTO). (1994). *Guide specifications for design and construction of segmental concrete bridges*, 1994 interim specifications. AASHTO, Washington, D.C.

American Concrete Institute-ASCE Committee 343 on Bridge Design. (1995). *Analysis and design of reinforced concrete bridge structures*. ACI 343R-95, Amer. Concrete Inst., Detroit, Mich.

DeHaven, Thomas A. (1995). "Chesapeake & Delaware Canal Bridge." *Aberdeen's Concrete Construction*, 40(9), 739-744.

Leonhardt, F. (1984). *Bridges*. MIT Press, Cambridge, Mass.

Menn, C. (1990). *Prestressed concrete bridges*. Birkhäuser, Boston, Mass.

Dynamic Behaviour of Masonry Church Bell Towers

Alan R. Selby[1] and John M. Wilson[2]

Abstract

Several masonry church bell towers situated in north-east England were forced to vibrate by tolling single bells. The fundamental natural frequencies, mode shapes and damping factors for the sway modes of the towers were deduced from the traces of the tower responses obtained using geophones and/or accelerometers. The natural frequencies in the north-south and east west directions and associated levels of damping are generally similar. The sway mode comprises components of rocking, shear and bending deformations all of which are generally significant. The shear component was found to be the most critical contribution to the response.

Using a computer program PAFEC based on the finite element method of structural analysis selected towers were modelled as monolithic structures using beam or brick finite elements. The periodic forces due to bell ringing were computed by assuming that the bells behaved as compound pendulums. Then the tower responses were computed using a numerical technique to solve the differential equations governing their forced vibration. To allow for masonry joints and other uncertainties concerning the structural fixity of the towers, the Young's modulus and density of the sandstone used in their construction were adjusted to provide the best agreement between the computed and measured natural frequencies, mode shapes and dynamic responses of the towers. It was found that the overall adjustment factors varied from 0.07 to 0.37 for the Young's modulus and from 0.33 to 1.09 for the density.

Introduction

Generally the older English churches are of stone construction and feature a masonry bell tower which is square in plan and fairly stocky. The relatively thick

[1]Reader [2]Lecturer, School of Engineering, University of Durham, Science Laboratories, South Road, Durham DH1 3LE, U. K.

tower walls are of sandwich construction, with ashlar inner and outer leaves and rubble infill as shown in Figure 1. The bells are housed in a bell frame situated in the bell chamber towards the top of the tower. The bell ringers stand in the ringing chamber below.

The bells are of bell metal (a bronze containing up to 20% tin) and are tuned to a particular scale. The lightest (treble) bells typically have masses from 100 to 200 kg and the largest (tenor) from a 500 to 3000 kg depending on the tower size.

In the English system of bell ringing the bells are swung full circle. Each bell is attached at its crown to a headstock with gudgeon bearings mounted on the bell frame which is built into the tower walls. Near one gudgeon is attached the rope wheel on which is wound the rope used by the bell ringer to control the rotation of the bell. Prior to ringing, each bell is raised until it rests against a stay with its mouth nearly uppermost in either handstroke or backstroke positions. From either position the bell can easily be set in motion and controlled to swing full circle by the bell ringer.

There are several modes of ringing. The most common are tolling (a single bell), firing (a number of bells sounded simultaneously), rounds (a number of bells sounding in a fixed repetitive sequence) and change ringing by methods which are complex sequences of ringing.

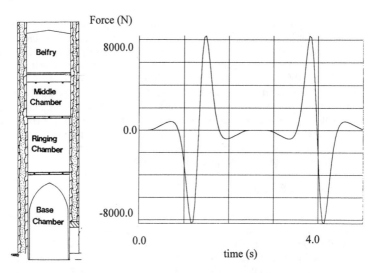

Figure 1. Sectional Elevation through Typical Tower

Figure 2. Typical Horizontal Force for a Whole Pull (Bell 8, St. Oswald's Church)

When swinging full circle in a regular manner, each bell behaves as a compound pendulum and transmits through its bearings to the tower walls periodic horizontal and vertical force components. The horizontal force acts in the direction of swing and contributes significantly to the tower sway response. Figure 2 shows a typical trace of the horizontal force component produced by a whole pull (a complete full circle oscillation of the bell starting and finishing on the same stroke). For a given bell undergoing a single oscillation the time dependent nature of these forces has been analysed (Wilson and Selby, 1993). By summing the separate effects of several bells, the forces imposed by any ringing mode can be found provided that due regard is taken of the phases and planes of the full circle oscillations. To aid the calculation of the forces a computer program BELL was written (Wilson, 1988).

When a structure is allowed to vibrate freely it does so in its natural modes each of which is associated with a characteristic natural frequency. However because the modes associated with higher frequencies are generally much more heavily damped than the lower modes, after a short time the fundamental mode usually predominates. This is particularly so if the natural frequencies are well spaced or there is little cross coupling between modes associated with similar frequencies. Natural frequencies and mode shapes are characteristic of a particular structure and can be used to identify the structure or to validate a given structural model.

The measurement of responses of many masonry church towers located in north-east England is described in this paper. In particular two churches (St. Brandon's, Brancepeth and St. Oswald's, Durham) and Durham Cathedral have been extensively studied both experimentally and by use of computer modelling (Selby and Wilson (1991), Wilson et al. (1993), Wilson and Selby (1993) and Lund et al. (1995)). The experimental results show that the towers vibrate primarily in their fundamental sway modes in north-south and east-west directions. Although shear deformation is significant in the tower vibration, some rocking of towers on their foundations may occur and also bending. Uncertainties regarding the foundations, wall construction, masonry joints and structural connection to the rest of the church make straightforward computer predictions difficult.

Measurement of Response

Two sets of instrumentation were used to obtain responses of towers to bell ringing. In the first method the responses of St. Brandon's, Brancepeth, St. Oswald's, Durham, St. Cuthbert's, Benfieldside and Durham Cathedral were obtained using geophones. Also the responses of one of these churches (St. Oswald's) and several other local churches were measured using accelerometers.

Geophones are instruments which measure velocity. They tend to be quite bulky but produce large enough voltage signals to enable direct recording onto a pen recorder capable of simultaneous plotting of two traces. Three geophones, two measuring horizontally and one vertically, were used. One of the horizontal geophones was used as a reference in a fixed location and the others were placed at different stations to enable mode shapes to be measured. The geophones were placed wherever possible on flat horizontal stone surfaces rather than a wooden floor or on the bell frame so that the tower response was measured rather than some local vibration. Stations were selected at several convenient heights and plan positions within the tower.

With the moveable geophones at a given station in each tower, the pen recorder was activated and the tower was excited separately in the north-south and east-west directions by tolling the largest bell acting in those directions for an interval of at least 30 seconds. This was considered sufficient to bring the response into a steady state. After this interval ringing ceased and the bell was allowed to rest against its stay. The tower then vibrated freely and the subsequent damped motion was recorded for a further interval of at least 10 seconds. Figure 3 shows part of a geophone trace of the horizontal velocity response for St. Oswald's Church at bell chamber level lasting over 5 seconds. The typical decay between each full circle swing can be seen.

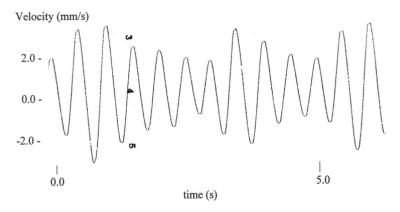

Figure 3. Geophone Trace for St. Oswald's Church, Durham

Accelerometers measure acceleration directly. A pair of accelerometers were used for the second investigation, one being used as a reference. Their signals have to be amplified by a charge amplifier before they are strong enough to drive the pen plotter. Otherwise the experimental method and the results obtained were similar to those for the geophones. Figure 4 shows part of an accelerometer trace for the horizontal response of the tower of Christchurch, Consett with excitation provided by bell 6. The damping after the cessation of ringing can be seen in the final portion of the trace.

Acceleration   (mm/s²)

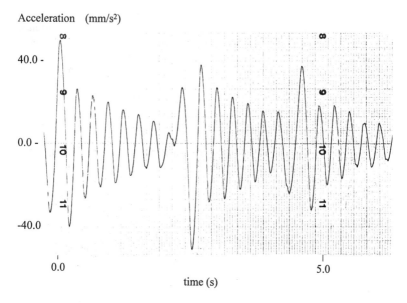

time (s)

Figure 4. Accelerometer Trace for Christchuch, Consett

<u>Experimental Results</u>

By examining all the traces for a given tower it was found that there was a predominant response frequency common to all traces, and that all pairs of traces recorded simultaneously were in phase. Furthermore the mode shape derived from both steady state and decaying portions of the traces were similar and showed that the towers were responding in a fundamental sway mode. The traces showed that the horizontal component of the response varied almost linearly with the height of the geophone station and was not influenced significantly by its plan position. This indicated significant rocking and/or shear deformation and a much smaller amount of bending. Generally there was some vertical motion which increased with the height of the measuring station. This confirmed that some rocking of the towers occurred on their foundations and that some bending deformation was present. Attempts were made to investigate whether there was a torsional component in the mode but nothing conclusive was found. The results from these measurements are shown in Table 1 where the peak response velocities obtained using the geophones have been converted to equivalent accelerations to render them consistent with the accelerometer results.

From the table it can be seen that the natural frequencies associated with the fundamental sway modes of a given tower are generally similar. This is because the majority of the towers considered are nearly square in plan. The levels of damping

Name of Church and Location	Bell				Natural Freq (Hz)	Damping Factor	Mode Percentage			Peak Accn (mm/s2)
	Number	Orientation	Mass (kg)	Tolling Period (s)			Rocking	Shear	Bending	
Durham	9	E-W	1096	2.52	*1.28	*0.016				*10.5
Cathedral	10	N-S	1425	2.46	*1.31	*0.016				*23.0
All Saints	6	E-W	432	2.40	2.56	0.026	16	72	12	84.8
Lanchester										
Christchurch	6	E-W	434	2.37	3.38	0.031	14	62	24	48.1
Consett	8	N-S	822	2.42	2.48	0.078	6	87	7	96.5
St. Andrew's	6	E-W	356	2.57	2.20	0.051	27	45	28	32.3
Bishop Auckland	8	N-S	829	2.65	2.51	0.065	26	44	30	28.5
St. Brandon's	7	N-S	498	2.57	*2.57	*0.032				*71.1
Brancepeth	8	E-W	693	2.66	*2.85	*0.020				*71.1
St. Cuthbert's	3	E-W	338	2.43	*2.25	*0.026				*22.6
Benfieldside	6	N-S	643	2.42	*2.27	*0.027				*31.4
St. Cuthbert's	8	E-W	966	3.30	1.59	0.028	20	36	44	129.8
Chester-le-Street										
St. Edmund's	4	E-W	469	2.46	2.20	0.017	35	46	19	40.9
Sedgefield	5	N-S	533	2.44	2.28	0.018	40	37	13	33.2
St. John's	7	N-S	589	3.05	3.68	0.028	20	26	54	15.8
Shildon	8	E-W	829	3.21	3.49	0.046	16	40	44	18.6
St. Margaret's	7	N-S	432	2.09	3.88	0.027	18	11	71	10.4
Tanfield	8	E-W	660	2.28	3.13	0.045	14	32	54	37.8
St. Mary's	7	N-S	408	3.51	3.04	0.029	29	32	39	17.4
Richmond	8	E-W	559	3.66	2.55	0.038	25	59	16	33.2
St. Mary's	6	N-S	210	2.50	2.36	0.018	21	20	59	11.1
Shincliffe										
St. Matthew's	7	E-W	1102	3.32	1.49	0.028	9	60	31	73.4
Newcastle	8	N-S	1588	3.29	1.92	0.026	22	60	18	57.5
St. Michael's	4	E-W	432	2.94	3.63	0.016	6	64	30	20.1
Heighington	6	N-S	782	3.08	3.16	0.023	9	64	27	34.8
St. Michael's	7	N-S	464	2.61	2.71	0.031	14	38	48	38.6
Houghton	8	E-W	610	2.7	2.98	0.049	19	43	38	36.3
St. Nicholas	1	E-W	203	2.31	1.82	0.014	12	39	49	31.7
Durham	6	N-S	516	2.55	1.38	0.014	6	34	60	46.8
St. Oswald's	7	N-S	447	2.60	2.05	0.025	22	35	43	44.6
Durham	8	E-W	651	2.65	2.01	0.025	30	23	47	59.8
Newcastle	12	N-S	1912	2.99	2.04	0.014	6	59	35	36.1
Cathedral										

* Measured by geophone

Table 1. Measured Church Tower Responses

associated with these modes are generally small, again generally similar and comparable with those found in many modern structures. Rocking, shear and bending deformations vary widely but all are generally significant.

Once the relative components of rocking, shear and bending in the mode shapes have been determined it is possible to identify which mode of failure is critical to excessive dynamic response. The critical conditions were assumed to be:

i)   rocking sufficient to reduce the vertical stress to zero in one outer edge of   the masonry at the base of the tower;

ii) shear stress in excess of 0.2 MPa, with insignificant coincident normal stress, such as occurs at bell frame level;

iii) bending stress just sufficient to exceed the vertical stress component due to self weight at the tower base.

The measured rocking, shear and bending deformation components were analysed with respect to these criteria. The risk of rocking failure was negligible. In only one case, St. Margaret's Church, Tanfield, was bending found to be most severe. In all other cases, shear was the critical mechanism of failure, with factors of safety typically of ten or more.

Computer Modelling of the Towers

The finite element computer program PAFEC-PC (Henshell,1984) was used to compute natural frequencies, mode shapes and responses to bell ringing of three selected towers. In each case a survey of the tower structure and bell frame had to be undertaken, some properties of the bells measured (Heyman and Threlfall, 1976) and also the material properties of the local sandstone used in the towers' construction found.

Detailed surveys of the bell towers at St. Brandon's Church, Brancepeth, St. Oswald's Church, Durham, and Durham Cathedral were undertaken. These towers date from the 12th to 15th centuries and are constructed from a local sandstone. St. Cuthbert's, Benfieldside was also surveyed but differs in many respects being comparatively modern and having a spire. For each tower measurements were made of the overall dimensions of the tower, wall thicknesses and the position of significant apertures. The positions of the bells within their frames and the heights of their bearing axes were also measured. Laboratory tests were undertaken to measure the density and elastic properties of locally quarried sandstone similar to that used for the tower construction.

The masses of the bells were found from foundry records. Other mechanical properties were deduced by measuring the period of small oscillations of the freely swinging bells and by hanging weights on the bell ropes and measuring the rotation of the bells (Heyman and Threlfall, 1976).

In producing structural models for the towers it was assumed that each tower behaved as an independent structure, that the foundation at ground level was rigid and that the elastic and inertial properties used for the model could be factored from those measured for the rock to allow for structural inhomogeneities such as the masonry joints and the rubble infill walls. Bell forces were calculated assuming frictionless bearings and that the effect of the bell ringers' actions was equivalent to impulses applied to the bells. Sensitivity studies were also undertaken to test the validity of these assumptions.

In the finite element method of structural analysis, the structure is divided up into a number of elements of finite size. These finite elements are of simple shape and their structural behaviour is either known or can be easily computed. The complete behaviour of the structure can be built up from the behaviour of the constituent elements.

Two types of finite element, a two-dimensional beam element and a three-dimensional brick element, available with PAFEC were used to form the tower structures. The beam element was capable of representing shear behaviour as well as bending. Although the brick element model is capable of more accurate modelling of the tower geometry, the beam elements produce simpler models requiring less computational effort. However the brick element models can be used to improve the accuracy of the beam models. Figure 5 shows a three-dimensional finite element brick model of the tower at St. Oswald's Church.

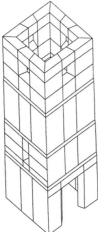

Figure 5. 3D Finite Element Brick   Figure 6. Fundamental Sway Mode
Model of St. Oswald's Tower   of Durham Cathedral Tower

From each tower survey a brick element model of the tower could be constructed and the natural frequencies and mode shapes computed using the PAFEC program. These could be forced into agreement with the measured values by adjusting the value for the Young's modulus of the rock used in the computation from that measured. From the surveys and the results from the brick models, the parameters required to produce each beam model could be calculated and adjusted so that the computed natural frequencies and mode shapes again agreed with the measured values. Figure 6 shows the fundamental sway mode of vibration for the tower of Durham Cathedral produced by PAFEC.

The response calculations were performed using the PAFEC program with the forces due to bellringing computed separately using the program BELL. As far as possible the input data for computing the bell forces corresponded to the actions of the bellringers in terms of the regularity and length of ringing. Using different techniques PAFEC allows both sinusoidal (steady state) and transient responses to be computed. Although the former is simpler its application imposes restrictions and therefore the latter method, the solution of which is based on the Newmark β method (Newmark, 1959) was used. Transient responses were produced with the level of damping set to the measured values, the value of β set to 0.25 to ensure computational stability and the time step chosen to be of the order of milliseconds to ensure accuracy in the computations. Figure 7 shows part of the velocity response of the tower of St. Oswald's Church computed using a beam model. By adjusting both the values of the Young's modulus (for the second time) and the density by the same factor, optimal agreement between the computed and measured responses could be obtained.

Velocity (mm/s)

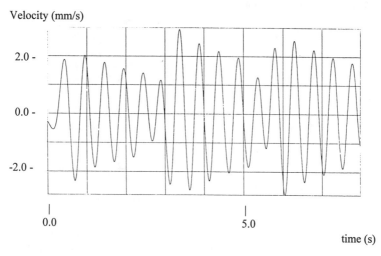

Figure 7. Computed Horizontal Velocity Response for St. Oswald's Church Tower

Computed Results

Table 2 shows a summary of the results obtained from the measurements using the geophones and those produced by the PAFEC program using brick and beam elements. The table shows the location of the tower, the number of the bell tolled, its mass, the direction in which it swings and the time for a whole oscillation of the bell. The table also shows the factors applied to the material properties of the sandstone, the measured and computed natural frequencies in each of the tower's fundamental sway modes, the damping factors measured from the decaying oscillations, the contribution to each mode from rocking, shear and bending deformation and the measured and computed peak responses, the latter obtained using beam models only.

Name of Church and Location	Bell				Material	Factors	Natural Freq (Hz)			Damping Factor	Peak Accn (mm/s2) Comp/ Meas
	Number	Orientation	Mass (kg)	Tolling Period (s)	Young's Modulus	Density	Measured	Computed			
								Beam	Brick		
Durham	9	E-W	1096	2.52	0.37	1.09	1.28	1.28	1.29	0.016	10.5
Cathedral	10	N-S	1425	2.46	0.15	0.46	1.31	1.3	1.31	0.016	23.0
St. Brandon's	7	N-S	498	2.57	0.08	0.47	2.57	2.56	2.55	0.032	71.1
Brancepeth	8	E-W	693	2.66	0.10	0.64	2.85	2.87	2.86	0.020	71.1
St. Cuthbert's	3	E-W	338	2.43	0.07	0.37	2.25	2.25	-	0.026	22.6
Benfieldside	6	N-S	643	2.42	0.07	0.41	2.27	2.27	-	0.027	31.4
St. Oswald's	7	N-S	447	2.60	0.07	0.46	2.08	2.08	2.09	0.025	43.1
Durham	8	E-W	651	2.65	0.06	0.32	2.01	2.02	2.01	0.025	75.6

Table 2. Measured and Computed Responses of Selected Towers

Discussion

The analysis of masonry structures is notoriously difficult because of the nature of and techniques used with the constuction material. Modelling masonry joints, whether formed with or without mortar, is uncertain due to cracking which produces non-linear behaviour. Furthermore because of uncertainty about the tower wall construction, particularly the thickness of the leaves, the arrangement of ties between the leaves and the integrity of the rubble infill, this problem is exacerbated. Also the nature and condition of the structural connections of the tower to other parts of the church can only be surmised. Thus the towers have been modelled as elastic, isolated, rigidly fixed, uniform, monolithic structures and discrepancies accounted for by applying reduction factors to the known properties of the rock.

Only rarely is it possible, for example when remedial work is being undertaken, to determine foundation conditions or the exact nature of the tower wall constuction. Extensive maintenaince and alterations over the centuries means that possibly little of the original structure remains even though old masonry is re-dressed and re-used.

Although computer resources needed to model structures elastically are readily available, the models so produced are necessarily restricted by the quality and accuracy of the input data. In church towers, the forced excitation, although not available as a pure harmonic form, does allow the possibility of estimation of the size and nature of the forces. The major inaccuracy here is the representation of the action of the bellringers and the assumptions of periodicity.

It would also be possible in the finite element model to account for foundation flexibility and for the stiffness of attached structures in a relatively simple manner if such data were available or more measurements undertaken. However the latter would be subject to the limitations and sensitivity of the instrumentation. Although foundation movement was recorded in some cases, such movement was generally slight and required confirmation.

Conclusions

The studies on church towers have generally shown that there are two dominant fundamental modes of vibration of the tower which involve uncoupled sway motions in either the north-south or the east-west directions. The natural frequencies associated with these modes are generally similar. By tolling a bell which swings in the appropriate direction only that mode will be excited. Higher modes, which tend to involve both sway and/or torsion, do not contribute significantly to the response.

The levels of damping found in the predominant modes are generally small, nearly equal and comparable with those found in many modern structures.

Because of uncertainties concerning the detailed tower structure, the method of factoring the rock properties offers a promising approach. Overall adjustment factors varied from 0.07 to 0.37 for the Young's modulus and from 0.33 to 1.09 for the density.

The components of the sway response modes were identified in terms of the proportions of rocking shearing and bending components. For nearly every tower, failure would be initiated in shear.

Acknowledgements

The authors would like to acknowledge the collaboration over several years of the many people involved in or contributing to this work including church authorities for permission to undertake the measurements, bell ringers for their willing labours and former students for their enthusiasm in carrying out final year projects.

References

Henshell, R. D. (ed.), 1984, PAFEC-PC-Data Preparation, Pre- and Post-processor and Theory Manuals, Pafec Ltd., Nottingham.

Heyman, J. and Threlfall, B. D., 1976, Inertia forces due to bell ringing, Int. J. Mech. Sci., Vol. 18, pp 161-164.

Lund, J. L., Selby, A.R. and Wilson, J. M., 1995, The dynamics of bell towers - a survey in northeast England, in STREMA4 [Proc. of 4th Int. Conf. on Structural Repair and Repair of Historic Buildings], (ed. C. A. Brebbia and B. Leftheris), Vol. 2, pp 45-52, Crete, 1995, Computational Mechanics Publications, Southampton.

Newmark, N. M., 1959, A method of computation for structural dynamics, Proc. Am. Soc. Civ. Engrs., J. Engrng. Mech. Div., Vol. 85 (EM3), pp 67-94.

Selby, A.R. and Wilson, J. M., 1991, The dynamic response of a church bell tower to bell ringing, in STREMA2 [Proc. of 2nd Int. Conf. on Structural Repair and Repair of Historic Buildings], (ed. C. A. Brebbia), Vol. 2, pp 3-16, Seville, 1991, Computatinal Mechanics Publications, Southampton.

Wilson, J. M., 1988, Periodic Forces on Bell Towers due to Bell Ringing, Internal Report, School of Engineering, University of Durham.

Wilson, J. M. and Selby, A.R., 1993, Durham Cathedral tower vibrations during bell-ringing, in Engineering a Cathedral [Proc. of Conf. on Engineering a Cathedral], (ed. M. Jackson), Vol. 3, pp 491-500, Bath, 1993, Computational Mechanics Publications, Southampton.

Wilson, J. M., Selby, A.R. and Ross, S. E., 1993, The dynamic behaviour of some bell bell towers during bell ringing, in STREMA3 [Proc. of 3rd Int. Conf. on Structural Repair and Repair of Historic Buildings], (ed. C. A. Brebbia and R. J. B. Frewer), Vol. 3, pp 491-500, Bath, 1993, Computational Mechanics Publications, Southampton.

Dynamic Response of Hagia Sophia

A.S. Cakmak*, C.L. Mullen*, and M.N. Natsis*

Abstract

Finite element studies of Hagia Sophia, a sixth century masonry edifice, in Istanbul, Turkey, provides insight to the structure's response to dynamic loads. The church contains four great brick arches springing from stone piers that offer primary support for a 31-meter diameter central dome and two semidomes. Stone and brick masonry material properties for the numerical model are adjusted to match system mode shapes and frequencies identified from measured response to a recent low-intensity earthquake. The calibrated model is used to predict the measured responses, and the effect of soil-structure interaction is demonstrated. Stresses under simulated severe earthquake loading are estimated at the critical locations in the arches.

Introduction

Begun in 532 as the principal church of the Eastern Roman Empire (and converted to a royal mosque after the fall of Constantinople in 1453), Hagia Sophia in Istanbul held the record as the world's largest domed building for some 800 years. In order to preserve this historical structure, it is necessary to understand its earthquake response in its present condition. This paper addresses aspects of the present day dynamic behavior of the primary dome support structure under recorded and likely earthquake excitation[10].

Early development of a numerical model for eigenvalue analysis of Hagia Sophia has been discussed by Cakmak et al. [1]. The first three mode shapes correspond to simple horizontal translation (modes 1 and 2) and a complex form of torsional rotation (mode 3) of the entire primary structural system. A very good match was attained in the first three natural frequencies measured during an ambient vibration survey of the actual structure (see Erdik et al. [2]). A low-level event of magnitude 4.8 occurred on March 22, 1992, with epicenter

---

*Department of Civil Engingeering and Operations Research, Princeton University, Princeton, NJ 08544

at Karabacey, Turkey, about 120 km south of Hagia Sophia. Strong motion acceleration time histories recorded for this event have been analyzed in both the time and frequency domains.

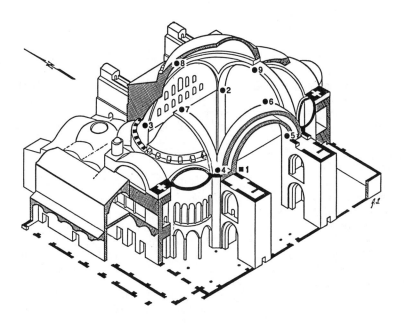

Figure 1. Main Dome support structure and strong motion instrument array.

The response of the numerical model with the material parameters obtained from a system identification procedure are compared to the measured experimental results and the model is modified to improve the accuracy of the predictions and provide damage estimates under severe earthquakes.

System Properties and Measured Response

The primary structure supporting the main dome of the Hagia Sophia and its orientation are illustrated in a cutaway view in Figure 1. The main dome is spherically shaped and rests on a square dome base. Major elements include the four main piers supporting the corners of the dome base and the four main arches that spring from these piers and support the edges of the dome base. The instrumentation array described by Erdik et al. [3] has been designed to capture the motion of the major elements comprising the main dome support structure during earthquake events.

Table 1: Comparison of strengths of mortar materials

Binder	Time to set (h)	Time to full strength	Tensile str. (MPa)	Compr. str. (MPa)
Lime	24	100 days-yr	0.3 - 0.7	9
Gypsum	0.5 - 1.0	0.5 - 1.0 h	4.6 - 5.0	46 - 50
Portland Cement	5-8	100-150 days	2 - 3	21 - 28
Pozzolanic	10 - 12	150 days - 1 yr	3.4 - 3.8	14 - 17

The main piers are comprised of stone masonry. The stone blocks are almost rigid, whereas the mortar is relatively compliant. The main arches and dome are comprised of brick masonry with a mass density of about 1500 kg/m^3, which is lighter than present day concrete. The mortar used in the brick masonry may be classified as a pozzolanic material. Pozzolanas generally contain phases that contribute soluble silica in the presence of lime. Burnt shale or brick dust could be considered as artificial pozzolanas, and there is evidence of the use of crushed brick as a pozzolana in ancient Roman masonry called 'coccio pesto'[4].

Such cementitious mortars have much higher tensile strengths than pure lime ones. Table 1 provides a comparison of strengths for various mortars[5].

Tensile strength of the mortar used in this study has been estimated from tests on samples taken from a 6th century rib in the Hagia Sophia main dome. Two samples were tested at Princeton University using adhesion test specimens and strengths of 0.4 - 0.7 MPa were estimated. Another sample was tested at the National Technical University of Athens (NTUA) using the scratch width method with a range of tensile strength estimated to be 0.5 - 1.2 MPa. Two split cylinder specimens approximately 40 mm long and 35 mm diameter have been taken from the SE buttress stairwell. Estimates of the split cylinder tensile strength are 0.7 and 1.0 MPa.

Elastic moduli have been estimated using non-destructive, in-situ, ultrasonic tests at various brick and mortar locations in Hagia Sophia including a main dome rib, the west arch, and the north arch[6]. The estimated dynamic elastic moduli are:

$$\begin{aligned}
\text{Brick:} & \quad E_b = 3.10 \ GPa; \\
\text{Mortar:} & \quad E_m = 0.66 \ GPa; \\
\text{Composite:} & \quad E_{bm} = 1.83 \ GPa.
\end{aligned}$$

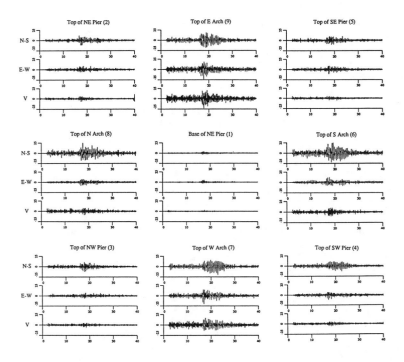

Figure 2. Measured accelerations at the locations shown in Figure 1 for March 22, 1992, Karacabey earthquake

Figure 2 shows the earthquake induced N-S (Y), E-W (X), and vertical (V or Z) acceleration component time histories recorded during the 1992 Karacabey event. The three acceleration components for each response location are displayed in a plan arrangement oriented to the view seen from the entrance to the basilica. In all cases the records indicate a nonstationary process acting with three approximately stationary time intervals. The first interval extends from 2 to 12 s and corresponds to the action of compressional waves; the second from 15 to 25 s corresponds to the arrival of shear waves; and, the third from 30 to 40 s corresponds to a period of decay in the energy of the input motion. From Figure 2 it is shown that the N-S motion tends to dominate the response motion with particularly intense N-S motions at locations 4 and 6. The peak displacement at location 4 is estimated as .071 cm which is double that of the locations 2, 3, and 5. Similarly, the peak displacement at location 6 is estimated to be 0.13 cm which is almost double that at location 8.

Table 2: Comparison of Measured and Modal Frequencies (in Hz)

	Translation East-West Mode 1	Translation North-South Mode 2	Torsion Mode 3
Ambient Vibration Study	1.8	2.1	2.4
March Earthquake			
Power Spectral Analysis	1.7	1.8	2.3
Transfer Function Analysis			
Interval 1	1.8	2.0	2.4
Interval 2	1.5	1.9	2.2
Interval 3	1.7	1.8	2.3
HSDYNTC9 Model	1.8	1.9	2.2

Figure 3 shows the experimental linear system transfer functions, H_{XX}, and H_{YY}, obtained for the first two modes using the procedure described by Cakmak et al. [1]. H_{XX} relates X (E-W) response motion to X (E-W) input motion at the floor level, and H_{YY}, the Y (N-S) response caused by Y (N-S) input at the floor level. Using the second interval, the observed mode 1 and mode 2 frequencies given in Table 2 are obtained.

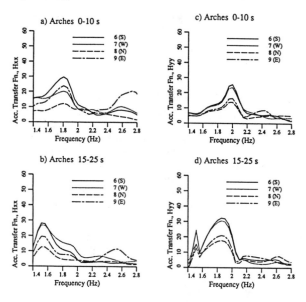

Figure 3. Transfer functions for first and second mode responses

Two characteristics of the measured response are not captured by the numerical models discussed in Cakmak et al. [1]. First, as seen in Figure 3 and Table 2, the system responds primarily at 1.9 Hz in the second interval, a 10 percent reduction in the mode 2 frequency relative to the ambient vibration frequency of 2.09 Hz. The first interval indicates mode 2 response at 2.1 Hz, a 5 percent reduction relative to the ambient vibration. Second, the SW main pier and the arches that spring from it respond at higher amplitudes in both modes 1 and 2 than the other piers and arches.

Table 3: Earthquake Calibration[1]

Mode	Observed	Simulation[2]		Dominant Motion
		Fixed Base	Soil Springs	
1	1.53	1.79	1.57	E-W (X-axis) translation
2	1.85	1.94	1.80	N-S (Y-axis) translation
3	2.15	2.16	2.02	Torsional (Z-axis) rotation

[1] All frequencies are in Hz

[2] Fixed Base = $hsdyntc9$, Soil Springs = $hsdynss$

Simulation

Two attempts at matching the recorded response using linear FE models are discussed here with particular attention paid to accounting for the two characteristics described above. The first model named $hsdyntc9$ has been constructed by Davidson[7]. It is similar to the linear FE model used by Cakmak et al. [1] for eigenvalue analysis but has more refinement of the mesh in the region of the arches. The elastic properties have been adjusted to match the observed frequency during the second interval of the earthquake response motion. The elastic moduli and density used in this model are the same as those for the eigenvalue model, except the Young's moduli, E, in surcharge and tension areas were not reduced and a ratio of 1.00:0.68 was used for E values of stone and brick masonries, respectively. The second model named $hsdynss$ has been constructed by Natsis[8], with the same geometry above the floor level as the $hsdyntc9$ model, however, the effect of soil-structure interaction has been incorporated by supporting the portions of the main piers below floor level with linear translational soil springs acting normal to the faces of the piers. Soil spring stiffnesses have been distributed in a manner reflective of expected variations in soil elastic moduli. Such patterns have been obtained from seismic tomography which measured compressional wave velocity of foundation material along a grid of horizontal and vertical plans in the area below and contained by the four main piers[9]. Figure 4 shows representative contours of equal velocity. The frequencies obtained by eigenvalue analysis of the $hsdyntc9$ and $hsdynss$ models are given in Table 3.

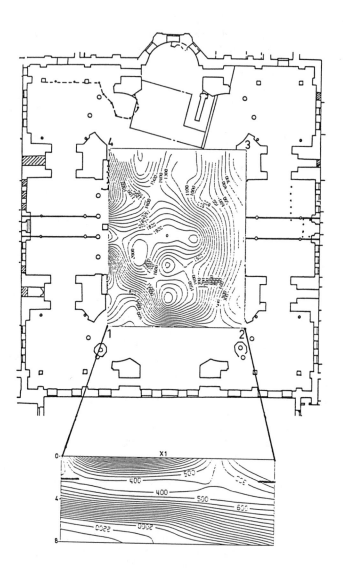

Figure 4. P-wave velocity contours obtained by seismic tomography.

Simulated response of the *hsdynss* model to the earthquake was calculated using the mode superposition method with input acceleration at all soil spring locations in the model identical to that measured in the corresponding component direction at the Kandilli seismographic station, a nearby bedrock free-field location. A pseudo-nonlinear response has been estimated by selecting different moduli in the first and second time intervals of the earthquake response motion. Figure 5 shows a comparison of some of the measured and simulated acceleration time histories.

Stress Analysis

Static dead load stresses in a model named 10*try*6 have been calculated using a pseudo-nonlinear procedure described in Davidson[7]. The procedure attempts to capture intermediate selfweight deformations experienced during a number of major stages of construction. The FE mesh for the 10*try*6 is essentially the same as that used in the *hsdyntc*9 model. Using magnitude $M = 6.5$ and $M = 7.5$ earthquake input accelerations generated at the site according to the procedure described by Findell *et al.* [10], response time histories for the *hsdyntc*9 model have been simulated. Maximum stresses in the critical crown region of the east and west arches corresponding to the static and dynamic loadings are summarized in Table 4. These results give benchmarks for the structure's response to severe earthquakes and highlight the importance of dynamic behavior to past earthquake and potential future failures of the primary support structure.

Table 4: Simulated Stresses[1] for Severe Earthquakes

Load Case	East Arch			West Arch	
	M=6.5		M=7.5	M=6.5	M=7.5
DL		0.80(3.30)		1.21(3.61)	
EQ	0.60(0.60)		1.72(1.72)	0.51(0.51)	1.48(1.48)
TOTAL	1.40(3.90)		2.52(5.02)	1.72(4.12)	2.73(5.09)

[1] All stresses are in MPa. Maximum tensions for *hsdyntc*9 are listed with maximum compressions given in parentheses.

Conclusion

Numerical modeling of the Hagia Sophia has been performed using calibrated linear finite element analyses. System identification from a recent low-intensity earthquake indicate a nonlinear behavior for the masonry structure even at very low response levels. The linear models provide reasonable estimates of overall dynamic characteristics including frequency and primary modes of response. Incorporation of nonlinear behavior and soil-structure interaction may be achieved in an approximate way and improves the low-intensity predictions. Numerical modeling provides an important means of monitoring the earthquake worthiness of Hagia Sophia. As larger intensities

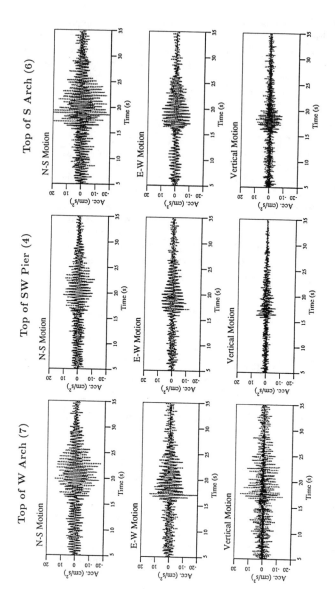

Figure 5. Simulated earthquake response predicted by soil-spring model.

are recorded and appreciable damage becomes evident, it will become increasingly important to accurately incorporate in such models the characteristic nonlinear behavior of the masonry construction.

References

1. Cakmak A.S., Davidson R., Mullen C.L. and Erdik M., Dynamic analysis and earthquake response of Hagia Sophia, in STREMA/93 (ed C.A. Brebbia and R.J.B. Frewer), pp. 67-84, *Proceedings of the 3rd Int. Conf. on Structural Studies, Repairs and Maintenance of Historical Buildings*, Bath, UK, 1993, CML Publications, Southampton, 1993.

2. Erdik M. and Cakti E., Istanbul Ayasofya muzesi yapisal sisteminin ve deprem guvenliginin saglanmasina yonelik tedbirlerin tespiti, 2nd Research Report, Earthquake Research Institute, Bogazici University, November, 1990.

3. Erdik M. and Cakti., Instrumentation of Aya Sofya and the analysis of the response of the structure to an earthquake of 4.8 magnitude, in STREMA/93 (ed C.A. Brebbia and R.J.B. Frewer), pp. 99-114, 1993.

4. Moropoulou A., Biscontin G., Bisbikou K., Bakolas A., Theoulakis P., Theodoraki A. and Tsiourva T. Physico-chemical study of adhesion mechanisms among binding material and brick fragments in 'Coccio Pesto', *Scienz E Beni Culturali IX*, 1993.

5. Lea F.M., *The Chemistry of Cement and Concrete*, Chemical Publishing Co., New York, pp. 419-35, 1971.

6. Cakmak A.S., Moropoulou A. and Mullen C.L., Interdisciplinary study of dynamic behavior and earthquake response of Hagia Sophia, *Soil Dynamics and Earthquake Engineering*,14, pp. 125-133, 1995.

7. Davidson R.A., The mother of all churches: a static and dynamic structural analysis of Hagia Sophia, unpublished senior thesis, Civil Engineering and Operations Research Dept., Princeton University, 1993.

8. Natsis M.N. The enigma of Hagia Sophia: A dynamic structural analysis of Justinian's great church, unpublished senior thesis, Dept. of Civil Engineering and Operations Research, Princeton University, 1994.

9. Gurbuz C., Deal M., Brouwer J., Bekler T., Cakmak A.S. and Erdik M., Ayasofya'nin Zemininde Yapilan Sismik Kirilma Calismalar i Konusunda Teknik Rapor, Bogazici University, Kandilli Observatory and Earthquake Research Institute, August, 1993.

10. Findell K., Koyluoglu H.U. and Cakmak A.S., Modeling and simulating earthquake accelerograms using strong motion data from the Istanbul, Turkey region, *Soil Dynamics and Earthquake Engineering*,12, pp. 51-59, 1993.

Seismic Resistance of Partially-Grouted Masonry Shear Walls

Arturo E. Schultz[†], A.M. ASCE

Abstract

In-plane cyclic load tests of partially-grouted masonry shear walls with bond beams are presented. Lateral force resistance mechanisms are discussed in conjunction with observed displacement response. Parameters relating to initial stiffness, shear strength, residual strength, energy dissipation and deformation capacity are highlighted, and predictions for the latter two are compared with measurements.

Introduction

The development of earthquake-resistant design procedures in the U.S.A. for masonry which is fully grouted and reinforced has been validated by experiment and observed performance during recent earthquakes. However, reinforced masonry has not gained widespread popularity in the eastern and central U.S.A. where seismic hazards are lower than those in the west coast. Cost-effectiveness and competitiveness with other structural materials has often placed reinforced masonry at a disadvantage. In partially-grouted masonry, vertical reinforcement is concentrated in fewer cells than in reinforced masonry, and only those vertical cells with rebar are grouted. Horizontal bars are concentrated in bond beams, or they are replaced altogether by bed joint reinforcement. Large savings in construction costs are expected from partially-grouted masonry (Fattal 1993a), yet, intuition and common sense are sufficient to render the seismic resistance of partially-grouted masonry inferior to that of reinforced masonry. The challenge lies in quantifying the seismic performance of partially-grouted masonry, as well as verifying the applicability of current analysis and design methods for reinforced masonry.

A research program was initiated in the Building and Fire Research Laboratory (BFRL) of the National Institute of Standards and Technology (NIST) to study the seismic performance of partially-grouted masonry shear walls (Fattal 1993a, Schultz 1994). An experimental study was designed to determine the influence of horizontal reinforcement ratio, type of horizontal reinforcement and height-to-length aspect ratio on the shear strength and behavior of partially-grouted masonry walls. A secondary goal of the study was to generate physical data on partially-grouted masonry shear walls for the verification of finite element modelling techniques.

The concept of partially-grouted masonry for lateral load resistance is not without precedent, as it shares similarities with a popular type of masonry construction in

----

†Associate Professor, Dept. of Civil Engineering, University of Minnesota, 238 Civil Engineering Bldg., 500 Pillsbury Drive S.E., Minneapolis, MN 55455

Latin America which is known as confined masonry (Casabbone 1994). Confined masonry comprises unreinforced masonry walls, usually made using clay masonry units, which are erected between vertical gaps used for tie columns. A "confinement" frame of lightly-reinforced tie columns and tie beams is cast after the masonry mortar has hardened. Shrinkage of the concrete frame serves to confine the unreinforced masonry panel, as well as to provide structural redundancy and more uniform distribution of lateral forces to the masonry. This system has been reasonably successful throughout Latin American in mitigating seismic damage to low-rise masonry structures (Schultz 1994).

The costs associated with the forming of tie beams and tie columns detracts from the appeal of confined masonry in the U.S.A. Yet, the function of the "confinement" frame is easily replaced by interconnected bond beams and grouted vertical cells. The placement of horizontal and vertical bars and grout necessitates the use of hollow masonry units with large openings, thus, the NIST program focuses on concrete block masonry. Furthermore, it is presumed that the performance of partially-grouted masonry walls is most sensitive to (1) amount of horizontal reinforcement, (2) wall height-to-length aspect ratio, and (3) magnitude of vertical compression stress. This paper discusses the first of two series of partially-grouted masonry shear wall tests in which specimens utilize bond beams for the placement of horizontal reinforcement, and horizontal reinforcement ratio (ρ_h) and height-to-length aspect ratio (H/L) are the principal experimental variables.

Experimental Program

A total of six partially-grouted masonry shear wall test specimens with bond beams were constructed and tested. An illustration of a typical specimen and the testing configuration is given in Fig. 1. Only the outermost vertical cells of the walls were reinforced vertically and grouted, and a single bond beam containing all horizontal reinforcement was placed at mid-height of the walls. The remainder of the masonry in the walls was not grouted. All walls were built from seven courses of masonry for a total height of masonry equal to 1422 mm (56 in.). The first four masonry courses, including the middle course of bond beam units, were placed in a single day. Bond beams and the lower portion of exterior vertical cells were grouted on the second day. The top three courses of masonry were placed on the third day, and the remainder of the exterior vertical cells were grouted on the fourth day. Wall lengths equal to 1422 mm (56 in.), 2032 mm (80 in.), and 2845 mm (112 in.) were used to define three different height-to-length aspect ratios, as noted in Table 1.

All six walls were reinforced with two #6 Grade 60 reinforcing bars in each exterior vertical cell, for a total area of vertical reinforcement equal to 1135 mm^2 (1.76 in.^{2}). This amount of reinforcement was deemed necessary to preclude flexural distress of the shear walls. Even though flexural yielding is a preferred mode of response for earthquake-resistant design, the purpose of this study requires that the specimens respond and fail in shear. Two horizontal reinforcement schemes were used for the bond beams (Table 1) to define horizontal reinforcement ratios, based on gross dimensions, that are equal to 0.05% and 0.12%, respectively. The first of these ratios represents the minimum value cited in most building codes and standards of practice for masonry in the U.S.A. (MSJC 1992, *UBC* 1994, *NEHRP* 1995).

The walls were constructed using two-cell concrete blocks with average length, thickness and height equal to 396 mm (15.6 in.), 195 mm (7.67 in.), and 194 mm

(7.63 in.), respectively. The units were 51% solid with a minimum face-shell thickness equal to 33.7 mm (1.33 in.). All masonry was face-shell bedded, except for the ends of the walls in which exterior webs were also bedded. The concrete block met the requirements for ASTM C90 (ASTM 1985), and the mortar was mixed using proportions that satisfy Type S in ASTM C270 (ASTM 1989). The grout mix was designed according to ASTM C476 requirements for coarse grout (ASTM 1983). The 28-day compression strengths for mortar and grout were 21.7 MPa (3140 psi) and 29.6 MPa (4300 psi), respectively, and the resulting masonry had 28-day compression strengths equal to 17.1 MPa (2480 psi) and 17.6 MPa (2550 psi), respectively, for ungrouted and grouted prisms.

The masonry panels were connected to precast concrete header and footer beams (Fig. 1) after the mortar hardened. Protruding ends of the vertical bars were inserted through sleeves into large pockets in the shape of a truncated pyramid in the header and footer beams, and the pockets were filled with high-strength grout. Vertical bars, the ends of which had been previously threaded, were anchored by means of steel plates and high-strength steel nuts. Horizontal bars were anchored with 180° hooks that engaged the interior vertical bars in each exterior jamb.

The cyclic load tests were conducted in the NIST Tri-Directional Testing Facility (Woodward and Rankin 1989), as shown in Fig. 1. The header and footer beams were attached to the TTF by means of high-strength post-tensioned steel rods. In-plane cyclic drift histories were applied to the header beam while the footer beam remained fixed. Since rotation of the header and footer beams was constrained, the panels were subjected to reversed bending. A summary of response forces and displacements from the cyclic load tests is given in Table 2.

The cyclic drift histories were patterned after the TCCMAR phased-sequential displacement procedure (Porter and Tremel 1987). In this procedure, groups of drift cycles are organized around peak amplitudes that are gradually increased to failure. Peak amplitudes after the initial elastic displacements are tied to the first-major event (FME). In the present study, the FME was associated with masonry cracking, and the loads and displacements at the FME (Table 2) were found to be very similar to those defining the limit of proportionality in the force-displacement relation. Because the TTF is essentially a displacement-controlled testing system vertical loads (Table 2) were maintained only approximately constant.

<u>Observed Behavior</u>

At early stages of the tests, all specimens developed cracks in the ungrouted masonry above the bond beam, but there were two distinct cracking modes. For all specimens except Wall No. 3, the first cracks were vertical and they formed at the top of the walls near the interface between the grouted vertical cell and the ungrouted masonry (Fig. 1). The cracks formed on the leading jamb of the wall (i.e. east jamb for east displacement and vice-versa), and, as the tests proceeded, they quickly propagated downward towards the joint between grouted vertical cell and bond beam. For Wall No. 3, the first cracks were inclined, they initiated at mid-length along the top edge of the wall, and they propagated towards the "bond beam"-"vertical cell" joints. Inclined cracks formed in the upper one-half of the masonry panels in Wall Nos. 1, 5, 7, 9 and 11, but this occurred after the vertical cracks along the jambs were well developed, and the width of the vertical cracks were considerably larger than those of the inclined cracks throughout the tests.

Initial vertical cracks in Wall Nos. 1, 5, 7, 9 and 11 occurred in response to the horizontal stress concentration between grouted and ungrouted masonry. It is not completely clear why Wall No. 3 did not exhibit this type of behavior, but it appears that differences in the actual anchorage conditions for vertical bars in Wall No. 3, as compared with the other specimens, may have played a role. It is clear, however, that the cracking mode played a significant role in the post-cracking behavior and resistance of the walls. For Wall Nos. 1, 5, 7, 9 and 11, the vertical cracks dominated the behavior of the walls, as few other cracks formed during the tests. As the header beam of the walls were displaced, the width of the vertical cracks on the leading jamb grew very wide, i.e. up to 6 mm (1/4 in.). These cracks eventually propagated into the "bond beam"-"vertical cell" joint, with the effect of disturbing the anchorage of the horizontal reinforcement. In addition, sliding displacements were evident between the masonry panels and the header and footer beams, and the mortar joints were damaged by abrasion from sliding.

Wall No. 3 responded in a manner similar to that of reinforced and fully-grouted masonry walls. Numerous inclined cracks formed following the initial cracks described earlier, and none of these well-distributed cracks opened as widely as did the vertical cracks in the other specimens. Furthermore, some of the inclined cracks propagated into the "bond beam"-"vertical cell" joint, but this occurred late in the test and does not appear to have disturbed the anchorage of horizontal bars as much as in Wall Nos. 1, 5, 7, 9, and 11. Horizontal reinforcement in Wall No. 3 achieved peak strains (Table 2) comparable to the nominal yield strain of the bar (0.002), whereas peak strains in the horizontal reinforcement of the other specimens were a small fraction of the nominal yield strain. It is noted that these strains were measured at mid-length of the horizontal bars, because inclined cracks were expected to traverse the bond beam at mid-length. However, this proved to be the case only for Wall No. 3, while the bond beams of the other walls cracked at the ends. It is possible that peak horizontal bar strains in Wall Nos. 1, 5, 7, 9, and 11 were much larger at the ends. In any case, it appears that the cracking mode for these walls decreased the effectiveness of horizontal reinforcement.

Force-Displacement Response

The response of all shear wall specimens to the cyclic drift histories was reasonably stable and well-behaved (Fig. 2 and 3). Initial response to load was linear and was characterized by large stiffnesses. As the peak amplitude of the drift cycles increased, the force-displacement hysteresis curves widened and peak resistance deteriorated. In addition, "pinching" behavior became evident, i.e. reduction in unloading stiffness at low lateral load levels. As noted in Fig. 2 for a stocky wall ($H/L = 0.5$) and Fig. 3 for a slender wall ($H/L = 1$), deterioration of post-peak strength and pinching of the hysteresis loops increased with increasing height-to-length aspect ratio. However, none of the specimens displayed sudden failure, and, peak resistance gradually deteriorated with load cycles. All specimens except Wall No. 1 were tested until the peak resistance for a given group of cycles was less than 75% of the largest horizontal force registered by that wall.

Envelopes of force-displacement response were constructed by sequentially scanning each history and collecting all force-displacement coordinate pairs for which displacement exceeded previous peak displacement. These envelopes were subsequently "smoothed" by applying an 11-point moving average to the force history. These smoothed envelopes, shown in Fig. 4 and 5, demonstrate that the response of the shear wall specimens is essentially bilinear. The envelopes have a marked limit of proportionality that strongly resembles a yielding system, even though there was no widespread yielding of reinforcement in these walls. These envelopes also illustrate the manner in which total lateral load resistance of the walls generally decreased with increasing aspect ratio (H/L).

The initial portion of the smoothed force-displacement envelopes were fitted with a linear least-squares trend, the slope of which is taken as initial stiffness. That portion of the envelopes for which the magnitude of force does not exceed 40% of that for the limit of proportionality was used in this calculation. Generally, the force-displacement data in this range demonstrated a strongly linear relation. These inferred stiffnesses are listed in Table 3 along with estimates based on elastic behavior. The calculated stiffnesses were obtained by combining and inverting flexural and shear flexibilities for an elastic panel, the thickness of which was taken as twice the minimum face-shell thickness of the block. The modulus of elasticity E_m was estimated as $600f'_m$, where f'_m is masonry compression strength (Atkinson and Yan 1990), and the shear modulus G was approximated as $0.4E_m$. In the direction of first loading, the inferred stiffnesses are found to be approximately 3/4 of the calculated stiffnesses. This is taken as good agreement because the stiffness of masonry walls decays rapidly with cracking, and, because slip was present between the masonry panels and the header and footer beams. In the negative loading direction, actual stiffnesses are roughly 3/5 of the calculated stiffnesses, as the panel has undergone cracking damage in the opposite loading direction.

Wall Toughness and Strength

The toughness of the walls can be quantified by means of their ability to dissipate energy. Cumulative energy dissipation was obtained by integrating each force-displacement loop and accumulating these areas. Because the specimens have different strengths and stiffnesses, cumulative dissipated energy was normalized by one-half of the product of force and displacement at the limit of proportionality to define an energy dissipation factor. This factor is given in Table 3 at instants when wall displacements correspond to maximum response (Δ_{max}) and drift equal to 0.54% of panel height (Δ_{054}). The amount of horizontal reinforcement does not seem to affect the ability of the walls to dissipate energy, but increasing height-to-length aspect ratio has a detrimental effect on wall toughness. Deformation capacity, defined as the displacement at which lateral load resistance decreases to 75% of the peak value, is another measure of toughness. These capacities, which were obtained from the smoothed envelopes, are given in Table 4 and they range from 0.33% to 1% of the height of masonry. These values appear to be low for regions of high seismic risk, but may be acceptable for regions of low seismicity.

In planning the experimental study, Schultz (1994) compared four different empirical formulas for estimating in-plane shear strength of masonry walls. Upon comparing measured peak loads with calculated loads in this study, one of these formulas was found to have less variation in the ratio of measured-to-calculated strengths. This formula is an adaptation by Fattal (1993b) of a formula developed by Matsumura (1987). Strengths calculated using this formula are listed in Table 4, along with the ratios of measured peak load to calculated strength. On the average, measured peak load is 3/4 of the shear strength calculated using Fattal's formula. It is also interesting to note that Wall No. 3, which displayed a different cracking mode and vastly larger peak horizontal bar strains, also displays a peak load which is essentially equal to the estimated shear strength. All other specimens were weaker than their estimated shear strengths.

Certain characteristics of the force-displacement relation for stocky walls (Fig. 2), such as steep unloading slopes, flat post-peak loading branches, and wide, stable loops, strongly suggest that the stocky walls relied on friction to resist horizontal forces. The mean values of the ratio of lateral force to vertical force in the post-peak branch are given in Table 4, and this ratio is equal to the coefficient of friction for an ideal sliding system. These ratios are smallest for the stocky walls ($H/L =$ 0.5), and a magnitude of 0.6 is not unreasonable for sliding friction between masonry, mortar and concrete. For walls with larger aspect ratios, larger values of this ratio suggest that other mechanisms contributed to lateral load resistance.

Influence of Experimental Parameters

The peak load resisted by the wall specimens, after averaging the values for both loading directions, were divided by net horizontal area to define ultimate shear stresses. The net horizontal area was taken as the product of twice the minimum face-shell thickness of the block and the total length of the wall. These ultimate shear stresses, except that for Wall No. 3, show a marked dependence on aspect ratio (Fig. 6). As aspect ratio doubles from the stocky walls to the slender walls, ultimate shear stress increases by 42% for walls with $\rho_h = 0.05\%$ and by 28% for walls with $\rho_h = 0.12\%$. Horizontal reinforcement ratio can be seen to have a modest beneficial influence on ultimate shear stress (Fig. 7), with the stocky walls ($H/L = 0.5$) and slender walls ($H/L = 0.5$) seeing 29% and 16% increases, respectively, as ρ_h increases from 0.05% to 0.12%. However, strength data for Wall No. 3 does not fit these trends. In fact, the ultimate shear stress for Wall No. 3 exceeds that of all other specimens. In view of this discrepancy, as well as those noted earlier (crack pattern, horizontal bar peak strain), it is concluded that the response of Wall No. 3 is out of character for this series of shear wall tests.

Conclusions

The data presented suggests that partially-grouted masonry is a viable lateral-load resisting system for regions of moderate and low seismic risk. Resistance to the drift histories is stable and features high initial stiffness and ample energy dissipation. The lateral load resisting mechanism is vastly different from that for reinforced masonry walls. Vertical cracks arising from stress concentrations between ungrouted and grouted masonry appear to dominate wall behavior. Height-to-length aspect ratio has a beneficial effect on ultimate shear stress, but a detrimental effect on toughness (strength deterioration, deformation capacity and energy dissipation capacity). Horizontal reinforcement ratio has a modest beneficial effect on ultimate shear stress, but it does not appear to affect toughness.

Acknowledgements

This work was conducted as part of the National Earthquake Hazard Reduction Program activities at NIST, of which the author was formerly Research Structural Engineer. The advice and support of the National Concrete Masonry Association, including Mr. Mark Hogan, Mr. Robert Thomas, and Mr. Larry Breeding, is gratefully acknowledged. The assistance and dedication of the technical staff in the BFRL Structures Division, including Mr. Frank Rankin, Mr. James Little, and Mr. Max Peltz is acknowledged, as are the contributions of Mr. Shawn McKee, graduate student at the University of Maryland at College Park, and Mr. José Ortiz, undergraduate student at the University of Puerto Rico at Mayagüez.

References

American Society for Testing and Materials, "Standard Specification for Grout for Masonry," ASTM C476-83, ASTM, Philadelphia, PA, 1983.

American Society for Testing and Materials, "Standard Specification for Hollow Load-Bearing Concrete Masonry Units," ASTM C90-85, ASTM, Philadelphia, PA, 1985.

American Society for Testing and Materials, "Standard Specification for Mortar for Unit Masonry," ASTM C270-89, ASTM, Philadelphia, PA, 1989.

Atkinson, R. H. and Yan, G. G., "Results of a Statistical Study of Masonry Deformability," *The Masonry Society Journal*, Vol. 9, No. 1, Aug. 1990, pp. 81-94.

Casabbone, C, "General Description of Systems and Construction Practices," in *Masonry in the Americas*, D. P. Abrams ed., SP-147, American Concrete Institute, Detroit, MI, 1994, pp. 21-55.

Fattal, S. G., "Research Plan for Masonry Shear Walls," NISTIR 5117, National Institute of Standards and Technology, Gaithersburg, MD, June 1993a, 33 pp.

Fattal, S. G., "Strength of Partially-Grouted Masonry Shear Walls Under Lateral Loads," NISTIR 5147, National Institute of Standards and Technology, Gaithersburg, MD, June 1993b, 66 pp.

International Conference of Building Officials, *Uniform Building Code*, Whittier, CA, 1994.

Masonry Standards Joint Committee, "Building Code Requirements for Masonry Structures," Publication No. ASCE 5-92, American Society of Civil Engineers, New York, 1992.

Matsumura, A., "Shear Strength of Reinforced Hollow Unit Masonry Walls," *Proceedings,* Fourth North American Masonry Conference, Paper No. 50, Los Angeles, CA, 1987.

NEHRP Recommended Provisions for Seismic Regulations for New Buildings, Part 1, Provisions, Federal Emergency Management Agency, FEMA 222A, May, 1995.

Porter, M. L. and Tremel, P. M., "Sequential Phased Displacement Procedure for TCCMAR Testing," Third Meeting of the Joint Technical Coordinating Committee on Masonry Research, U.S.-Japan Coordinated Earthquake Research Program, Sapporo, Japan, October 1987.

Schultz, A. E., "Performance of Masonry Structures under Extreme Lateral Loading Events," in *Masonry in the Americas*, D. P. Abrams ed., SP-147, American Concrete Institute, Detroit, MI, 1994, pp. 85-125.

Woodward, K., and Rankin, F., "The NBS Tri-Directional Test Facility," NBSIR 84-2879, National Bureau of Standards, Gaithersburg, MD, May 1984, 44 pp.

Table 1 Schedule of Shear Wall Specimens with Bond Beams

Wall Specimen		Length	Aspect Ratio, H/L	Horizontal Reinforcement	ρ_h (%)
No.	Designation	mm[1]			
1	CO-R05-B05-Q10	2845	0.5	2-#3	0.05
3	CO-R07-B05-Q10	2032	0.7	2-#3	0.05
5	CO-R10-B05-Q10	1422	1	2-#3	0.05
7	CO-R05-B12-Q10	2845	0.5	1-#4 & 1-#5	0.12
9	CO-R07-B12-Q10	2032	0.7	1-#4 & 1-#5	0.12
11	CO-R10-B12-Q10	1422	1	1-#4 & 1-#5	0.12

[1] 1 mm = 0.03937 in.

Table 2 Summary of Cyclic Drift Tests

Wall No.	Vertical Load[1], kN[2]		Lateral Load[1], kN[2]		Displacement[1], mm[3]		Horizontal Bar Peak Strain
	Mean	Std. Dev.	FME	Peak	FME	Peak	
1	267	14.0	136	187	0.84	13.61	0.00003
3	191	8.2	183	245	1.41	10.74	0.00226
5	133	4.8	115	133	1.27	10.41	0.00004
7	266	6.4	223	240	0.95	10.44	0.00011
9	177	5.2	151	192	0.88	7.29	0.00015
11	132	5.9	162	154	1.60	4.88	0.00007

[1]Average of both loading directions   [2]1 kN = 0.2248 kips   [3]1 mm = 0.03937 in.

Table 3 Initial Stiffnesses and Energy Dissipation Factors

Wall No.	Initial Stiffness, kN/mm[1]			Inferred-to-Calculated Stiffness Ratio		Energy Dissipation Factor	
	Calcu-lated	Inferred		positive	negative	@ Δ_{054}	@ Δ_{max}
		positive	negative				
1	431	330	235	0.76	0.55	103	374
3	287	175	149	0.61	0.52	55	73
5	175	166	160	0.95	0.91	81	119
7	431	348	241	0.81	0.56	111	221
9	287	202	168	0.70	0.58	66	61
11	175	129	109	0.74	0.62	49	62

[1] 1 kN/mm = 5.710 kips/in.

Table 4 Summary of Strength and Deformation Estimates

Wall No.	Peak Lateral Load[1], kN[2]		Measured-to-Calculated Peak Lateral Load Ratio	Mean Lateral Load-to-Vertical Force Ratio	Deformation Capacity[1], mm[3]
	Measured	Calculated			
1	187	305	0.611	0.567	13.61[4]
3	245	229	1.066	0.917	5.46
5	133	169	0.783	0.642	5.00
7	240	340	0.708	0.676	9.98
9	192	254	0.755	0.832	4.70
11	154	186	0.825	0.806	5.21

[1]Average of both loading directions   [2]1 kN = 0.2248 kips   [3]1 mm = 0.03937 in.
[4]Deformation capacity was not reached, peak displacement is shown

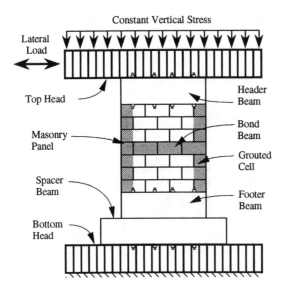

Fig. 1 Test Setup for Partially-Grouted Masonry Walls

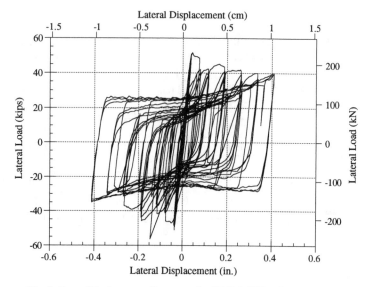

Fig. 2 Force-Displacement Response for Wall 7 ($H/L = 0.5$, $\rho_h = 0.12\%$)

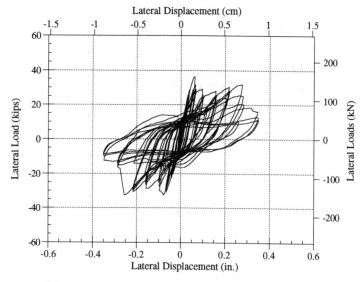

Fig. 3 Force-Displacement Response for Wall 11 ($H/L = 1$, $\rho_h = 0.12\%$)

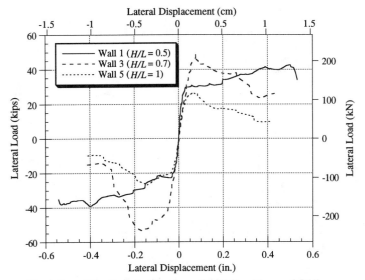

Fig. 4 Force-Displacement Envelopes for Walls with $\rho_h = 0.05\%$

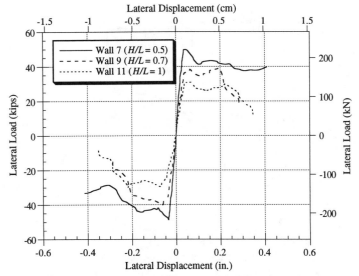

Fig. 5 Force-Displacement Envelopes for Walls with $\rho_h = 0.12\%$

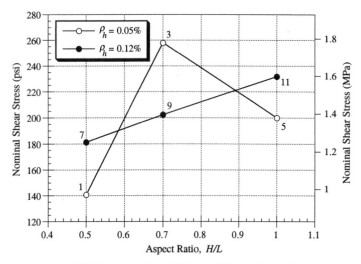

Fig. 6 Influence of Aspect Ratio on Ultimate Shear Stress

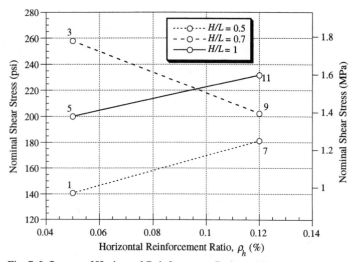

Fig. 7 Influence of Horizontal Reinforcement Ratio on Ultimate Shear Stress

RESPONSE OF LIME MORTAR JOINT ARCHES TO MOVING LOADS

Barry T. Rosson[1], Member ASCE
Thomas E. Boothby[2], Member ASCE
Ketil Søyland[3]

Abstract

The response of masonry arch bridges to moving wheel loads was simulated in the laboratory by constructing four half-scale arches with a span length of 1.22 m, using 17 voussoirs and lime mortar joints. The moving loads varied from 115 kg to 910 kg and were simulated by hanging steel weights from the center-of-gravity of the voussoirs. The finite element program ADINA was used to model the arch ring using 13,312 8-node isoparametric elements. Experimental results show the lime mortar joints exhibit significant plastic deformation accumulations with the first few cycles of loading, then diminish during subsequent load cycles. When the Drucker-Prager material model is used for the lime mortar, the FEM results indicate the plastic accumulations do not occur after the first load cycle. It is believed that sliding occurs between the voussoirs and the mortar, producing small cyclical deformations under moving loads. The four arches were also monotonically loaded to collapse in the laboratory, and the FEM stress variations normal to the mortar joints are illustrated for this loading condition.

Introduction

The original mortars used in the construction of historic masonry arch bridges were composed of lime, sand, and water. The use of hydraulic cement mortars, either naturally hydraulic or portland cement, is relatively recent. The use of portland cement in the U.S. did not begin until after 1873 (Grimmer 1990). It has previously been inferred that the weaker lime mortar joints undergo

---

[1] Assist. Professor, Civil Engr. Dept., Univ. of Nebraska, Lincoln, NE 68588.
[2] Assist. Professor, Arch. Engr. Dept., Penn State Univ., Univ. Park, PA 16802.
[3] Grad. Res. Asst., Civil Engr. Dept., Univ. of Nebraska, Lincoln, NE 68588.

larger deformations prior to failure and significant energy is absorbed under cyclic loading (Boothby 1995). Recent experimental observations have shown that a lime mortar arch bridge subjected to the passing of an overload truck develops irreversible deformations along the arch ring. However, under subsequent load passes of the same truck, the arch appears to behave in an elastic manner (Rosson & Boothby 1995). To gain a better understanding of this phenomenon, four half-scale arches were constructed in the laboratory and tested under varying magnitudes of moving load to investigate the elastic-plastic behavior of the arch ring. This research is intended to contribute to the rational evaluation of the safety of older masonry arch bridges on the basis of the mortar properties used in their construction.

    A nonlinear finite element model of the arches was also used to determine the stresses in the mortar joints and to investigate the plastic deformations in the mortar joints under the same loading conditions. The Drucker-Prager material model was used to simulate the nonlinear behavior of the lime mortar.

    The wheel loads from a vehicle traversing a masonry arch bridge were simulated in the laboratory by moving suspended weights back and forth across the arch ring. The weights were hung from the center-of-gravity of selected voussoirs symmetrically placed about the keystone. The behavior of the arch ring was observed by measuring the displacements normal to the extrados.

Experimental Program

    Each arch was designed to span 1.22 m (4 ft) with a rise of 30.5 cm (1 ft). As shown in Figure 1, each arch consisted of 17 voussoirs, each with a height of 203 mm (8 in) and thickness of 305 mm (12 in). The width of the voussoir varied from 98.4 mm (3.875 in) along the extrados to 76.2 mm (3 in) along the intrados. The voussoirs were made from portland cement concrete with a compressive strength of approximately 31 MPa. The typical weight of a voussoir

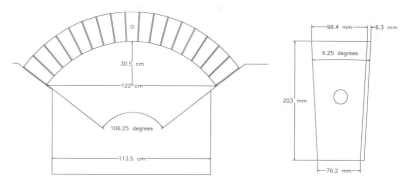

Figure 1. Dimensions of the Arch and Voussoirs.

was 11.8 kg (26 lbs). The voussoir identification scheme consisted, from left to right, of 1L, 2L, 3L, 4L, 5L, 6L, 7L, 8L, K, 8R, 7R, 6R, 5R, 4R, 3R, 2R, 1R.

A U-shaped foundation, with a length of 3.7 m (12 ft), width of 1.8 m (6 ft), and height of 1.1 m (3.5 ft), was constructed to support all four arches simultaneously. The foundation was conservatively designed to minimize the lateral displacements of the arches during load application. Measurements showed that the lateral displacements at the top of the foundation walls were limited to 0.009 mm under the heaviest loading condition. The lateral deflections were determined to be insignificant and were neglected in the analysis of the displacement data.

The lime mortar joints were 6 mm (1/4 in) thick. Arches 1 and 4 were constructed with a mortar that had a 1:3 lime to sand ratio by volume; Arches 2 and 3 had a mortar with a 1:2 ratio (1:6 and 1:4 by weight). The compressive strength of the 1:3 mortar was 1.3 MPa and the 1:2 mortar was 1.1 MPa. The mortar was allowed to dry for approximately six weeks before applying the loads.

To detect the deformations under cyclic loading, displacement data were recorded at 16 locations around the extrados of each arch. As shown in Figure 2, LVDTs were used to measure the displacements normal to the extrados. All of the voussoirs were instrumented — except for 2L and 2R. An LVDT was placed at each edge of the keystone to verify the existence of any twisting of the arch during loading. The data acquisition system typically collected 60 scans per load application. Deformations before, during and after loading were recorded during a fifteen second interval — comprising approximately five seconds of scanning before, during, and afterwards. Figure 3 illustrates a typical load application vs. time plot. D_1, D_2 and D_3 represent the displacement at each load stage (before, during, and after). The displacement δ_e (D_1-D_2) represents the total displacement during the load application, and δ_p (D_1-D_3) represents the permanent displacement remaining after unloading. Before removing the plywood centering, each arch was fully instrumented to determine the existence of initial strains in the mortar joints prior to loading. The data acquisition system did not detect displacements of the voussoirs during removal of the centering. The arches were also examined visually during each load application to check for any cracks in the mortar.

Figure 2. Photograph of arch ring, instrumentation, and loading apparatus.

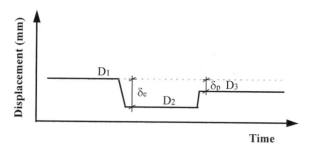

Figure 3. Schematic of LVDT Displacement vs. Time.

A 13 mm (1/2 in) diameter PVC pipe was placed at the center-of-gravity of each voussoir before pouring the concrete in the voussoir forms. This pipe provided for a convenient way to load the arch ring. By inserting a steel rod through the PVC pipe and hanging the load from it, the load produced very limited twisting of the ring and the data acquisition system was not interrupted when moving the load. The load frame was hung from voussoirs 5L, 7L, K, 7R, and 5R. This was conveniently accomplished by moving the load frame with an overhead crane. The load cycles were conducted with loads that varied from a minimum magnitude of 115 kg (250 lbs) to a maximum of 910 kg (2,000 lbs).

Laboratory Results

Figures 4 and 5 illustrate the accumulation of permanent deformations in Arch 2 due to the moving loads. The displacement histories of the five voussoirs from which the load was hung are shown in Figure 4. Arch 2 was tested through 14 load cycles; three cycles with the 460 kg (1,000 lbs) load; four cycles with the 690 kg (1,500 lbs) load; and seven cycles with the 910 kg (2,000 lbs) load. The first five load applications in the initial cycle produced a sudden accumulation of permanent deformations. This drop was more significant in the mid-section of the arch — from 7L to 7R. The data indicates a cyclical downward trend, and as expected, the amplitude is largest with heavier loads. The figure shows a decreasing trend through the first seven load cycles, but as the load was increased to 910 kg (2,000 lbs), the accumulated deformations began to stabilize. The permanent deformations of each instrumented voussoir at the end of selected load cycles are presented in Figure 5. It is noted that the accumulations are the greatest in the mid-section of the arch, and the first seven load cycles produce the most accumulation, although they were obtained with magnitudes of load that were less than that used for the last seven load cycles. Similar plots were obtained for the other three arches but are not included for brevity, refer to the paper by Soyland et al. (1995) for those results.

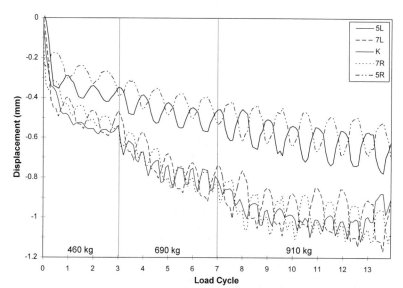

Figure 4. Cumulative Permanent Deformations (δ_p) of Arch 2.

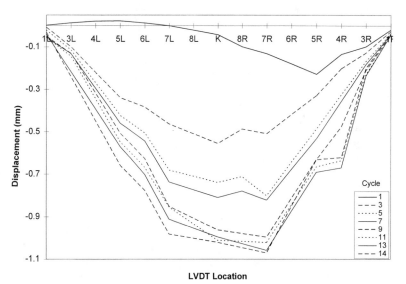

Figure 5. Deformation Profile of Permanent Deformations (δ_p) Across Arch 2.

Finite Element Method Results

The deformations of the voussoirs and the stresses normal to the mortar joints were assessed using a high-resolution finite element method (FEM) model. The model used in the study consisted of 13,312 8-node isoparametric finite elements (17 voussoirs: 8 columns × 64 rows of elements, and 18 mortar joints: 4 columns × 64 rows of elements). The lime mortar joints were modeled using a nonlinear constitutive relationship based on the Drucker–Prager yield function (1952), with further modifications developed by Bathe *et al.* (1980) to include a hardening cap and tension cut-off (ADINA 1995). A linear elastic material model was used to model the concrete voussoirs because the strains in the voussoirs were much smaller than those in the mortar joints.

A validation effort was conducted to determine the appropriateness of using the Drucker–Prager material model to represent the behavior of the lime mortar joints. A batch of the lime mortar was mixed and placed in 50 mm (2 in) I.D. steel cylinders, with mortar thicknesses of 6 mm and 13 mm (1/4 and 1/2 in). The confined mortar specimens were subjected to a monotonically increasing load and then unloaded. A typical stress vs. strain plot is presented in Figure 6. The plot shows that the uniaxial strain test using the Drucker-Prager material model, and a simple FEM model (2 columns × 2 rows of 8-node axisymmetric elements), can accurately simulate the plastic behavior of the lime mortar. The constants used for the material model were: $D=-7.25165\times10^{-7}$, $\nu=.25$, $\alpha=.1$, $E=6.2055\times10^{7}$, $k=3.4475\times10^{5}$, $W=-.1$, $T=1.379\times10^{5}$, $^{0}I_{1}=0$.

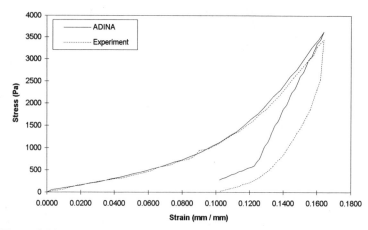

Figure 6. Lime Mortar Uniaxial Strain Comparisons.

This lime mortar material model was then used in the FEM arch ring model in order simulate the behavior of Arch 2 under the same moving load

conditions. Figure 7 shows the accumulated permanent displacements of the same
five voussoirs found in Figure 4. It is noticed that the displacements do not
accumulate after the first load pass (loads 1 through 5) using the nonlinear FEM
model. The model was constructed without contact surfaces between the mortar
and voussoirs; therefore sliding action between these surfaces was prohibited.
Because of this result, it is believed that sliding between the joints produces the
permanent deformations shown in Figure 4, and that the permanent compressive
strains in the mortar develop as a result of the initial hardening done by the first
load pass and by wedging of the voussoirs under the sliding action of the
subsequent load passes. Further laboratory testing with newly constructed arches
is being conducted to study this behavior more closely.

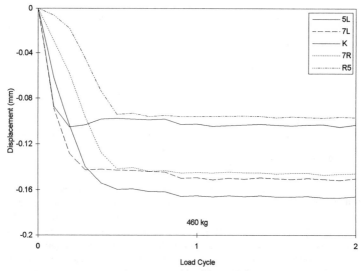

Figure 7. FEM Model Cumulative Permanent Deformations (δ_p) of Arch 2.

In addition to the moving load study, the four arches were monotonically
loaded to collapse in the laboratory. Each arch was loaded at 5R, and the
displacements normal to the extrados were recorded up to and after the collapse
load was achieved. Figure 8 illustrates the displacement vs. load responses at 5R,
K, and 5L of Arch 1. The collapse loads for the four arches were: Arch 1 - 1,970
kg (4,300 lbs); Arch 2 - 1,500 kg (3,300 lbs); Arch 3 - 1,360 kg (2,970 lbs); and
Arch 4 - 1,050 kg (2,300 lbs). It is interesting to note that although Arches 1 and
4 were constructed to the same specifications and with the same mortar, the
collapse load of Arch 1 is almost double that of Arch 4. The only significant
difference between the two arches was the loading history prior to the collapse
load being applied. Arch 1 was loaded through six load cycles with a maximum

load of 690 kg (1,500 lbs), and Arch 4 was loaded through two load cycles with a maximum load of 910 kg (2,000 lbs). Although the hardening of the mortar in each arch was not the same, this may not fully explain the wide differences found during testing. Significant sliding between the joints preceded the collapse load condition in Arches 1, 3, and 4, whereas only Arch 2 developed a traditional four hinge mechanism at collapse.

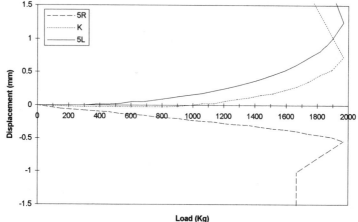

Figure 8. Displacement vs. Load to Collapse of Arch 1.

The FEM model normal stresses perpendicular to the mortar joints were obtained at two magnitudes of monotonic load. Figure 9 shows the normal stress variations throughout the depth of each mortar joint for magnitudes of load 353 kg (770 lbs) and 707 kg (1,540 lbs). The Drucker-Prager material model with a tension cut-off was used so that large tensile stresses could not develop in the model. The plots show an almost linear variation of the compressive stresses at each mortar joint, regardless the load. Also, the location of the stress resultants move closer to the compressive side of the arch ring when the load is increased, much like a thrust line would move under a similar loading condition.

Conclusions

In this study, a comprehensive laboratory investigation was conducted to examine the behavior of four arches subjected to moving loads. The results indicated the lime mortar joints developed irrecoverable deformations that accumulated throughout the load cycles. The accumulations were cyclical with an increasing average, but as the load and the number of cycles increased, the deformations began to stabilize about a given magnitude. The results suggest the possibility of a shakedown phenomenon, by which the mortar hardens under the

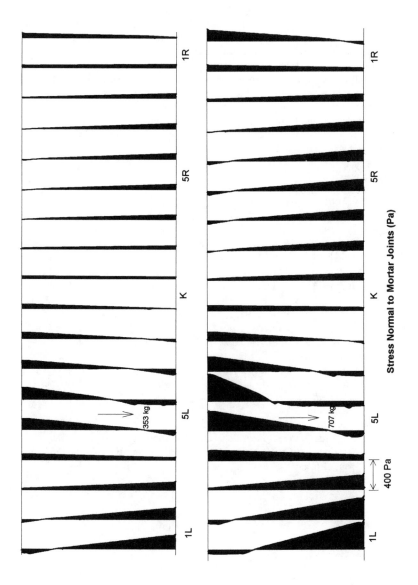

Figure 9. Normal Stress Distributions Along the Depth of the Mortar Joints.

compressive stress and wedging action, then reverts to an elastic response after several cycles of loading.

As a result of this study it is suggested that further computer modeling be conducted with additional loading cycles and magnitudes of load using contact surfaces to evaluate further the existence and extent of the sliding between the mortar and the voussoirs. Further work is also needed to investigate the possibility of using the permanent deformation responses produced from the passing of an overload truck to detect the loading history of historic stone arch bridges. If the displacement measurements are only cyclical about a given magnitude, it could be reasonable to assume the bridge has been previously loaded with at least the same vehicle weight, thus yielding limited but useful loading history information.

References

ADINA (1995). A finite element program for Automatic Dynamic Incremental Nonlinear Analysis, ADINA R & D, Inc., Watertown, MA.

Bathe, K.J., Snyder, M.D., Cimento, A.P., Rolph, W.D. (1980). " On Some Current Procedures and Difficulties in Finite Element Analysis of Elasto-Plastic Response." *Computers and Structures*, Vol. 12, pp. 607-624.

Boothby, T.E. (1995). "The Interpretation of Ancient Masonry Structures by Elastic Plastic Stability Criteria." *International Conference on Structural Stability, ICSS-95*, PSG College of Technology, Coimbatore, India, June.

Drucker, D.C., Prager, W. (1952). "Soil Mechanics and Plastic Analysis or Limit State." *Quarterly Applied Mathematics*, Vol. 10, pp. 157-165.

Grimmer, A. (1990). *Preservation Briefs 22: The Preservation and Repair of Historic Stucco*, U.S. Department of the Interior. National Park Service, Preservation Assistance Division, Washington.

Rosson, B.T., Boothby, T.E. (1995). "Shakedown of Masonry Arch Bridges." *Proceedings of ASCE Structures Congress XIII*, pp. 1727-1730, Boston, MA, April 2-5.

Soyland, K., Rosson, B.T., Boothby, T.E. (1995). "The Influence of Mortar Properties on the System Behavior of Masonry Arch Bridges." *First International Conference on Arch Bridges*, Bolton, England, pp. 355-364, Sept. 3-6.

Design of Seismic Resistant Concrete Columns for Confinement

Murat Saatcioglu,[1] Member, ASCE

Abstract

Design of confinement reinforcement is discussed for earthquake resistant concrete columns. The mechanism of confinement is presented and the significance of design parameters are illustrated. An analytical procedure is outlined to compute inelastic force-displacement relationship of a reinforced concrete column. Current design requirements for column confinement are examined and possible improvements are discussed. A displacement based design approach is presented as a rational approach for column confinement, with sample design aids.

Introduction

Performance of reinforced concrete structures during past earthquakes demonstrated that poor column behavior was responsible for significant structural damage. Although the current seismic design approach is based on the strong-column weak-beam concept, inelasticity in columns is difficult to prevent during a strong earthquake, especially at the first story level. Therefore, current building codes (Building Code 1989) call for confinement of concrete columns in seismically active regions. Confinement of concrete by properly designed and detailed reinforcement improves strength and deformability of the core, which results in improvements on overall column behavior.

The confinement provisions of ACI 318 (Building Code 1989) are based on research conducted by Richart et al. (1928) on cylinders confined by hydrostatic pressure and circular spirals. Although modified somewhat since then, the primary design parameters and design philosophy remain essentially unchanged in

[1]Professor, Department of Civil Engineering, University of Ottawa, Ottawa, Ontario, K1N 6N5, CANADA.

233

spite of significant research conducted during the last two decades. It is the objective of this paper to review the current design approach and make recommendations based on recent experimental and analytical research, as well as lessons learned from past earthquakes.

Mechanism of Confinement

Plain concrete subjected to concentric compression is in a state of uniaxial compression. The applied axial compression results in transverse tensile strains due to the Poisson's effect which may produce longitudinal splitting cracks. As concrete reaches its uniaxial capacity the transverse strains and longitudinal cracks reach their limiting values. For a linearly elastic and isotropic material, the limiting value of lateral strain can be expressed as:

$$\epsilon_{ul} = \frac{\mu f_{u2}}{E} \tag{1}$$

where f_{u2} is the maximum axial stress that can be sustained by the material. If transverse tensile strains are counteracted by lateral pressure, however, the failure is delayed while axial stresses and strains continue increasing. For a linearly elastic and isotropic material the lateral strain associated with confinement pressure f_{tl} can be determined from Hook's law.

$$\epsilon_{tl} = \frac{f_{tl} - \mu(f_{t2} + f_{tl})}{E} \tag{2}$$

Assuming failure under uniaxial and triaxial stress conditions occur at the same transverse strain level, i.e., $\epsilon_{ul} = \epsilon_{tl}$, Eqs. 1 and 2 can be set equal to each other to derive the following expression:

$$f_{t2} = f_{u2} + k_1' f_{tl} \tag{3}$$

where;

$$k_1' = \frac{(1-\mu)}{\mu} \tag{4}$$

As μ increases, coefficient k_1' decreases. Eq. 3 can be re-written for concrete in terms of confined (triaxial) strength f'_{cc} and unconfined (uniaxial) strength f'_{co};

$$f'_{cc} = f'_{co} + k_1 f_l \tag{5}$$

where, f_l represents lateral confinement pressure and k_1 represents k'_1 for concrete. Coefficient k_1 can best be determined from experimental data because of material nonlinearity, lack of isotropy, and internal cracking. However, k_1, like k_1', decreases with increasing μ which is accompanied by an increase in passive confinement pressure f_l in laterally expanding concrete. This was observed experimentally by Richart et al. (1928), and is reflected in Eq. 6, which was recommended by Saatcioglu and Razvi (1992).

$$k_1 = 6.7(f_l)^{-0.17} \qquad (6)$$

Closely spaced circular spirals and circular hoops produce near-uniform lateral pressure. Therefore, Eq. 5 is directly applicable to circular columns where the uniform lateral pressure f_l can be computed from hoop tension. While the lateral pressure provided by circular reinforcement may be represented by uniform pressure, the same may not be true for square and rectangular columns confined with rectilinear reinforcement. The resistance provided by transverse legs of reinforcement becomes maximum at hoop corners and tie hooks, while resistances between these nodal points are generated by flexural resistance of tie reinforcement, which is usually very small. Therefore, the lateral resistance provided by rectilinear reinforcement becomes non-uniform as illustrated in Fig. 1. Eq. 5 may be applied to columns with rectilinear reinforcement provided that an equivalent uniform pressure f_{le} is used in place of f_l. An empirical equation was developed for f_{le} by Saatcioglu and Razvi (1992) from a large volume of test data, and was later generalized to include high-strength concretes (Razvi and Saatcioglu 1995). The equivalent pressure is obtained by reducing the average pressure f_l through coefficient k_2 so that the degree of non-uniformity in lateral pressure, resulting from different arrangements of rectilinear reinforcement, is reflected in pressure calculations.

$$f'_{cc} = f'_{co} + k_1 f_{le} \qquad (7)$$

$$f_{le} = k_2 f_l \qquad (8)$$

$$k_2 = 0.15 \sqrt{(\frac{b_c}{s})(\frac{b_c}{s_l})} \leq 1.0 \qquad (9)$$

Coefficient $k_2 = 1.0$ for circular columns with closely spaced circular spirals or hoops. It may also be equal to 1.0 for square and rectangular columns with well distributed and laterally supported longitudinal reinforcement. The equivalent uniform pressure may be computed to be different in two cross-sectional directions, especially for rectangular columns. An average value, weighted with respect to cross-sectional dimensions, may be used to find the value of f_{le} to be used in Eq. 7.

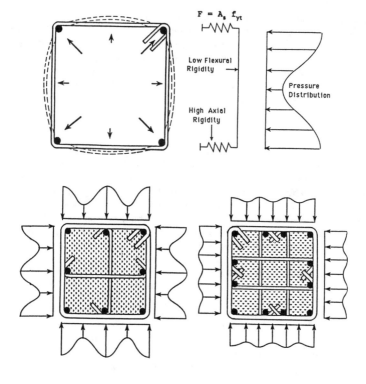

Fig. 1 Lateral Pressure in Square Columns

The mechanism of confinement described above illustrates two important aspects of concrete confinement. First; the importance of the magnitude of average lateral pressure, which is a function of the volumetric ratio as well as the grade of transverse reinforcement, and second; the importance of the efficiency of lateral pressure as dictated by the arrangement and distribution of reinforcement. Both the magnitude and efficiency of lateral pressure are important for proper design of confinement reinforcement. It is also important to recognize the potential trade off that exists between the parameters of confinement to achieve the most suitable design in terms of economy, constructibility and performance. For example, the volumetric ratio of transverse reinforcement may be reduced if a superior tie arrangement is selected with well distributed longitudinal reinforcement and overlapping hoops or cross ties. Similarly, the spacing requirements may be relaxed if reinforcement arrangement is favorable. However, this type of adjustment to design parameters should be done with caution and without exceeding the range of applicability of each parameter. The spacing limitations stated in ACI (Building Code 1989) may be used as limits within which the parameters of confinement can be adjusted to achieve the desired lateral pressure and efficiency (uniformity) of this pressure. Furthermore, the volumetric ratio of transverse reinforcement plays a dominant role on ductility of concrete and should not be reduced to adversely affect the post-peak behavior of concrete.

The magnitude and efficiency of confinement pressure also play important roles on axial strain of confined concrete at peak stress. The following expression provides a good estimate of this strain value (Balmer 1949, Mander et al. 1988, Saatcioglu and Razvi 1992).

$$\epsilon_1 = \epsilon_{01}(1+5K) \tag{10}$$

$$K = \frac{k_1 f_{le}}{f_{co}} \tag{11}$$

Deformability of confined concrete beyond the peak stress is significantly affected by the behavior of longitudinal reinforcement. Spalling of cover, at approximately the peak stress, makes the longitudinal reinforcement susceptible to buckling. The stability of reinforcement cage is essential for confinement of core concrete. The stability of longitudinal reinforcement between the ties is ensured by providing sufficiently high lateral reinforcement relative to the unsupported length of longitudinal reinforcement. Therefore, the amount of lateral reinforcement, as expressed in terms of reinforcement ratio ρ_c plays a major role on post-peak characteristics of confined concrete. Eq.12, proposed by Saatcioglu and Razvi (1992), may be used to estimate the strain corresponding to 85% of peak stress during a linear descending branch.

$$\epsilon_{85} = 260\rho_c\epsilon_1 + \epsilon_{085} \tag{12}$$

$$\rho_c = \frac{\sum A_s}{s(b_{cx}+b_{cy})} \tag{13}$$

The area ratio, ρ_c, is equal to one half the volumetric ratio of transverse reinforcement (ρ_s) for circular and symmetrically reinforced square sections.

The above stress and strain values can be used to construct an analytical model with a parabolic ascending branch, followed by a linear descending branch. The analytical model can then be used to compute inelastic column behavior for design of earthquake resistant columns.

Effect of Concrete Confinement on Column Response

The primary purpose of confining concrete in columns is to increase the deformability of column as a member to sustain seismic induced inelastic deformations. Therefore, it is important to understand the effects of concrete confinement on member behavior.

Reinforced concrete columns under seismic loading develop three components of inelastic deformation caused by flexure, shear and anchorage slip. Shear deformations are usually limited to a small fraction of total displacement unless the column has a low aspect ratio (as in the case of short columns). Shear dominant short columns are beyond the scope of the confinement design requirements discussed in this paper. The remaining two components can be equally important depending on the level of axial compression. Anchorage slip is essentially caused by the extension of longitudinal reinforcement in an adjoining member, due to yield penetration. This deformation component can be very significant if the reinforcement is strained to strain hardening.

Seismic behavior of a concrete column is influenced considerably by the level of axial compression. Columns under low levels of axial compression behave similar to beams and can exhibit ductile behavior with little concrete confinement. In a typical under-reinforced member the behavior is dominated by yielding of longitudinal reinforcement much before the crushing of concrete. Penetration of yielding into the adjoining member results in extension of reinforcement in the adjoining member with potential slippage depending on the embedment length. Lateral displacement caused by anchorage slip can be as high as that due to flexure. As the level of axial compression increases, concrete plays a more dominant role on member behavior. Confinement of concrete also becomes important. Columns under high axial compression may fail in a brittle manner if the core concrete is not properly confined. Deformations due to anchorage slip becomes less important, and may be ignored. Most columns in practice, however, are subjected to moderate levels of axial compression, ranging between 20% to 40% of their concentric capacities. Both anchorage slip and concrete confinement may be important in these columns. The lateral drift consists of displacement components due to flexure and anchorage slip. Furthermore, these two modes of behavior are closely related through sectional strains and stresses at the interface with the adjoining member.

Columns in braced frames are not expected to experience significant inelasticity during a seismic action. Therefore, these columns may not have to be designed for high lateral drift. Columns that are expected to undergo significant lateral drift, on the other hand, must be confined to develop the expected level of inelastic deformability. Therefore, computation of inelastic displacement capacity of a column becomes important for proper design of confinement reinforcement.

Computation of Column Drift Capacity

Flexural Displacements: Inelastic displacements due to flexure can be computed starting from sectional analysis. The curvature distribution can be established along the height, from which member rotations and displacements can be computed. Although this involves a standard procedure, two aspects remain to be crucial for reliability of results. One is the selection of material models, including those for confined concrete and steel with strain hardening, and the other is the treatment of hinging region and progression of plastification.

Sectional behavior can be determined through plane section analysis. Parameters of confinement can be introduced by means of the confined concrete model. It is important to select a model that includes the amount, grade, spacing and arrangement of reinforcement as parameters for different cross-sectional shapes. The strain hardening of longitudinal reinforcement must be considered as this range is inevitably reached when concrete is confined. The curvatures increase with applied load during the initial ascending branch of the force-displacement relationship. Moment of the area under the curvature diagram gives inelastic flexural displacement at desired location. Cover spalling and concrete crushing at or shortly after the peak load lead to reductions in load resistance. As moments decrease the curvatures continue increasing in the hinging region. This has to be modelled carefully in member analysis. An algorithm was developed by Razvi and Saatcioglu (1994) to establish curvatures in column hinging region and progression of hinging during descending and ascending branches of force-displacement relationship. This procedure can be used in computing inelastic displacements caused by flexure. An alternative approach may be to use an idealized moment-curvature relationship with an assumed fixed hinge length.

Displacement due to Anchorage Slip: Any reinforced concrete member that has a critical section at the end, where it frames into another member, develops some degree of anchorage slip due to the straining of longitudinal reinforcement within the adjoining member. The extension of reinforcement in the adjacent member results in a rigid-body rotation which may give rise to significant member displacement. The extension of reinforcement is sometimes accompanied by slippage, depending on the embedment length. Analytical procedures have been developed for computation of these deformations, ranging from crude simplifications to involved numerical procedures (Otani and Sozen 1972, Soleimani et al. 1979, Bannon et al. 1981, Fillippou et al. 1983, Morita and Kaku 1984, Fillippou and Issa 1988, Saatcioglu et al. 1992). Anchorage slip becomes important in columns only if the level of axial compression is limited to moderate levels, and the longitudinal column reinforcement develops strain hardening.

Design for Confinement Reinforcement

The ACI 318-89 Design Approach: The current ACI 318 (Building Code 1989) requirements for confinement reinforcement are established on the basis of a performance criterion on axial deformability of columns. The amount of confinement reinforcement is determined such that the column concentric capacity is maintained after the spalling of cover with reduced cross-sectional area but increased concrete strength due to confinement. This indicates a certain level of axial deformability while the strength is maintained in the inelastic load range. The increase in concrete strength is computed on the basis of constant coefficient $k_1 = 4.1$, which after the simplifications of design expression becomes 3.8, as opposed to that suggested in Eq. 6. This results in underestimation of confined concrete strength since k_1 can be as high as 6.0 for columns typically used in practice. The following expression can be derived for volumetric ratio of circular spirals for axial deformability implied in the code.

$$\rho_s = 0.45(\frac{A_g}{A_c}-1)\frac{f_c'}{f_{yt}}$$ (14)

A lower bound for ρ_s is provided to safeguard against unsafe values of volumetric ratio that may result for columns with large cross-sections where A_g/A_c ratio approaches 1.0. This limiting value is obtained by substituting $A_g/A_c=1.2$ into Eq. 14.

$$\rho_s = 0.12\frac{f_c'}{f_{yh}}$$ (15)

ACI 318 (Building Code 1989) adopted Eq. 14 to square and rectangular columns with the premise that rectilinear reinforcement was 3/4 as effective as circular spirals. This implies that 1/3 more confinement steel is needed to confine square and rectangular columns. The lower limit of confinement reinforcement was established to produce 50% more steel than that specified in Eq. 15 for circular spirals. The following expressions give the confinement steel requirements for rectilinear ties in terms of the area of steel in each cross-sectional direction.

$$A_{sh} = 0.3 \ (sh_c\frac{f_c'}{f_{yh}}) \ [(\frac{A_g}{A_{ch}})-1]$$ (16)

$$A_{sh} = 0.09 \ sh_c\frac{f_c'}{f_{yh}}$$ (17)

Eq. 17 is associated with confined core concrete strength of;

$$f_{cc}' = 0.85 f_c' + 2.8 f_l \tag{18}$$

where f_l is the average lateral pressure computed in the same manner as for circular columns. The above equations are based on a constant value of $k_1 k_2 = 2.8$, as indicated in Eq. 18, instead of the values given in Eqs. 6 and 9 for coefficients k_1 and k_2, respectively, which properly reflect the relationship between equivalent uniform lateral pressure and strength enhancement, as well as the distribution of lateral pressure associated with different reinforcement arrangements. Consequently, a square column designed on the basis of ACI 318 would require the same amount of confinement steel irrespective of whether it is confined with perimeter hoops only or well distributed longitudinal reinforcement laterally supported by overlapping hoops and crossties.

The ACI-318 performance criterion for confinement reinforcement is based on the axial deformability of columns with implications that columns that have axial deformability also have lateral deformability under seismic loading. Lateral drift capacity of columns as affected by confinement reinforcement is not addressed in the code. Furthermore, the effect of axial compression on lateral deformability is not recognized since the approach followed does not lend it self to such treatment. These shortcoming of the current design approach can be overcome by a displacement based design procedure.

Displacement Based Design for Column Confinement: Columns of a multistory building are subjected to combined axial and lateral load reversals during an earthquake. Columns that are susceptible to inelastic deformation reversals, either due to high seismic forces and/or lack of adequate lateral bracing, need to be confined for improved deformability. Column confinement increases lateral drift capacity. Hence, a rational design approach should recognize the close relationship that exists between concrete confinement and lateral drift capacity. While columns with low drift demands may not require significant confinement, others expected to experience high lateral drift may have to be confined stringently to meet their drift demands.

A displacement based design requires computation of lateral column displacement and a criterion for maximum usable lateral drift beyond which the column can be declared to have exhausted its capacity. It may be reasonable to assume that up to 20% of decay can be tolerated in lateral load capacity of a column due to potential redundancies in the system and possible redistribution of resisting forces among other components. With the usable column capacity defined as such, and the procedure discussed above for computation of inelastic drift capacity, a design approach can be devised for column confinement. The parameters of confinement can be introduced through the confinement model. The design process involves determination of these parameters such that the required drift is attained before the drift capacity is reached. The application of the procedure may require a computer software to carry out the iterations involved in selecting the parameters of confinement. Alternatively, design charts and tables can be established relating drift capacity to confinement parameters. One such procedure was employed by Razvi and Saatcioglu (1994) and design aids were developed for column confinement. Fig. 2 illustrates sample design charts that relate the area ratio ρ_c for confinement reinforcement to lateral drift for different levels of axial load, section geometry and material strengths. These charts may be used to determine the required amount of confinement steel for an expected level of lateral drift.

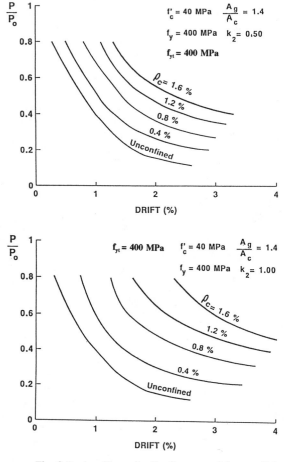

Fig. 2 Design Charts for Confinement of Square Columns

Summary and Conclusions

Confinement design requirements for reinforced concrete columns are discussed. The importance of the magnitude and efficiency of passive confinement pressure is illustrated. It is concluded that the current Building Code (1989) requirements for column confinement have shortcomings in terms of the performance criterion used, and design parameters considered. Confinement of concrete should be done to improve lateral drift capacity of columns which is not addressed explicitly in the code. Similarly, the effects of the arrangement of rectilinear reinforcement on column confinement are not recognized, although the beneficial effects of overlapping hoops and crossties, supporting well distributed longitudinal reinforcement, have been shown both experimentally and analytically.

A procedure for computation of column drift capacity is outlined with a discussion of a displacement based design procedure for column confinement. It is concluded that a design procedure based on the computation of drift capacity, properly reflecting all the relevant parameters of confinement, provides a rational alternative to some of the shortcomings of the current code approach.

References

Balmer, G. G. (1949). "Shearing strength of concrete under high triaxial stress computation of Mohr's envelope as a curve." Structural Research Lab. Report No. SP-23, U.S. Bureau of Reclamation, Denver, Co.

Bannon, H., Biggs, J., and Irvin, H. (1981). "Seismic damage in reinforced concrete frames." J. Struct. Div., ASCE, 107(9), 1713-1729.

Building code requirements for reinforced concrete and commentary. American Concrete Inst., Detroit, Mich., 353.

Fillippou, F. C., Popov, E. P., and Bertero, V. V. (1983). "Effect of bond deterioration on hysteretic behavior of reinforced concrete joints." Report No. EERC 83/19, Univ. of California, Berkeley, Calif., 184.

Fillippou, F. C., and Issa, A. (1988). "Nonlinear analysis of reinforced concrete frames under cyclic load reversals." Report No. EERC-88/12, Univ. of California, Berkeley, Calif., 120.

Mander, J. B., Priestley, M. J. N., and Park, R. (1988). "Observed stress-strain behavior of confined concrete." J. Struct. Engrg., ASCE, 114(8), 1827-1849.

Morita, S., and Kaku, T. (1984). "Slippage of reinforcement in beam-column joint of reinforced concrete frame," Proc., 8th World Conf. on Earthquake Engineering, Prentice-Hall, Inc., 477-484.

Otani, S., and Sozen, M. (1972). Behavior of multistory reinforce concrete frames during earthquakes." Struct. Res. Series No.392, Univ. of Illinois, Urbana, Ill., 551.

Razvi, S. R., and Saatcioglu, M. (1995). "Confinement model for normal-strength and high-strength concrete." Research Report, Ottawa-Carleton Earthquake Engineering Research Center, Dept. of Civil Engrg, Univ. of Ottawa, Ottawa, Ont., Canada.

Razvi, S. R., and Saatcioglu, M. (1994). "Design of reinforced concrete columns for confinement based on lateral drift." Research Report, Ottawa-Carleton Earthquake Engineering Research Center, Dept. of Civil Engrg., Univ. of Ottawa, Ottawa, Ont.

Richart, F. E., Brandtzaeg, A., and Brown, R. L. (1928). "A study of the failure of concrete under combined compressive stresses." Bulletin No. 185, Univ. of Illinois Engrg. Experimental Station, Urbana, Ill.

Saatcioglu, M., and Razvi, S. R. (1992). "Strength and Ductility of Confined Concrete." J. of Struct. Engrg., ASCE, 118(6), 1079-1102.

Saatcioglu, M., Alsiwat, J. M., and Ozcebe, G. (1992). "J. of Struct. Engrg., ASCE, 118(9), 2429-2458.

Soleimani, D., Popov, E. P., and Bertero, V. V. (1979). "Hysteretic behavior of reinforced concrete beam-column subassemblages." ACI J., 76(11), 1179-1195.

Notations

A_c, A_{ch} : Core area measured to outside of spiral or perimeter hoop.

A_g : Gross cross-sectional area of concrete.

A_s : Area of a single leg of transverse reinforcement.

b_c, h_c : Confined core dimension, center-to-center of perimeter hoop.

b_{cx}, b_{cy} : Confined core dimension in x and y directions.

E : Elastic modulus of concrete.

f'_c : Concrete cylinder strength.

f'_{co} : In-place strength of unconfined concrete in column.

f'_{cc} : Confined strength of core concrete.

f_l : Average uniform lateral pressure in one direction $= \sum A_s f_{yt}/b_c s$

f_{le} : Equivalent uniform pressure defined in Eq. 8.

f_{u2} : Longitudinal stress under uniaxial stress condition.

f_{t1} : Transverse stress under triaxial stress condition.

f_{t2} : Longitudinal stress under triaxial stress condition.

K : Factor defined in Eq. 11.

k_1 : Coefficient defined in Eq. 6.

k'_1 : Coefficient defined in Eq. 4.

k_2 : Coefficient defined in Eq. 9.

s : Tie spacing.

s_l : Spacing of laterally supported longitudinal reinforcement.

f_y : Yield stress of longitudinal reinforcement.

f_{yh}, f_{yt} : Yield stress of transverse reinforcement.

ε_{u1} : Transverse strain under uniaxial stress conditions.

ε_{t1} : Transverse strain under triaxial stress conditions.

ε_1 : Strain at peak stress of confined concrete.

ε_{01} : Strain at peak stress of unconfined concrete.

ε_{85} : Post-peak strain at 85% of peak stress of confined concrete.

ε_{085} : Post-peak strain at 85% of peak stress of unconfined concrete.

ρ_c : Area ratio of confinement steel, defined in Eq. 13.

ρ_s : Volumetric ratio of confinement reinforcement.

μ : Poisson's ratio.

BEHAVIOR OF REINFORCED CONCRETE BUILDINGS:
DEFORMATIONS & DEFORMATION COMPATIBILITY

John W. Wallace [1]

Abstract

New U.S. code documents have placed a greater emphasis on the role of deformations in the design process.   This emphasis has elevated the role that deformation calculations play in the design process; therefore, it is critical that modeling and analysis limitations and their influence on deformations be appropriately considered. Given this need, general issues relating to deformation demands and deformation compatibility are discussed to assess design issues for structural walls and "gravity load-carrying" columns. Experimental studies of reinforced concrete (RC) structural walls and "non-ductile" RC columns are presented to highlight some of these issues.

Introduction

New U.S. code documents such as ACI 318-95 (Building, 1995) and UBC-1994 (Uniform, 1994) have placed a greater emphasis on the role of deformations in the design process.   UBC-94 design requirements for detailing of reinforced concrete structural wall boundaries are based on using displacement response to evaluate detailing requirements at wall boundaries, and more stringent detailing requirements for "gravity load columns" were incorporated into ACI 318-95 following the Northridge earthquake. These changes, as well as current code trends (Vision, 1995; Guidelines, 1994) have elevated the role that deformation calculations play in the design process; therefore, it is critical that modeling and analysis limitations and there influence on design be appropriately considered.

[1]Assoc. Prof., Dept. of Civ. and Envir. Engrg., Box 5710, Clarkson Univ., Potsdam, NY, 13699-5710

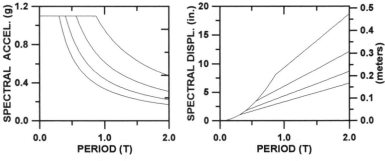

Figure 1   Code Aceleration Spectra   Figure 2   Code Displacement Spectra

Design Requirements for RC Buildings

Seismic design requirements for RC buildings are based on the use of an equivalent code spectrum. For example, the 1994 Uniform Building Code (Uniform, 1994) incorporates spectra to account for four general soil conditions. These spectra are commonly represented in the form of spectral acceleration S_a versus building period T (Fig. 1); however, it is important to note that they can also be represented as spectral displacement S_d versus building period T if it is noted that $S_d = T^2 S_a/4\pi^2$ (Fig 2).

Current code requirements (Uniform, 1994) are organized such that they emphasize the spectral acceleration relation, that is, the emphasis is on forces and not displacements. The design forces used to proportion the structural elements are commonly based on elastic structural analysis techniques with stiffness properties of the structural members modeled using gross-section concrete values. This approach is widely practiced because it is efficient and it tends to lead to conservative design force levels (because the effects of concrete cracking are not accounted for, a lower fundamental period is computed, leading to higher design force levels). Design force levels are not those expected for elastic response for a given equivalent code design spectrum, but are reduced by the factor R_w to account for a variety of effects (overstrength, system ductility, and damping). Once members are proportioned, current codes use primarily prescriptive detailing provisions. For example, if a column cross section remains unchanged over the building height, the transverse reinforcement provided is the same for the roof column as it is for the column at the base of the building. Prescriptive detailing rules are used because they are simple to apply and the expected response of the system has not been evaluated from which to assess performance-based detailing requirements.

An analysis such as described in the preceding paragraph is likely to underestimated displacement response for reasons that include: (1) actual fundamental periods are typically underestimated due to modeling assumptions, (2) force reduction factors are used to determine design level forces, (3) diaphragm/collector deformations may not be adequately incorporated, and (4) foundation deformations may not be considered. Each of these effects is discussed in the following paragraphs.

A low estimate of fundamental period will underestimate displacement response because displacement response generally increases with increasing period (Fig. 2). Therefore, to obtain more realistic estimates of displacement response, the effects of load-induced concrete cracking should be considered.   For simplified (SDOF) analyses, the fundamental period of the "cracked" system can be approximated as $\sqrt{2}$ times the fundamental period obtained using gross-section stiffness values (for 5% damping).   For more detailed analyses, stiffness assumptions can be made for individual elements based on moment-curvature analyses, or typical values can be assumed (e.g., Paulay, 1986; Guidelines, 1994). Stiffness modification is most easily accomplish for elastic analysis by changing the modulus of elasticity. Assuming that the stiffness of the structural system is reduced to one-half of the stiffness values based on concrete gross-section values and that the period of the structure falls in the linear range on a displacement spectrum, displacement response for an elastic, cracked structure will be approximately 40% greater than that for an analysis of an elastic, uncracked structure.

The use of a "force reduction factor," R_w, also reduces computed displacement response of an elastic system by R_w (Fig. 2). Therefore, to accurately assess maximum displacement response of a building system, the influence of the "force reduction factor" on the response of the building must be considered. Current code provisions (Uniform, 1991; 1994) amplify the displacements obtained from an elastic analysis for reduced forces by $3(R_w/8)$. As a result, the integrity of the vertical load-carry elements and detailing requirements at the wall boundaries are evaluated for a displacement response equal to 3/8 of the maximum value expected for elastic response, assuming stiffness values can be represented by gross-section properties. Therefore, the performance evaluation of the structural elements is intrinsically tied to the validity of using this displacement level for design.

Considerable research has been conducted on displacement response of elastic and inelastic systems.   In general, this research shows that estimating maximum displacement response as 3/8 of the elastic displacements based on a gross-section model will significantly underestimate the maximum displacement response in the short

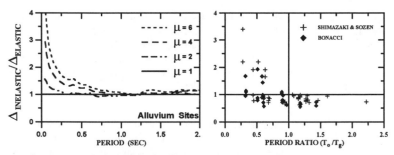

Figure 3   Analytical SDOF Studies   Figure 4   Earthquake Simulator Studies

period range and, at best, provide a lower bound estimate in the long period range. This research is outlined in the following paragraph.

Miranda (1993) studied the average displacement response of SDOF elastic and bilinear systems for a large number of ground motions. For ground motions obtained on rock or alluvium, the average maximum displacement response of an inelastic system is approximately equal to the maximum displacement response of an elastic system if the initial period of the building exceeds approximately 0.75 seconds (Fig. 3). This is commonly referred to as the "equal displacement rule." For shorter periods, the average maximum displacement response of an inelastic system is greater than the average maximum displacement response of an elastic system, and depends on the strength and period of the building (Fig. 3). Similar trends to those reported by Miranda are reported for SDOF and MDOF systems tested on earthquake simulators (Shimazaki and Sozen, 1984; Bonacci, 1994; Fig. 4). For earthquake simulator tests, if the initial period of the building model exceeded the characteristic ground period T_g (approximately 0.5 seconds for the results reported in Fig. 4), then the maximum displacement response for an inelastic system can be reasonably estimated by the maximum displacement response of an elastic system. For initial periods less than T_g, then maximum displacement for inelastic response is greater than that for elastic response.

Based on the overview provided in the previous paragraph, current code provisions will significantly underestimate the maximum displacement response, especially for short periods. This very substantial underestimation of expected displacement response is likely to have a significant impact on structural element performance.

Design Issues for Structural Walls

Considerable research has been conducted in recent years on developing analytical modeling techniques and performance-based design guidelines for structural walls (e.g., Wallace and Moehle, 1992; Sittipunt and Wood, 1993; Wallace, 1994a; 1994b;

1995; Wallace and Thomsen, 1995). In addition, recent experimental studies of rectangular and T-shaped wall cross-sections (Thomsen and Wallace, 1994; 1995) have established the validity of using a displacement-based design approach for structural walls. Therefore, considerable information exists to evaluate new wall design and to address the shortcomings of current code requirements. Evaluation of "older" wall construction is complicated by the lack of experimental evidence on the performance of walls with widely spaced transverse reinforcement at the wall boundaries, premature termination of longitudinal reinforcement, and short compression splices; however, with available tools, it should be possible to devise reasonably economical retrofit strategies for the structural walls until further information is available.

Design Issues for Gravity Load-Carrying Columns

Current code provisions (Uniform, 1994, Section 1621.2.4; Uniform, 1991, Section 2337(b)4) for gravity load-carrying columns not intended to participate in the lateral force resisting system require that the vertical load-carrying capacity of a column be verified for the lateral displacements imposed on the column by the movement of the lateral force resisting system. This provision is based on the concept that even though the vertical load carrying system is not intended to resist lateral forces, due to kinematic constraints, it must drift with the lateral force resisting system (due to the presence of a floor diaphragm). Thus, a critical component in evaluating the behavior of a "gravity load-carrying system" is evaluating the role of deformation compatibility between the lateral force-resisting and gravity force-resisting systems.

The performance of gravity load-carrying columns may also be influenced by the diaphragm deformations. These deformations may result from the use of relatively few structural walls that are widely spaced, such that the columns located midway between the walls are subjected to greater displacement demands than those adjacent to the walls. This situation is shown in Fig. 5. Current code provisions (Uniform, 1994, Sec. 1631.2.9; Uniform, 1991, Sec. 2337(b)9A) consider this effect, and require that diaphragm deformations be limited so that they do not "exceed the permissible deflection of the attached elements." Typically, simple beam formulas are used to model in-plane diaphragm displacements between structural walls. In special cases, an elastic finite element analysis may be conducted.

A potentially significant contribution to diaphragm deformations may result from the use of collectors, which are needed to provide a load path to the lateral force resisting elements (e.g., structural walls). If the collectors are not adequately designed, the forces cannot be delivered to the resisting elements and the system will not behave as designed. The need for collector elements is discussed in Uniform (1994) Section 1631.2.6 (Uniform, 1991, Sec. 2337(b)6); "**Collector elements.** *Collector elements shall be provided which are capable of transferring the seismic forces originating in other portions of the building to the element providing the resistance to those forces.*" This provision is consistent with current code philosophy of focusing on

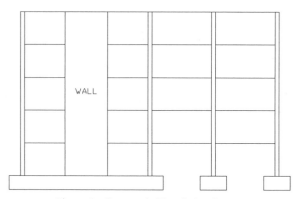

Figure 5   Structural - Foundation System

strength requirements.   No mention is made of the potential impact of collector deformations on the behavior of the structural system.

The deformation of collector elements may have a very significant affect on the behavior of gravity load-carrying columns, particularly if the collector is long.   For example, consider a 25 m (82 ft) long collector (long collectors are not unusual, especially where walls are centrally located) in which the strains in the reinforcing steel approach yield (0.002).   Extension of the collector would impose a displacement of 50 mm (1.97 in.) on exterior columns at a given level, or 1.5 to 2% of the story height.   It is important to note that to accurately assess the impact of collector deformations on column behavior, the interstory displacement of the column must be evaluated.   This requires that the influence of diaphragm/collector deformations be evaluated at all levels of the building (in addition to the displacements of the lateral force resisting system).   It is clear that collector deformations may influence the imposed interstory displacements on a column as much, or more than the displacements of the lateral force resisting system.

Modeling of collectors cannot be accomplished within an elastic analysis environment although simple hand calculations could be used to conservatively estimate deformations.   New modeling options for collectors should be incorporated into widely used inelastic analysis programs (for example, in DRAIN-2DX, Prakash, 1992; or IDARC, Kunnath et al., 1992) so that detailed evaluations can be conducted for buildings damaged in the Northridge and Kobe earthquakes.   These studies should focus on the realistic estimation of column deformation demands and to establish whether specific damage can be traced to the inadequacy of current U.S. code provisions.   In addition, detailed parametric studies of "generic" buildings should be conducted to better evaluate the role of diaphragm and collector deformations in the design of gravity load-carrying columns.

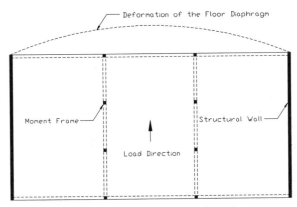

Figure 6   In-Plane Diaphragm Deformations

The type of foundation system used may also influence the deformations imposed on the column system. To resist the wall overturning forces, the foundation system for the wall may have to "reach out" to engage several columns (Fig 6.); therefore, the wall and the columns supported on the "wall" footing will rotate as a unit. However, if adjacent columns are supported on individual footings (Fig. 6), these adjacent columns may be subjected to large interstory deformations.   Case and parametric studies are needed to assess how foundation system influences deformation demands on the structural components. In addition, it is likely that improved element modeling options for soils and piles should be incorporated into inelastic analysis programs such as DRAIN-2DX.

Based on observations from the Northridge earthquake, "emergency" changes were introduced into Building (1995) Chapter 21 for frame members not proportioned to resist forces induced by earthquake motions.   Section 21.7 of ACI 318 (Building, 1992) was replaced on an interim basis with new provisions. The new provisions are intended to provide "gravity-load columns" capable or sustaining vertical load-carrying capacity under moderate (Section 21.7.1) and significant (Section 21.7.3) excursions into the inelastic range. The reason cited in the commentary for the change is stated as: "actual displacements resulting from earthquake forces may be several times the displacements calculated using the code-specified design forces and commonly used analysis models. The new provisions adopted in ACI 318 (Building, 1995) recognize that detailed studies of both demand and capacity issues for "gravity load-carrying" columns need to be conducted to develop more comprehensive design guidelines. The ultimate objective of these studies should be the development of performance-based design guidelines.

Experimental Studies: Deformation Issues

Structural Walls: Recent experimental studies of structural walls with rectangular and T-shaped cross-sections (Thomsen and Wallace, 1994; 1995) were conducted to investigate the behavior of unsymmetrical walls as well as to validate the use of a displacement-based design approach. The walls were approximately 1/4 to 1/3 scale and cyclic lateral loads were applied at the top of the walls (Fig. 7). A constant axial load of approximately $0.10A_g f'_c$ was maintained for the duration of the testing. Test results for two wall specimens are discussed in the following paragraphs.

Reinforcing details at the base of the walls are presented in Thomsen and Wallace (1995). Transverse reinforcement at the wall boundaries was selected assuming a design drift level of 1.5% of the wall height using a displacement-based design procedure (Wallace, 1995). A rectangular cross-section with eight U.S. #3 (9.53 mm) deformed bars at each wall boundary was used for specimen RW1. Web reinforcement consisted of deformed U.S. #2 (6.35 mm) bar and transverse boundary reinforcement consisted of 3/16 in. (4.76 mm) diameter hoops and two cross-ties spaced at $8d_b$ (76.2 mm, 3 in.). Specimen TW2 consisted of a T-shaped cross section. Design requirements were evaluated based on the expected behavior of the T-shaped wall cross-section using a displacement-based design approach (Wallace, 1995). The evaluation for TW2 indicated that more stringent detailing was required at the boundary of the wall web, as well, additional shear reinforcement was needed. However, details were relaxed at the web-flange intersection and at the flange boundary (note that only uniaxial loads in the plane of the wall web were applied to the specimens).

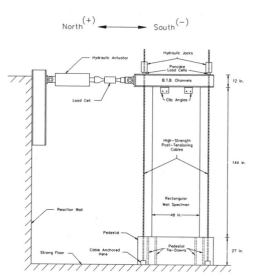

Lateral load versus lateral displacement at the top of the wall are presented in Fig. 7 for both specimens. Good performance was observed for specimen RW1, as the design drift was 1.5% and the wall was subjected to two cycles of drift at 2% before failing on the first cycle to approximately 2.5% drift. The failure occurred due to buckling of the longitudinal boundary bars, which was expected based on the provided spacing of transverse

Figure 7   Test Setup

reinforcement ($s = 8d_b$). Results for specimen TW2 also indicate good performance. The wall was subjected to a large number of displacement cycles up to 2.5% lateral drift before failure occurred due to lateral web instability. Very large spacing of transverse reinforcement was used at the web-flange intersection and in the flange (101.6 mm, 4 inches, or $10.67d_b$, where d_b is the diameter of the boundary vertical reinforcement). Even though the flange reinforcement was subjected to significant tensile strains, the experimental results indicate that special detailing requirements for the wall flange and the web-flange intersection are not required. Maximum flange compression strains of 0.003 and 0.004 were measured at 1.5% and 2.5% drift. An evaluation of the test results (Thomsen and Wallace, 1995) indicates that a displacement-based approach can be used to predict both drift capacities and failure modes of the wall specimens. The studies indicate that wall shear stresses as high as $6\sqrt{f'_c}\ psi$ ($0.5\sqrt{f'_c\ MPa}$) can be resisting without affecting the flexural capacity and ductility of the wall significantly.

The studies also indicate the importance of properly evaluating unsymmetrical wall cross-sections. A critical aspect of this evaluation is estimating the effective flange width of the wall. UBC (Uniform, 1994) specifies that the effective flange width not exceed ten percent of the wall height, which is a relatively low value (Paulay, 1986; Sittipunt and Wood, 1993; Thomsen and Wallace, 1995). This provision appears to be based on the premiss that it is conservative to underestimate the effective flange width since it leads to a low estimate of the wall flexural strength. However, a low estimate of the effective flange width could lead to inadequate detailing of the wall web opposite the flange in tension, inadequate shear reinforcement, and potential fracture of tension reinforcement. Recent studies (Thomsen and Wallace, 1995) indicate that ACI requirements for selecting effective flange widths for T-beams (Section 8.10) can be extrapolated to determine a single effective flange width that can be used to evaluate strength, stiffness, and detailing requirements for T-shaped walls. Use of these requirements was found to provide a reasonable upper bound estimate of the effective flange width for T-shaped walls; however, for L-shaped walls, using one-half of the effective flange width for a T-shaped wall was recommended versus extrapolating the values for L-shaped

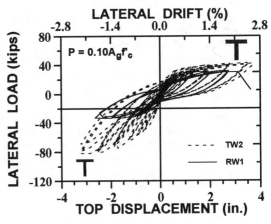

Figure 8   Load - Deformation Behavior

beams (ACI 318, Section 8.10). Extrapolating ACI 318 values for L-shaped beams to L-shaped or C-shaped walls was found to yield slightly unconservative values (Sittipunt and Wood, 1993).

Gravity Load-Carrying Columns: An additional concern with regard to the design of "nonductile" columns and walls is the degradation of shear strength with imposed deformation. Recent studies have noted the deterioration of column shear strength with increasing displacement ductility demand (Aschheim and Moehle, 1992), as well as the influence of axial stress and splices on column deformation capacity. Therefore, the accurate assessment of expected displacement response, including the influence of diaphragm/collector deformations, is essential for evaluating column behavior. Experimental and analytical studies of columns with "moderate details" may be needed to develop comprehensive guidelines. These studies should include columns with details that reflect columns designed to meet the new ACI 318 Chapter 21, Section 21.7 provisions.

Summary and Conclusions

An overview of issues relating to the design of structural walls and gravity load-carrying columns was presented. Based on this overview, significant shortcomings in current U.S. code requirements were noted and areas were additional development are needed were discussed. Based on this overview, the following conclusions are reached:

(1) Recent research and current code development efforts are elevating the role that deformations play in the design process. Given this trend, care must be exercised to assure that realistic deformation estimates are obtained under the design ground motions.

(2) By using a multiplier of $3R_w/8$, UBC (1994) provisions significantly underestimate the expected displacement response, especially for short period structures. As a result, insufficient transverse reinforcement may be provided at wall boundaries and vertical load-carrying capacity may be compromised for columns not intended to participate in the lateral force resisting system. Analytical and experimental studies have been conducted that suggest a multiplier of R_w should be used for "long period" structures, and a larger multiplier is needed for "short period" structures.

(3) Diaphragm and collector deformations may play a significant role in the deformation demands imposed on gravity load-carrying columns. Studies are needed to develop and verify appropriate modeling techniques for determining deformation demands on gravity load-carrying columns. As well, reasonable deformation estimates are needed to gauge column shear strength since shear strength degrades with increasing deformation demand.

(4) Experimental studies of RC structural walls indicate that a displacement-based design methodology provides a robust tool to evaluate wall failure modes and detailing requirements. If wall shear stress is limited to $6\sqrt{f'_c} \, psi$ ($0.5\sqrt{f'_c} \, MPa$), then shear does not significantly affect wall flexural capacity and ductility.

Acknowledgements

The work presented in this paper was supported by funds from the National Science Foundation under Grants No. BCS-9112962 and CMS-9416487. This financial support, as well as the support of Program Director, Dr. Shih-Chi Liu, is gratefully acknowledged. The assistance of Dr. John H. Thomsen IV in evaluating the wall test results and in preparation of figures for this paper is appreciated. Opinions, findings, conclusions, and recommendations in this paper are those of the author, and do not necessarily represent those of the sponsor.

Appendix A: References

Aschheim, M.; Moehle, J. P. (1992). "Shear Strength and Deformability of RC Bridge Columns Subjected to Inelastic Cyclic Displacements." *Rep. No. UCB/EERC-92/04*, Earthquake Engineering Research Center, University of California, Berkeley, Calif.

Bonacci, J. F. (1994). "Design Forces for Drift and Damage Control: A Second Look at the Substitute Structure Approaches," *Earthquake Spectra*, 10(2), 319-332.

"Building Code Requirements for Reinforced Concrete." (1992, 1995). American Concrete Institute (ACI), Detroit, Mich.

"Guidelines and Commentary for the Seismic Rehabilitation of Buildings." (1994). *ATC-33.02 50% Submittal*, Applied Technology Council, Redwood City, Calif.

Kunnath, S. K.; Reinhorn, A. M.; Lobo, R. F. (1992). "IDARC Version 3.0: Inelastic Damage Analysis of Reinforced Concrete Structures." *Rep. No. NCEER-92-0022*, National Center for Earthquake Engineering Research, State University of New York, Buffalo, NY

Miranda, E. (1993). "Evaluation of Site-Dependent Inelastic Seismic Design Spectra," *J. Struct. Engrg.*, ASCE, 119(5), 1319-1338.

Paulay, T. (1986). "The Design of Ductile Reinforced Concrete Structural Walls for Earthquake Resistance." *Earthquake Spectra*, 2(4), 783-823.

Prakash, V. (1992). "Dynamic Response Analysis of Inelastic Building Structures: The DRAIN Series of Computer Programs." *Rep. No. UCB/SEMM-92/28*, Department of Civil Engineering, University of California, Berkeley, Calif.

Shimazaki K., Sozen, M. A. (1984). "Seismic Drift of Reinforced Concrete Structures." *Research Reports*, Hazama-Gumi, Ltd., Tokyo.

Sittipunt, C.; and Wood, S. L. (1993). "Finite Element Analysis of Reinforced Concrete Shear Walls." *Struct. Res. Series Rep. No. 584*, Civil Engineering Studies, University of Illinois, Urbana, Ill.

Thomsen IV, J. H., and Wallace, J. W. (1994). "T-Shaped Shear Walls: Design Requirements and Preliminary Results of Cyclic Lateral Load Testing." *Proc., 5th U.S. Nat. Conf. on Earthquake Engrg.*, Vol. II, Earthquake Engineering Research Institute, Oakland, Calif., 891-900.

Thomsen IV, J. H., and Wallace, J. W. (1995). "Displacement-Based Design of RC Structural Walls: An Experimental Investigation of Walls with Rectangular and T-Shaped Cross-Sections." *Rep. No. CU/CEE-96/06*, Department of Civil and Environmental Engineering, Clarkson University, Potsdam, NY

"Uniform Building Code." (1991, 1994). International Conference of Building Officials (ICBO)," Whittier, Calif.

"Vision 2000." (1995). Structural Engineers Association of California (SEAOC), Sacramento, Calif.

Wallace, J. W. (1994a). "New Methodology for Seismic Design of RC Shear Walls." *J. Struct. Engrg.* 120(3), 863-884.

Wallace, J. W. (1994b). "Displacement-Based Design of RC Structural Walls." *Proc., 5th U.S. Nat. Conf. on Earthquake Engrg.*, Vol. II, Earthquake Engineering Research Institute, Oakland, Calif., 191-200.

Wallace, J. W. (1995). "Seismic Design of RC Structural Walls. Part I: New Code Format." *J. Struct. Engrg.*, 121(1), 75-87.

Wallace, J. W., and Moehle, J. P. (1992). "Ductility and Detailing Requirements of Bearing Wall Buildings." *J. Struct. Engrg.*, 118(6), 1625-1644.

Wallace, J. W., and Thomsen IV, J. H. (1995). "Seismic Design of RC Structural Walls. Part II: Applications." *J. Struct. Engrg.*, 121(1), 88-101.

Analysis Requirements for Performance-Based Design
of Beam-Column Joints

By John F. Bonacci,[1] Member ASCE

Abstract

Modeling requirements for performance-based earthquake-resistant design of reinforced concrete beam-column joints are outlined. It is shown that, in order to base joint detailing and design requirements on local demands, it is necessary to determine the portion of total story drift that results from joint deformation. It is further argued that the model of joint behavior should be sensitive to all design variables if design adjustments are to be considered for the sake of improved seismic performance. A number of modeling approaches for beam-column joints are briefly summarized and evaluated for potential application in performance-based evaluation of joints. It is concluded that a modeling approach as similar as possible to that which is familiar for beam flexure would provide the best compromise between effort in application and capability to furnish needed results.

Performance-Based Design and Detailing

Current seismic design procedures are based on strength of an overall structure and on the ordering of the relative strength of its various members. Strength, then, is the main attribute that would be used to distinguish one prospective performance from another. Displacements are compared to a fixed limit, much as they would be in designing for wind loads. For most structures designed for severe earthquake loading, it is expected that the lateral-load strength provided will be reached and that resistance beyond this point will rely on hysteretic action to dissipate the energy absorbed from the ground. Under these circumstances, strength and internal forces become much weaker indicators of the likely performance than other aspects of inelastic response such as the maximum displacement response, the pattern of

---

[1]Assoc. Prof., Dept. of Civ. Engrg., Univ. of Toronto, Toronto, ON M5S 1A4, Canada.

damage sites that are contributing to overall hysteresis, or the local deformations at sites of concentrated inelasticity.

In a general sense, performance-based design is an attempt to base the design and detailing of a system on the most tangible consequences of a post-yield response. The individual steps in this kind of design evaluation can be summarized, in very general terms, as follows:

1. Initial proportioning and reinforcement of members: This is accomplished from consideration of other load effects (gravity, occupancy, wind, etc.). In addition, conventional seismic design provisions can be applied to establish a starting point for lateral-load resistance.

2. Evaluation of maximum inelastic story drift response: This should be done in a manner that accounts for member softening and the relationship of the softened-structure period to the frequency content of the earthquake motion.

3. Decomposition of total story drift into member contributions: For the example of a frame, this would involve determining contributions from beams, columns, joint panels, and anchorages on the basis of their relative flexibilities at the appropriate distortion levels.

4. Member detailing on the basis of local deformation demand: Member behavioral models are applied to determine detailing measures that are required for the member to sustain its local deformation demand. If the demand is judged to be too large, the system must be reproportioned to increase stiffness or redistribute story drifts (step #1). If member detailing measures change the structural period, the overall lateral-load strength, or the relative flexibility of story members, then it may be necessary to repeat steps #2-#4.

Clearly, the outline touches on a number of indefinite issues which would call for in-depth consideration before these steps can be applied in design. This paper will focus only on what the above steps imply about the requirements for modeling of beam-column joint behavior.

Prescriptions for Joint Modeling

The objective of a performance-based design is much the same as any other kind: make sure that demand does not exceed supply. But from the outline of steps in the previous section, one distinguishing feature is that displacements and deformations, rather than forces, are used to characterize the inelastic structural behavior. Ideally then, models of member behavior should be useful in both the translation of story drift to local member demand (step #3 in the outline) and for establishing the capacity of the member to sustain the deformation demands (step #4 in the outline).

In order to satisfy these objectives an analytical model must provide information about member flexibility, stress and strain in concrete and reinforcement, response milestones (e.g. cracking, yield, crushing), maximum deformability, strength, failure mode, and the sensitivity of each of these characteristics to variables of the member design and detailing. For a beam-column joint, variables that are known to influence behavior include: dimensions, load history, framing conditions, concrete strength, column axial force, amounts of reinforcement (joint hoops, beam bars, column bars), reinforcement stress-strain characteristics, and detailing of reinforcement.

Beam flexure is an example for which conventional analytical procedures are ideally suited for performance-based evaluations. Results available from routine beam sectional analyses that are vital to steps #3 and #4 of the performance-based design outline include:

- Moment-curvature relationship.

- Milestones in sectional response, such as: cracking, yield of reinforcement layers, and concrete crushing.

- Distribution of strain along the depth at the critical section, which can be used to detect failures at the material level.

- Conditions at failure, including: failure mode, maximum deformability, and strength.

- Sensitivity to design variables, including: geometry of section, concrete strength, longitudinal reinforcement (amount, layout, and material properties), and axial force.

With little added sophistication, the analysis can be made to account for the confinement of the compression zone by transverse reinforcement and the spalling of unconfined cover concrete. Integration of curvature over the beam axis will furnish a moment-rotation relationship, which provides the basis for linking local demand to story drift (Bonacci and Wight in press).

Because of the extensive information it provides from application of familiar and relatively simple principles, beam analysis can be considered a bench mark for the evaluation of modeling schemes for other member types. In the following section, a variety of models for beam-column joints are considered from the viewpoints of usefulness in a performance-based design and the amount of effort required in application.

Modeling of Beam-Column Joints

Empirical Model: ACI-ASCE Committee 352 Recommendations

ACI-ASCE Committee 352 recommendations (1985) for earthquake-resisting (Type 2) joints are based on an extensive summary of data available from laboratory structural tests of beam-column joint assemblies. The recommendations can be used to determine reliable horizontal joint shear strength as a function of concrete strength, and confinement provided by the combination of joint reinforcement and adjacent members that frame into the joint. The Committee 352 approach is based on the premise that joint reinforcement must be provided to confine the concrete core. The potential to function as stirrups (resisting diagonal tension associated with joint shear) is not considered. The prescribed joint reinforcement amount is therefore insensitive to applied forces. The Committee takes the view that it is better to provide a well-confined core with moderate shear stress than to attempt to carry higher stresses by providing proportionately higher ratios of joint reinforcement.

Using the Committee 352 model, calculated horizontal joint shear strength does not depend on column reinforcement ratio, column reinforcement properties, or the column axial force. No information is provided about joint flexibility, strain in joint reinforcement and concrete, predicted joint failure mode, or joint deformation. A statement in the commentary, that the Committee's joint shear provisions are "intended for limited displacement and rotation levels", is the only reference to deformation capacity.

The empirical approach of the Committee 352 recommendations was developed to provide a simple design criterion for promoting reliable cyclic behavior of beam-column connection assemblies. In considering experimental evidence, it was the behavior of overall test specimens (i.e. beams, columns, and anchorages in addition to the joint) that was judged in establishing tolerable joint shear stress levels. Consequently, it is not the objective of the 352 provisions to model joint behavior, per se. So, while it considers a number of the joint design variables and is simple to apply, the empirical model would not meet all of the specifications for analysis as part of a performance-based design.

Equilibrium Models

There are two prominent examples of equilibrium models, one which forms the basis for New Zealand design standards (*Code of practice* 1982) and the other for Japanese design standards (*Design Guidelines* 1990). Both of them postulate the existence of two separate force paths (Paulay et al 1978), a self-equilibrating diagonal compression strut and a truss mechanism. For the first mechanism, joint reinforcement is important for confinement of the compressed concrete diagonal. For the second mechanism, joint reinforcement functions as vertical and horizontal truss tension members that are required to equilibrate compression forces in diagonal concrete sub-struts (assumed to exist between diagonal tension cracks). Both design models consider the two mechanisms to be of equal importance for early stages of loading. Where they differ is in their views of which of the two mechanisms becomes more prevalent at advanced stages of inelastic response. Based on the premises that cyclic flexural cracks in beams at the joint face become increasingly more difficult to close and that beam bar bond resistance migrates to the joint centerline, the New Zealand viewpoint is that the diagonal strut mechanism tends to become less pronounced while the truss mechanism becomes more so. The Japanese viewpoint (Kitayama et al 1991) is that, at usual design drift levels (2% of story height) or beam-end rotation of 4 times yield, experiments demonstrate that a certain degree of bond deterioration is inevitable. Deterioration of bond along beam bars passing through a joint makes it progressively more difficult to engage the truss mechanism and therefore more likely that the forces will migrate to the diagonal compression strut.

As a result of these conflicting interpretations of the two postulated force paths, the New Zealand and Japanese joint design requirements are quite different. In the New Zealand standards, unless column axial load is relatively high, the contribution of the concrete strut to joint shear resistance is neglected and the entire force must therefore be resisted by reinforcement. Reinforcement requirements in both horizontal and vertical directions are therefore proportional to the shear carried by the joint. An upper limit on joint shear strength, which is proportional to concrete strength, is provided to preclude diagonal compression failure in sub-strut elements. In the Japanese standard, joint reinforcement is considered to function as confinement for the concrete diagonal, much like the premise of the ACI-ASCE Committee 352 recommendations. Minimum joint reinforcement and joint shear strength prescriptions are based on values determined from experiments to promote beam hinging without joint inelasticity.

Because both models postulate two distinct force paths, equilibrium considerations alone are not sufficient to divide the total joint shear between them. It is the required additional premises for each viewpoint, which are empirical in nature, that lead to the disparities in design concepts.

Neither of the design models considers joint flexibility or deformation capacity. By neglecting (typically) the participation of the diagonal strut mechanism, the New Zealand approach dictates the relationship between stress in joint reinforcement (horizontal and vertical) and applied loading. In the New Zealand model, when column axial stress is significant, it is deemed to have some effect on the distribution of the total shear among the two resistance mechanisms (i.e. the diagonal strut contribution is non-zero). Even though both load-carrying mechanisms involve diagonal compression, both design models place limits on joint shear stress rather than on the principal normal stress in concrete along the diagonal direction.

Equilibrium models, particularly the New Zealand design model, demonstrate slightly more sensitivity to variables in beam-column joint design than the ACI-ASCE Committee 352 approach. But the added sensitivity of the New Zealand model is due to the prescribed distribution of total resistance between the two postulated force paths rather than to equilibrium itself. That this is true is emphasized by the fact that the Japanese design model turns out to resemble the Committee 352 model even though it is based on consideration of the same two force paths as the New Zealand approach. Given the nature of the added sensitivity of the New Zealand model, and considering that neither approach models joint deformation, it is concluded that neither of the equilibrium-based design approaches would be sufficient to meet the stated specifications for performance-based evaluation of joints.

Joint Panel Model

In formulating this 2-D joint model (Pantazopoulou and Bonacci 1992), it was recognized from the outset that what distinguished beam flexural modeling from most empirical models was the ability to characterize, in a convenient manner (i.e. the "plane sections" idealization), the pattern of deformation at the section of interest. In the joint panel formulation (Pantazopoulou and Bonacci 1992), which is summarized only very briefly here, joint deformations were modeled with three average strain quantities (Fig. 1): horizontal strain, vertical strain, and shear strain. Equilibrium requirements for horizontal and vertical directions relate average stresses in concrete and steel to applied normal stresses at the joint center. Concrete stress-strain is modeled as parabolic, with consideration of the effects of confinement and softening caused by tension perpendicular to the direction of principal compression. Concrete tensile strength is neglected. Steel stress-strain is modeled as elasto-plastic. At yield, hoop reinforcement is assumed to become ineffective for confinement. Directions of principal stress and strain are assumed to coincide.

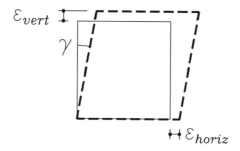

Figure 1. Idealized Joint Deformations

The resulting algebraic model is analogous to sectional analysis for beam flexure. For any selected value of strain in horizontal reinforcement, the direction and magnitude of principal stresses and strain are found, as are joint shear stress and strain. The analysis algorithm requires an initial guess for concrete modulus with iteration required to converge on the value that is compatible with the resultant state of strain. The complexity of this iteration is of the same order as that required to locate the neutral axis for beam bending when rectangular stress block idealizations are not used. Each converged solution provides the complete state of average stress and strain in the joint for a specified hoop strain. With these values, it is possible to make direct checks against the material models for detection of reinforcement yield or concrete crushing. Because, at each point, the shear stress and shear strain are determined, the shear flexibility is also known.

Conditions that establish maximum joint shear strength in the formulation are: (1) yield of horizontal joint reinforcement followed by concrete crushing along the principal diagonal; (2) yield of column reinforcement followed by crushing; or (3) yield of reinforcement in both horizontal and vertical directions. Design equations for each of these failure modes are provided for strength and failure mode determination from the governing minimum capacity.

Full application of the panel model requires considerably more effort than any of the models discussed so far which form the basis of published design standards. Once the algebraic expressions are entered in an electronic spreadsheet, application of the model to compute joint response at a single stage of loading requires very little effort. It is seldom necessary to compute the full force-deformation response of a joint for a performance-based evaluation. Experimental studies (Uzumeri 1977, Otani et al 1984, Otani et al 1985) have shown a correlation of poor cyclic response of joints with the detection of hoop yield. It has also been shown, for joints with conventional reinforcement ratios and concrete strengths, that: (1) the shear modulus computed from the 2-D panel model can be considered constant after cracking (which is not modeled) and before the first yield of joint reinforcement; and (2) iteration to convergence in the value of concrete modulus frequently causes only minor changes in the computed stress and strain states. Thus, perhaps the most useful application of this model-- and at the same time one of the simplest-- is to determine the conditions at yield of joint reinforcement and how they depend on design variables. Extensive details for this practical subset of the full formulation are given in a paper by Bonacci and Wight.

Finite-Element Analysis

Numerous possibilities exist for finite-element analysis of beam-column joints. Joints can be modeled as elements in a complete frame, as part of an isolated connection assembly (resembling structural test specimens), or as an isolated member with boundary conditions in place of adjacent members. In the first two approaches, the joint can be modeled with a discrete spring controlling the angle between connected bending elements or as a mesh of finite elements. Materials can be treated as linear or nonlinear. Load application can be monotonic or cyclic. Reinforcement can be modeled as discrete elements, joined to concrete with springs to model bond resistance, or it can be treated in smeared fashion by modifying concrete material properties. The characteristics of the model will depend on the objectives of the analysis.

In a performance-based evaluation of frame connections, it is important that the modeling technique can account for joint flexibility, material nonlinearity, axial forces, strain conditions in concrete along the path of diagonal compression, and strain for individual reinforcement types (joints hoops, beam bars and column bars). The options to analyze an isolated joint region or treating joints as discrete hysteretic springs in a complete frame analysis may not serve the objectives of performance-based design (though it is conceivable that a combination of these two approaches would).

The principal advantage of finite-element approaches for a performance-based evaluation are the detailed information that they provide. Though examples of finite-element approaches for modeling joint behavior can be found in literature (e.g. Noguchi and Watanabe 1987, Jordan 1991, Atrach 1992), they are still not incorporated as options in standard structural analysis programs. The highly specialized, labor-intensive and computationally demanding nature of this kind of analysis would make it impractical for application in most design offices.

Conclusion

Performance-based design considerations for beam-column joints place special requirements on the modeling technique to be employed for studying joint design and detailing measures. This paper briefly reviewed a number of known approaches for modeling the behavior of beam-column joints. Not suprisingly, it was shown that added sophistication goes hand in hand with added effort in model application. The purpose of this review was to demonstrate that the objectives of performance-based design, being different than traditional strength-based design, necessitate reconsideration of the type of modeling approach that should be followed. Models providing the basis for traditional strength-based code provisions were found to be lacking with respect to a number of the specifications for use in performance-based evaluations-- particularly important is the inability to model joint shear stiffness. It was also argued that nonlinear finite-element modeling is likely to be too laborious in practical applications to compensate for the added detail about performance and sensitivity to joint variables.

It is concluded that an approach like the 2-D joint panel model provides a reasonable compromise between added complexity and added capability. While the formulation itself may not be very familiar to practicing engineers, its basic premises are analogous to those used for modeling beam flexure. That it is relatively more complex than beam analysis has more to do with the planar nature of joints, and the fact that several distinct sets of reinforcement are engaged, than with the formulation strategy.

References

Atrach, O.M. (1992). *Behavior of interior and exterior beam-column joints under earthquake conditions.* M.A.Sc. thesis, Dept. of Civ. Engrg., Univ. of Toronto.

Bonacci, J.F., and Wight, J.K. (in press). "Displacement-based assessment of RC frames in earthquakes," ACI Special Publ., *Proceedings of the Mete A. Sozen Symposium*, Tampa, FL, Oct. 1994.

Code of practice for the design of concrete structures, NZS 3101 Part 1. (1982). Standards Assoc. of New Zealand, Wellington.

Design guidelines for earthquake resistant reinforced concrete buildings based on ultimate strength concept. (1990). Architectural Inst. of Japan, 337 pp., in Japanese.

Jordan, R.M. (1991). *Evaluation of strengthening schemes for reinforced concrete moment-resisting frame structures subjected to seismic loads.* Ph.D. dissertation, Dep. of Civ. Engrg., The Univ. of Texas, Austin.

Kitayama, K., Otani, S., and Aoyama, H. (1991). "Development of design criteria for RC interior beam-column joints." *Design of beam-column joints for seismic resistance*, Publ. SP-123, Amer. Concrete Inst., Detroit, MI, 97-123.

Noguchi, H. and Watanabe, K. (1987). "Application of FEM to the analysis of shear resistance mechanism of RC beam-column joints under reversed cyclic loading." *Proc. of the 3rd U.S-N.Z.-Japan seminar on design of RC beam-column joints*, Christchurch, N.Z.

Otani, S., Kitayama, K., and Aoyama, H. (1985). "Beam bar bond stress and behavior of RC interior beam-column connections." *Proc. of the 2nd U.S.-N.Z.-Japan seminar on design of RC beam-column joints*, Tokyo, Japan.

Otani, S., Kobayashi, Y., and Aoyama, H. (1984). "Reinforced concrete beam-column joints under simulated earthquake loading." *Proc. of the U.S.-N.Z.-Japan seminar on design of RC beam-column joints*, Monterey, CA.

Pantazopoulou, S.J. and Bonacci, J.F. (1992). "Consideration of questions about beam-column joints." *ACI Struct. J.* 89(1), 27-36.

Paulay, P., Park, R., and Priestley, M.J.N. (1978). "Reinforced concrete beam-column joints under seismic actions." *J. of the Amer. Concrete Inst., Proc.*, 75(11), 585-593.

Recommendations for design of beam-column joints in monolithic reinforced concrete structures, ACI-ASCE 352. (1985). Amer. Concrete Inst., Detroit, MI.

Uzumeri, S.M. (1977). "Strength and ductility of cast-in-place beam-column joints." *Reinforced concrete structures in seismic zones*, Publ. SP-53, Amer. Concrete Inst., Detroit, MI, 293-350.

Hybrid Moment Resisting Precast Beam-Column Connections.

John Stanton[1]

ABSTRACT

A precast concrete framing system for resisting earthquake loads is described. It uses both unbonded post-tensioned reinforcement and bonded mild steel bar reinforcement.

INTRODUCTION

Precast concrete does not enjoy a reputation as a good seismic-resistant framing system and the 1994 Northridge earthquake did nothing to improve it. The main difficulties lie in making connections that are tough enough to endure cyclic loading.

Prestressing is effectively banned by the UBC for use in resisting seismic loads on the basis that it offers little energy dissipation [Uniform 1994]. Unbonded prestressing is regarded by many engineers as particularly unsuitable for seismic applications, because it offers even fewer opportunities for energy dissipation and because of the consequences of possible anchorage failure. It would therefore appear perverse to try to design an unbonded post-tensioned precast concrete frame and to expect it to perform well during an earthquake, but such a system has been developed and tested, and the results are most encouraging. A comparable monolithic, cast-in-place frame was built and subjected to the same test program for reference. The new precast frame proved at least equal to the monolithic one in all respects, and superior in most.

This paper describes the underlying principles on which the design of the new system is based, outlines the test program and explains the basis for design. Detailed information on the tests can be found in Stone et al (1995). The design method is treated in detail by Cheok et al. (1995).

DESCRIPTION OF THE SYSTEM

The system is shown in Fig. 1. The beams and columns are precast and no permanent corbels are necessary. The column is assumed to be a

---

[1]Prof., Dept of Civ. Eng. University of Washington, Seattle WA 98195.

single, multi-story unit, in order to save the costs of splicing. This is possible up to a maximum of about six stories. The beam has a solid rectangular cross-section at its ends and, in its central region, it has a trough at the top and bottom. The beam-column connection is made with grouted reinforcing bars and unbonded post-tensioning (PT). (Partial bonding may be used instead, as discussed later). The rebars are at the top and bottom of the beam, and the PT tendon is straight and is located at the beam centroid.

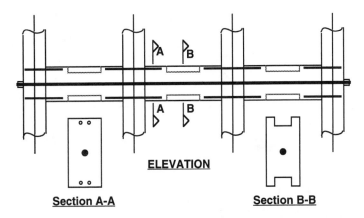

ELEVATION

Section A-A   **Section B-B**

Fig. 1.   Essential components of the framing system.

The construction sequence is expected to be that the columns are erected first, and are equipped with temporary steel corbels. The beams are set on them and the reinforcing bars are dropped into the trough in the beam and are slid through the ducts in the solid ends of the beam, which line up with matching ducts in the column. The gap between beam and column is then grouted with a fiber-reinforced grout. The ducts containing the reinforcing bars may be grouted at the same time. When a line of beams is in place and the grout has gained strength, the PT is installed and stressed. The tendon runs the full length of the structure and uses a multi-strand anchor at each end. The floor system is then installed, and the temporary corbels may be removed.

Shear resistance between each beam and column is provided by the PT, which clamps the two components and generates frictional resistance at all times. During an earthquake, the beams rotate relative to the columns by rocking and opening a crack at top or bottom of the grout joint. The unbonded PT strand remains elastic because the extension caused by the crack opening is distributed over a length that is great enough that the resulting change in strain does not cause yield. When the load reverses the crack first closes, thanks to the elastic PT force, then a crack opens on the other side. When the ground motion stops, the PT force closes all cracks and returns the structure to its undeformed position. The corners of the beams are protected from crushing by steel armor angles.

A joint that contains unbonded PT alone is clearly possible, and such a system has been tested recently [Priestley and MacRae, 1994]. It has excellent displacement capacity and suffers little damage during cyclic loading to large displacements. However, it dissipates little energy because no steel yields. The present system seeks to take the best features of unbonded PT and bonded mild steel, and contains reinforcing bars that act both as energy dissipators and sources of strength. They are placed at the top and bottom of the beams, because there they undergo the greatest strains for a given story drift and thus dissipate the maximum possible energy through cyclic yielding. The PT is placed at mid-height, in order to minimize the increase in strain caused by earthquake motion and thereby to protect the tendon against yielding.

In most conventional seismic designs, causing all the steel to yield and dissipate energy is considered desirable. Prestressing steel is essentially banned by the UBC because it offers little opportunity for energy dissipation, even if it is bonded. If a strand were to yield in tension, load reversal would cause buckling rather than yielding in compression, so yielding would occur only on the first extension.

In the new system, the strand is specifically designed not to yield, because its function is to maintain the clamping force between the elements and to provide a reliable and permanent restoring force, which it can only do if it remains elastic. The energy dissipation is provided separately by the rebars.

This division of the functions of dissipating energy and providing restoring force capability is an essential part of the concept, and it is clear that use of a single type of reinforcement could not meet the design objectives. For example, high strength bars alone would lose their restoring force if they yielded and would dissipate no energy if they did not. The simultaneous use of unbonded post-tensioning and bonded mild steel is referred to as hybrid reinforcement.

EXPECTED SEISMIC RESPONSE

Cast-in-place concrete frames reinforced with conventional bars ("rebar frames") are the de facto standard against which other seismic concrete framing systems are presently measured. It is therefore worth comparing their seismic properties to those of typical precast structures. First, the prevailing goal in a rebar frame is to ensure the maximum energy dissipation through yielding of the reinforcement. However, the bars must be bonded to the concrete to ensure yield and, in so doing, they inevitably cause damage to the concrete. That damage near the beam column joint may also lead to shear failure.   The structure is thus protected by the dubious philosophy of deliberately inflicting damage on it. This design approach appears paradoxical, but has been used for many years for lack of anything better.

A design that meets the goal of widespread yielding leaves few of the critical elements elastic and, after a moderate to severe earthquake, the structure will almost certainly display permanent drift and member damage. Straightening and repairing the structure may incur significant costs caused by physical repairs, down-time and interruption to business.

Precast frames may be constructed either with "wet joints", in which the members are connected by interlocking reinforcement in a cast-in-place joint or with "dry joints" which avoid the site-cast concrete by using bolts, welds, mechanical connectors, grouted sleeves, etc. to complete the connections.   The former are used almost universally in Japan and New Zealand, and usually constitute "emulation designs" which attempt to mimic the behavior of cast-in-place reinforced concrete. Dry joints are preferred for contractual reasons in the USA.

Typical dry-jointed precast frames are considered vulnerable to seismic forces largely because their connections represent concentrations of flexibility and weakness, at which inelastic deformation inevitably accumulate. However, this property also has the advantage of protecting the members themselves from damage.   Such protection becomes attractive if the connections can be constructed so that they survive an earthquake without serious loss of strength or stiffness. This philosophy has parallels with seismic isolation, wherein the structure itself is protected from damage at the expense of concentrating the deformations in the isolators, which are designed to accommodate the large displacements.

In a hybrid reinforced frame the connections are relatively resistant to damage because the unbonded post-tensioned connection possesses the capacity for enormous deformation after the crack opens.   The consequence is that jointed precast structures offer the possibility of seismic behavior that is philosophically different from, and arguably better than, that of cast-in-place frames. To take advantages of such behavior, rather than struggling to emulate cast-in-place construction, is a rational goal.

The hybrid reinforced precast frame possesses a number of potentially useful behavioral attributes. First, the post-tensioning leads to a high stiffness prior to cracking. After cracking, it gives the joint a positive material stiffness even if the rebars do not strain harden. A positive material stiffness is essential to counteract the negative component of stiffness caused by the "P-Δ effect", or geometric stiffness, as shown in Fig. 2. The need for a positive stiffness after yielding is the reason that underlies the UBC requirement that $f_u \geq 1.5 f_y$ for reinforcing bar steel. If the total stiffness, consisting of the sum of the material and geometric components, were to become negative, the onset of collapse would be only a matter of time [Newmark and Rosenblueth, 1971]. Second, the fact that the PT closes the cracks after the ground motions stops minimizes damage due to residual drift and consequent down-time.

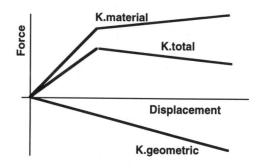

Fig. 2.   Material, geometric and total stiffnesses.

The fact that only some of the flexural strength of the beams is derived from rebars suggests that less energy might be dissipated in a hybrid frame than in a monolithic one. However the tests on the system [Stone et al., 1995] showed that, at drift angles less than about 1.5%, the hybrid frame actually dissipated more energy than the monolithic one. This behavior is explained partly by the fact that the post-tensioning stiffens the members, which therefore deform much less than do conventionally reinforced members. The system therefore behaves approximately as a collection of rigid beams connected by deformable joints. The concentration of the deformation at the connections means that, at low drifts, the local deformations are larger, and more yielding occurs, than would be the case in a comparable rebar frame. A second reason is that the bars were anchored very effectively in the grouted tubes, so they yielded with less slip than that which occurs in a cast-in-place frame.

The role of energy dissipation is also subject to question. Energy dissipation helps to inhibit the build-up of large displacements, but it is not the only issue that determines them. The relationship between the period of the input motion and the natural period of the structure is also important. An elastic system has a unique period, as does a sinusoidal ground motion. If these two periods are identical, resonance results and displacements build up without bound in the absence of damping. By contrast, most earthquakes are composed of a spectrum of frequencies with limited energy at each, and a yielding or a nonlinear elastic structure has no unique period, because even the "effective period" based on a secant stiffness varies with the amplitude of vibration. In this case, the lack of a true resonance by virtue of detuning plays a larger role in inhibiting displacements, so damping plays a smaller one.

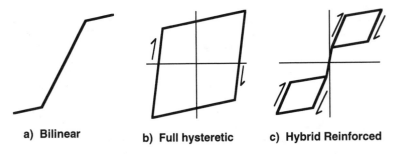

a) Bilinear   **b) Full hysteretic**   **c) Hybrid Reinforced**

Fig. 3.   Idealized load-deflection curves for bilinear elastic, full hysteretic and Hybrid systems.

Alternatively, damping may be thought of as a means of dissipating the energy after it has entered the structure, whereas detuning is a way of limiting the amount of energy that enters the structure in the first place. Studies on SDOF systems representing prestressed systems (e.g. Priestley and Rao (1993), Mole and Stanton (1994)) have shown that the displacements of a bilinear elastic system (Fig. 3) are larger than those of an idealized hysteretic system, but not by a substantial margin. They also showed that the ground motion chosen had much more influence on displacements than did the type of hysteresis loop and that a moderate amount of hysteresis, such as might be provided by a hybrid reinforced system, is very nearly as effective as larger amounts of hysteretic energy dissipation. This finding casts doubt on the conventional wisdom that a full, fat hysteresis loop is the only acceptable paradigm, so it merits further study. It also suggests that the dynamic response of the hybrid system should be satisfactory.

EXPERIMENTAL RESULTS
        Tests on 1/3 sale models of the hybrid connection have been conducted at NIST [Stone et al., 1994]. An earlier, three-phase, investigation there had concentrated on precast, prestressed joints made from bonded prestressing alone, so a fourth phase of that testing program was initiated to investigate the Hybrid system. Several different configurations were studied in Phase IVa, and the best one, shown in Fig. 1, was selected for more detailed study in Phase IVb. A conventional cast-in-place specimen was tested in the first phase and served as a reference.

Full details of the Phase IVb tests are presented elsewhere [Stone et al. 1994]. The hybrid specimens in general performed well. The PT remained elastic up to a drift of approximately 3.5%, which is approximately twice the value anticipated by the UBC design methodology for a maximum credible earthquake. One specimen was taken through two cycles at 6% drift (the test apparatus limit), after which its strength was still approximately 75% of the largest strength displayed in any previous cycle and all cracks in the beams and column had closed completely. At no point in any test was there any slip of the beam relative to the column, despite the presence of a permanent vertical load to simulate dead and live load shear.

Failure was defined as the point when the strength dropped to 80% of its peak value, and this generally coincided with fracture or pullout of the rebars. Pullout occurred in the first specimen, in which the bond to the bars was unintentionally reduced by waterproofing on the strain gages. It also occurred in a later specimen in which stainless steel bars were used for their high strain capacity. However lugs had to be made by machining deformations onto a plain bar, and the pattern was selected more for ease of machining than for good bond. Subsequent changes of the lug pattern machined into the stainless steel bars solved the debonding problem. The last specimen failed (i.e. the strength dropped below 80% of the peak value) when the bars fractured. The drift at which fracture occurs can be selected by debonding the bars locally for a short distance either side of the beam-column interface.

DESIGN

The design procedure contained in the UBC for reinforced concrete frames was developed for systems without prestressing that display relatively full hysteresis loops. It relies on Newmark's observation that the maximum displacement of an elastic system and a yielding system with the same initial period will be approximately the same [Veletsos and Newmark, 1960]. This leads to the concept of obtaining design loads by calculating elastic forces and then dividing them by a factor proportional to the ductility capacity of the system. This procedure is not particularly appropriate for the hybrid reinforced system for a number of reasons. Most important, the computational model used by Newmark to develop his hypothesis was a simple yielding one, in which the definition of ductility is unique, whereas in the Hybrid system there are two significant changes in slope in the moment-curvature plot (one when the crack opens and the next when the rebar yields) so the definition of ductility is not unique and a clear rationale for selecting a force reduction factor is lacking. A displacement-based, rather than a force-based, design approach would reflect the system behavior better, but the realities of obtaining approval from Building Officials require that, at least for the present, design must be carried out within the force-based UBC framework.

Based on the behavior observed in the tests, an R_w value of 12 is proposed. This is the same factor as is used for cast-in-placed rebar frames. The hybrid system demonstrated considerably more deformation capacity than the conventional cast-in-place rebar system, and comparable energy dissipation, so a case could be made for a higher value. However, obtaining approval for a higher R_w would be fraught with difficulties, and the cost savings in construction might not be significant. Too large an R_w would also lead to smaller members, a more flexible system and higher drift.

Two issues dominate the flexural design. The first is how to define and calculate M_n, the nominal moment resistance. In a rebar frame it is taken as the moment corresponding to a maximum concrete compression strain of 0.003 in/in, at which time the tension steel has yielded. In an under-reinforced beam this differs little from the moment at first yield. In design of non-seismic partially prestressed concrete, which contains both strand and rebar, M_n is also computed using the 0.003 in/in criterion, but it results in strand stresses that are often close to ultimate (270 ksi) and bar strains well in excess of ε_y. In the hybrid system, the strains at the beam end are clearly not linearly distributed, so use of a limiting compression strain is relatively meaningless. One possibility is therefore to use the moment corresponding to first yield of the rebars but that would be very conservative.

A compromise for M_n is proposed whereby the rebar is assumed to be strained to the onset of strain hardening (taken as 0.010 in/in, or $5\varepsilon_y$, in the absence of better information) and the corresponding strand stress is obtained by considering compatibility of deformations. The plastic moment capacity, M_p, is calculated separately as the moment corresponding to fracture of the rebars, where once again the strand stress is obtained using strain compatibility. M_p is used to define the largest possible moment that could exist, and is needed for the shear design.

For any given beam size, the required flexural strength can be satisfied by any one of an infinite number of combinations of strand and rebar. The selection of a suitable arrangement is influenced by the fact that the tension in the PT must be able to overcome the resistance of the bars at compressive yield so as to generate sufficient clamping force across the beam-column interface. The quantity of rebar used will also influence the amount of energy dissipated. In general, the displacement response is smaller if more energy is dissipated, and this is especially true if the input motion is nearly periodic. Thus, unless circumstances dictate otherwise, the amount of rebar steel should be maximized subject to the constraint of the clamping force requirement.

The selection of the prestress level in the PT steel is also important. If it is too high, the PT will yield at an unacceptably low drift. If it is too low, an uneconomically large area of strand will be necessary to provide the required clamping force. If the beam itself is assumed not to deform, the strain increase due to earthquake drift can be shown to be

$$\Delta\varepsilon = \theta\, h/L_{beam} \tag{1}$$

where

θ = drift angle

h = beam depth

L_{beam} = clear length of beam

In practice, local deformations of the grout joint or the concrete in the beam end will reduce the true strain change to a somewhat smaller value. Equation 1 allows the prestress level to be selected so as to ensure that the PT steel remains elastic at the expected drift. However, if some yielding occurs, perhaps during an event larger than anticipated, slip will not necessarily occur at the column face. During the test program, the system was taken to approximately twice the drift to cause yield, but some 25% of the initial prestress remained after unloading and no slip whatsoever occurred.

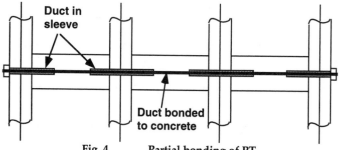

Fig. 4.   Partial bonding of PT.

In the simplest version of the system, the PT is unbonded and the stress is developed by mechanical anchorages. Supplementary anchorage by local bonding is not necessary but may be incorporated without detracting from the functioning of the system. Even if the PT tendon is unbonded it does not slip relative to the concrete at midspan of each beam, because this is a point of symmetry. Thus the PT could be bonded at midspan and unbonded elsewhere as shown in Fig. 4, so the relative slip could still occur where it is needed at the column face. By this means, back-up protection against anchorage failure could be obtained on a bay-by-bay basis. One way of achieving the partial bonding would be to place continuous duct in the beam, but to debond it (perhaps by placing it inside a sleeve) at the beam ends. At midspan it would be embedded directly in, and bonded to, the concrete. The duct could then be grouted along its full length, which would be a relatively simple operation on site and would provide the additional benefit of extra corrosion protection.

Shear reinforcement requirements in both the beam and the beam-column joint of the Hybrid system are significantly lower than in a cast-in-place rebar frame. Not only does the reduction in tie steel save material, but it significantly simplifies the task of rebar placement and thereby offers potential labor cost savings.

In a system containing only unbonded PT reinforcement, the beam end moment at any displacement is

$$M = T_p (1-\alpha)h/2 \qquad (2)$$

where

T_p = instantaneous tension in PT steel

αh = depth of compressed concrete, $\approx 0.1h$

The shear force demand acting at the same time is

$$V_{dem} = 2M/L_{beam} + wL_{beam}/2 \qquad (3)$$

and the shear resistance is

$$V_{cap} = \mu T_p \qquad (4)$$

where

μ = coefficient of friction between grout and beam.

Slip will not occur provided that

$$V_{cap} \geq V_{dem} \qquad (5)$$

or

$$\mu \geq \frac{(1-\alpha)h}{L_{beam}} + \frac{wL}{2T_p} \qquad (6)$$

Most seismic framing is placed in the perimeter frame, where gravity loads are relatively small, so the second term in Equation 6 is unlikely to play a large role. If it is neglected, μ is taken as 0.7 (UBC) and α as 0.1, the requirement becomes a function of the beam's span to depth ratio:

$$\frac{L_{beam}}{h} \geq \frac{(1-\alpha)}{\mu} \approx \frac{0.9}{0.7} = 1.3 \qquad (7)$$

The clear span to depth ratio of the beams in a seismic frame usually lies in the range of 4 to 8, so the requirement is satisfied with a large margin of safety. It should be noticed that the use of a load factor on the seismic component of the shear demand is inappropriate, because a larger shear force can only occur if T_p increases, in which the shear capacity and demand increase by the same amount. If some dead load is present, or if some of the moment resistance is supplied by nonprestressed rebar, the beam span/depth ratio must be larger than 1.3 to prevent slip. The exact value will depend on beam details.

For the same case of a beam reinforced only with unbonded PT and subject only to seismic shear, the shear can be examined from a strut-and-tie perspective as illustrated in Fig. 5. The shear force is introduced by a single diagonal force at one corner of the beam. Its angle of application is defined by

$$\psi = V_{dem}/T_p = \frac{(1-\alpha)h}{L_{beam}} \tag{8}$$

so the entire shear is theoretically be carried by a single strut running approximately between opposite corners of the beam. Thus no reinforcement would be necessary to resist shear. Some tie reinforcement would be needed to confine the compressed concrete at the beam ends and to hold the cage together during casting, so in practice some stirrups would be used. In a hybrid system that carried gravity load, some reinforcement might be required for shear, but the quantity would be significantly less than that required in a rebar frame.

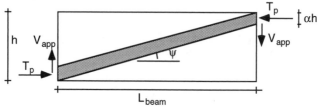

Fig. 5.   **Diagonal strut in beam.**

Joint shear strength in the beam-column joint often dictates the column size in a cast-in-place frame. Priestley and MacRae (1994) have shown by the use of strut and tie models that the required joint shear reinforcement in a purely unbonded PT frame is less than in a rebar frame. The reasoning is similar to that for the beam steel: the joint forces are introduced as external diagonal compression forces, rather than as bond forces from bars passing through the joint, so they can be largely carried by a single diagonal compression strut through the joint. The limitation on this behavior appears to be development of the vertical component of the diagonal force, which depends on bond to the vertical column bars. Only part of the benefit would be gained in a Hybrid frame, but any reduction in column size would reduce the column weight and facilitate erection.

CONCLUSIONS

This study has shown that:

1. Precast concrete has unique properties that can offer advantages over cast-in-place concrete for seismic resistance. Precast concrete should be viewed as an individual construction system and should not be relegated to mimicking cast-in-place technology.
2. The hybrid reinforcing concept is a valid one that offers the possibility of excellent seismic resistance.
3. The hybrid system minimizes expected post-earthquake down-time for repairs by limiting damage.

4. The system offers clean, economical architectural expression without the need for corbels.

ACKNOWLEDGEMENTS
Funding for this project was provided by Charles Pankow Builders, Ltd, by the American Concrete Institute, and by the National Institute of Standards and Technology. Their support is gratefully acknowledged. The testing was conducted at NIST by Ms. G. Cheok and Dr. W.C. Stone, and assistance with the design methodology was provided by Ms. S.D. Nakaki. Support and advice and encouragement was supplied throughout the project by the Project Advisory Committee, consisting of: Cathy French, S.K. Ghosh, Jacob Grossman, Grant Halvorsen, Paul Johal, Bob Mast, Courtney Phillips, Nigel Priestley and Norm Scott.

REFERENCES
Stone, W.C., Cheok, G.S. and Stanton, J.F. (1995). "Performance of Hybrid Moment-Resisting Precast Beam-Column Concrete Connections Subjected to Cyclic Loading." *ACI, Str. Jo.*, 92(2), March-April, pp. 229-249.

MacRae, G.A. and Priestley, M.J.N. (1994). "Precast Post-tensioned Ungrouted Concrete Beam-Column Subassemblage Tests." *Report No. SSRP 94/10*, Dept. of Applied Mechanics and Engineering Sciences, University of California, San Diego. March, 124 p.

Mole, A. and Stanton, J.F. (1994). "A Hybrid Precast Prestressed Concrete Frame System." Fourth meeting of the US-Japan Technical Co-ordinating Committee on Precast Seismic Structural Systems (JTCC-PRESSS), Tsukuba, Japan, May, 24 p.

Cheok, G.S. , Stone, W.C., and Nakaki, S.D. (1995). "Simplified Design Procedure for Hybrid Precast Concrete Connections." *NISTIR Report no* , Building and Fire Research Laboratory, National Institute of Standards and Technology, Gaithersburg, MD. 76 p.

Newmark, N.M. and Rosenblueth, E. (1971) "Fundamentals of Earthquake Engineering." Prentice Hall, 640 p.

Priestley, M.J.N. and Rao, J-T. (1993) "Seismic Response of Precast Prestressed Concrete Frames with Partially Debonded Tendons." *PCI Jo.* 38(1), Jan-Feb, pp. 58-69.

Uniform Building Code, (1994). International Conference of Building Officials, Whittier, CA.

Veletsos A.N. and Newmark, N.M. (1960) "Effect of Inelastic Behavior on the Response of Simple Systems to Earthquake Motions." *Proc. 2WCEE*, Tokyo and Kyoto, pp. 895-912.

Analysis and Design of the
Ponce Coliseum in 1969 and 1996

Alex C. Scordelis[1], Hon.M.ASCE, Pere Roca[2],
and Antonio R. Mari[3]

Abstract

The original analysis, design, and construction in 1969 of an 82m (271 ft) span, reinforced and prestressed concrete, hyperbolic paraboloid (HP) shell roof for the Ponce Coliseum in Puerto Rico is described. As a comparison, a discussion is presented on how the analysis and design might be done in 1996 using analytical methods and computer programs available today.

Introduction

The Ponce Coliseum was originally analyzed and designed in 1968-69 [Lo and Scordelis (1969)]. Construction began in late 1969 and was completed in 1971 [Scordelis et al. (1971)]. The Coliseum has had extensive use for sports events and other activities from 1971 to the present time, 1996.

The conceptual design was constrained by a limited budget, and since air conditioning the Coliseum would have cost at least three times the budget, it was decided at the outset to have only natural ventilation. The climate of Ponce is tropical with temperatures ranging from 24° to 32° C. (75° to 90° F.) throughout the day and night all year long. Thus, basically what was needed was a roof which would act as a giant shade from the hot sun and as an umbrella in case of rain, but would still permit the trade winds to blow in to cool off the spectators. This resulted in the selection of the HP roof, shown in Figs. 1, 2, and 3, which has only four support piers and is completely open on the sides to permit an

---
[1]Nishkian Professor Emeritus of Structural Engineering, 721 Davis Hall, University of California, Berkeley, CA 94720, U.S.A.
[2,3]Professor of Civil Engineering, Technical University of Catalonia, 08034 Barcelona, Spain.

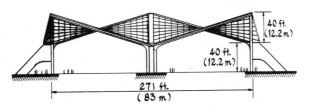

1- FRONT ELEVATION VIEW

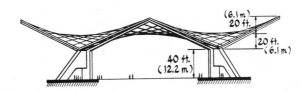

2-DIAGONAL ELEVATION VIEW

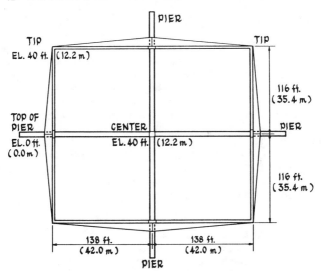

3 - PLAN DIMENSIONS

unobstructed air flow from one side of the Coliseum to the other.

Description of Shell Structure

The shell structure, Figs. 1, 2, and 3, has overall plan dimensions of 84.0m (276 ft) by 70.8m (232 ft). The complete roof is made up of four similar 102mm (4 in.) thick saddle-type shells connected to interior and edge beams to form a structure supported by four piers at the low points located at the centers of the four exterior edges. The high points of the shell, which rise 12.2m (40 ft) above the low points, are at the four corner tips and at the center of the shell. The clear spans between opposite support piers in the two directions are 82.6m (271 ft) and 69.2m (227 ft).

The cantilever edge beams are supported only at the piers and have a constant width of 762mm (30 in.) throughout their entire length. The depth of the 42.0m (138 ft) cantilever edge beams varies linearly from 457mm (18 in.) at the corner tips to 1346mm (53 in.) at a distance of 5.2m (17 ft) from the abutment and thence increases more rapidly to a maximum depth of 2388mm (94 in.) at the abutment. Similar depths for the 35.4m (116 ft) cantilever edge beams are 457, 1118, and 2210mm (18, 44, and 87 in.). The interior beams are 1524mm (60 in.) wide. The depth of the 42.0m (138 ft) long interior beams varies linearly from 475mm (18 in.) at the center of the shell to 1194mm (47 in.) at a distance of 3.1m (10 ft) from the abutment and thence increases more rapidly to a maximum depth of 1829mm (72 in.) at the abutment. Similar depths for the 35.4m (116 ft.) long interior beams are 457, 1016, and 1524mm (18, 40, and 60 in.).

The cantilever edge beams contain both normal reinforcing steel and prestressing steel. They are prestressed to help balance dead load stresses and deflections and to reduce the required size of these beams. The interior beams contain only normal reinforcing steel and are not prestressed.

The normal 102mm (4 in.) thick shell is gradually thickened over a 1.5m (5 ft) wide zone adjacent to each beam to a 152mm (6 in.) thickness where it joins the beams. Reinforcing for the shell consists of reinforcing bars in the top and bottom surfaces throughout the shell with additional reinforcement being added in the zones of high shell bending adjacent to the beams. In addition, to minimize cracking in the shell, prestressing is provided by tendons in the shell parallel to the straight line generators.

The thrusts from the beams are delivered to large pier-type abutments resting on pile foundations. Prestressed tie beams at the foundation level are used to connect opposite piers and carry the unbalanced horizontal thrust coming from the interior beams. Smaller tie beams running diagonally between abutments are also utilized to insure that no relative motion between the abutments will occur in case of ground motion due to earthquakes.

Original Analysis and Design in 1969

The structure was designed to insure that it had adequate strength, stiffness, and stability under all possible load conditions. The construction sequence was also carefully considered in the design process.

A preliminary analysis and design was first conducted for the shell roof using simple membrane theory to determine the major forces being transmitted through the shell, edge, and interior beams, piers, foundations, and ties in the structural system to insure statical control.

It was apparent from the beginning that the control of the deflections and the stresses in the cantilever edge beams was critical to the success of the design. A concept of load balancing by means of proper prestressing of these edge beams suggested itself; however, it was recognized that an accurate analysis was needed to determine the interaction of the shell and edge beam under the dead load of the shell and the beams and prestress in the edge beams.

A computer program developed at the University of California in 1965 which was based on the finite element method was used to perform a large number of analyses under different load conditions. The program used a linear elastic analysis and a direct stiffness solution to determine nodal point displacements and element forces in a general three-dimensional structure composed of a system of one-dimensional beam-type elements and two-dimensional triangular plane stress finite elements. To simulate the shell, a quadrilateral mesh was selected over the structure having two triangular elements in each quadrilateral and beam members forming a two-way grid along the two approximate normal lines of the nodal points. The plane stress triangular finite elements were used to represent the membrane stiffness of the shell, while the two-way grid of beam members, which were assigned only a flexural stiffness equivalent to the shell thickness, were used to represent the bending stiffness of the shell. In 1965 reliable shell elements which could account for shell membrane and bending action were not available.

The interior and edge beams were represented by beam-type members having axial, bending, and torsional stiffnesses.

Input into the program consisted of the nodal point coordinates, the uncracked concrete material properties of each beam element and each triangular finite element, and the loading and the boundary conditions. Output included the nodal point displacements; principal membrane stresses and bending moments in the shell; and axial forces, torques, and bending moments in the beams.

Using the computer output, the reinforcement and prestressing were designed initially using an allowable stress procedure for various combinations of dead load of the shell plus beams (D), prestress (P), live load (L) = 1.44 N/m^2 (30 psf), and wind loads (W). Ultimate strength design checks and adjustments of the reinforcement were then made under the critical factored load case of 1.5D + 1.0P + 1.8L.

The Ponce roof, designed in 1969, was the first large-span reinforced and prestressed concrete roof shell for which the final design was based on a detailed computer linear elastic finite element analysis.

Construction Sequence

The construction sequence, begun in late 1969, was as follows (Figs. 1, 2, and 3): (1) construct all four piers; (2) construct shell quadrant I with four surrounding beams; (3) stress shell tendons parallel to beams in sequence from tip towards pier; (4) pour shell slab cantilevering from edge beams in quadrant I; (5) repeat steps 2 through 4 for shell quadrants II, III, and then IV; (6) stress the foundation tie beams; (7) gradually release the vertical supports for the forms under the 4 in. shells working from the center of each of the four quadrants, simultaneously towards the four surrounding beams; (8) gradually stress the edge beam tendons in 4-to-6 stages, and at each stage work around the perimeter of the shell and stress at least 2 tendons at one time such that the forces are symmetrical about each pier at all times; and (9) gradually remove the supports of the edge and interior beams; for the edge beams start symmetrically from all 4 tips, and for the interior beams from the center towards the piers. Upon completion of this step the shell is self-supporting.

Step 9 was completed in August, 1971, at which time the average deflection of the four tips was measured to be 76mm (3.0 in.), which compared favorably with the calculated value of 89mm (3.5 in.).

Developments in Analysis and Computer Programs 1969-1996

During the period 1969-1996, there have been tremendous developments in analysis and computer program capabilities for concrete shells contributed by many researchers and designers summarized in papers by Scordelis [Scordelis (1986, 1990, 1993)]. It is estimated that today several hundred general purpose finite element programs exist for the linear elastic analysis of structures, many of which can analyze shells, using pre- and postprocessing programs and interactive graphics. A much smaller number of these programs have the capability to do nonlinear analyses of various types and only a very few can be applied to nonlinearities involved in the design of reinforced and prestressed concrete shells. One such program, NASHL, developed by Chan [Chan (1982)] at the University of California, Berkeley, can trace the complete structural response of reinforced concrete shells with edge beams throughout their service load history and under increasing loads up to ultimate failure. The time-dependent effects of load and temperature history and creep and shrinkage of the concrete can be taken into account. Working at Berkeley and at the Technical University of Catalonia, Barcelona, Roca [Roca (1988)] added prestressing elements to the shell and edge beams into the NASHL analytical model and program, and gave it the name NASHL1. Detailed descriptions of the theoretical development; analytical models; shell and beam elements; material properties for concrete, reinforcing, and prestressing steel; solution techniques for nonlinear material, geometric, and time-dependent effects; numerous numerical examples; availability; and input-output for the computer programs NASHL and NASHL1 can be found in the references by Chan [Chan (1982)], Roca [Roca (1988)], and Scordelis [Scordelis (1990)]. A brief description of the analytical models and elements is given below.

In the computer program, the analytical model consists of a series of joints interconnected by shell and beam elements. The shell element is a nine-node, two-dimensional curved isoparametric element with 5 DOF at each node. The element thickness is divided into concrete and reinforcing steel layers, and each layer is assumed to be under a two-dimensional stress state. The steel reinforcement can be placed in any layer and in several directions if desired. Cracking and nonlinear material response is traced layer-by-layer under increasing load and the time-dependent effects of creep and shrinkage. An updated Lagrangian formulation is used to take the effects of changing structural geometry into account.

By adding two straight, one-dimensional beam elements to the side of a curved shell element, thin shells with edge beams can be analyzed. Each beam element with 12 DOF is prismatic, but has an arbitrary cross-section made up of discrete numbers of concrete and reinforcing steel filaments for which the uniaxial strains and stresses are monitored.

In order to simulate elastic supports, obtain support reactions, or enforce irregular boundary conditions for the global structure, linear elastic spring elements can be placed in any specified direction at discrete nodal points.

Shell prestressing tendon elements are input individually as arbitrary spatial curves contained in the shell thickness using analytical parametric expressions for the definition of their geometry.

Distributed tangential and normal interactive forces between concrete and prestressing tendons are calculated taking into account frictional and anchor slip forces, and the end tendon segment forces.

Bonded or unbonded post-tensioned, as well as pretensioned tendons, may be considered.

Each edge beam prestressing tendon element is divided into a number of segments, each of which is straight, spans a single beam element, and is assumed to have a constant force. Using vector algebra to define the tendon geometry-nodal forces, strains and stresses can then be monitored throughout the analysis.

Analysis and Design of Ponce Coliseum in 1996

It is of interest to compare how an analysis and design for the Ponce Coliseum would be done in 1996 using analytical methods and computer programs available today as compared to the original analysis and design done in 1969.

The three major steps in the design of a shell roof structural system of the size and complexity of the Ponce Coliseum are (1) conceptual design, (2) preliminary analysis and design, and (3) final analysis and design. This is as true in 1996 as it was in the original design of 1969. Steps 1 and 2 would be essentially the same in 1996 as in 1969. The conceptual design, Step 1, is based on the structure's size, function, esthetics, location, environment, climate, possible methods of construction, etc. and to a high degree the designer's knowledge, experience, bias, and to a large extent the available budget. The preliminary analysis and design, Step 2, would probably again use a membrane analysis to insure statical control of the total structural system. A designer skilled in computer graphics today might find it useful for Steps 1 and 2.

The final analysis and design, Step 3, should be extended to two steps, 3a and 3b. In Step 3a, linear elastic analysis, with uncracked concrete section properties, because of its greater simplicity and possibility to use superposition of results, will remain the primary method for strength design for factored loads in determining design internal stress resultant to be resisted at ultimate load and for design of the reinforcement and prestressing. Improved shell and beam finite elements such as those in NASHL and NASHL1 can be used together with pre- and postprocessing units and computer graphics to greatly aid in this process.

Step 3b should be considered a verification step of the final design dimensions, reinforcement, and prestressing selected in Step 3a in which these are used to form the analytical model and input for a nonlinear material, geometric, and time-dependent analysis program such as NASHL1. The analytical model is then subjected to dead load (1.0D) + prestress (1.0P) + increasing multiples α of the design live load (αL) until failure.

One such analysis was made using NASHL1 [Roca, Molins, and Scordelis (1993); Scordelis (1993)], including material and geometric nonlinearities, which indicated an adequate margin of safety of $\alpha = 3.87$ for the original design in 1969 of the Ponce Coliseum shell roof. In addition, the NASHL1 output gave much valuable information on cracking, deflections, stresses in the concrete, and reinforcing and prestressing steel as the structure was loaded incrementally through its elastic, cracking, inelastic, and ultimate ranges. Especially important was the output on sequence and mode of failure. All of these give the designer an opportunity to study, understand, and check the structural behavior given by the computer solution.

Conclusions

The development and improvement, between 1969 and 1996, in methods of linear and nonlinear analysis, computer programs, pre- and post-processing, and computer graphics provide powerful tools to aid in the analysis and design of reinforced and prestressed concrete shell systems. However, conceptual design and verification of the final design are still the ultimate responsibility of an engineer who understands the true behavior of these fascinating structures.

Acknowledgements

The structural design of the Ponce Coliseum in 1969 was by joint venture of T. Y. Lin International, San Francisco, California and R. Watson, Engineer, and Sanchez, Davila and Suarez, Engineers, of San Juan, Puerto Rico. The architect for the project was V. Monsanto and Associates, Ponce, Puerto Rico. The contractor was Gabriel Alvarez and Associates, Ponce, Puerto Rico. The senior author of the present paper was a consultant on the detailed analysis and design of the shell roof structure.

This paper is a synthesis of excerpts taken from earlier papers by the senior author given in the References.

References

Chan, E.C. (1982). *Nonlinear Geometric, Material and Time Dependent Analysis of Reinforced Concrete Shells and Edge Beams*, Ph.D. Dissertation, UC-SESM 82-8, University of California, Berkeley, 361 pp.

Lo, K.S. and Scordelis, A.C. (1969). "Design of a 271-ft Span Prestressed HP Shell," *Proceedings, IASS International Congress*, Madrid, Spain, September, 1969, 13 pp.

Roca, P. (1988). *A Numerical Model for the Nonlinear Analysis of Prestressed Concrete Shells* (in Spanish), Ph.D. Dissertation, Technical University of Catalonia, Barcelona, 343 pp.

Roca, P., Molins, C., and Scordelis, A.C. (1993). "Nonlinear Analysis of an Existing Prestressed Shell," *Proceedings, IABSE Colloquium*, Copenhagen, pp. 401-408.

Scordelis, A.C., Lo, K.S., Lin, T.Y., Yang, Y.C., and Kulka, F. (1972). "Ponce

Coliseum Shell Roof in Puerto Rico," *Proceedings, IASS International Symposium*, Calgary, Canada, July, 1972, pp. 277-284.

Scordelis, A.C. (1986). "Computer Analysis of Reinforced Concrete Shells," *IASS Bulletin*, Vol. 27, No. 90, pp. 47-53.

Scordelis, A.C. (1990). "Nonlinear Material, Geometric and Time Dependent Analysis of Reinforced and Prestressed Concrete Shells," *IASS Bulletin*, Vol. 31-1,2, Nos. 102 and 103, pp. 57-70.

Scordelis, A.C. (1993). "Present Status of Nonlinear Analysis in the Design of Concrete Shell Structures," *IASS Bulletin*, Vol. 34, No. 2, pp. 67-80.

NONLINEAR BEHAVIOR OF RC COOLING TOWERS AND ITS EFFECTS ON STRAINS, STRESSES, REINFORCEMENT AND CRACKS

by Udo Wittek,[1] and Anmin Ji,[2]

SUMMARY

The physical nonlinear behavior of RC cooling towers will be compared with a classical approach based on linear elasticity theory.

A complex set of nonlinear elastic constitutive laws is used to analyze high-rise cooling towers. The nonlinear effects compared with those based on linear theory will be presented with respect to the overall behavior of the shell considering stress-, strain-, reinforcement-, and crack-distribution.

In order to meet the demands of durability, safety and economic aspects, the nonlinear behavior of the structure should be taken into account for the design of the RC cooling towers.

1. INTRODUCTION

It is known that a cooling tower shell usually exists with cracks under the normal working state, it indicating strong nonlinear behavior. The classical linear elasticity theory can not describe such actual nonlinear behavior of the reinforced concrete structure. Only with a nonlinear analysis method based on the real constitutive laws for concrete and steel, can the complete nonlinear material behavior of a structure be predicted until collapse. To effectively reflect the nonlinear behavior of a reinforced concrete cooling tower shell structure, a realistic numerical model is necessary.

----

[1]Prof. Dr.-Ing., Lehrstuhl für Baustatik, Universität Kaiserslautern, Postfach 3049, D- 67653 Kaiserslautern, Germany
[2]Dr.-Ing., Res. Assistant, Lehrstuhl für Baustatik, Universität Kaiserslautern, Postfach 3049, D-67653 Kaiserslautern, Germany

The aim of this paper is to depict the physical nonlinear material behavior of reinforced concrete cooling tower shells. The typical appearance of the nonlinear stress redistribution and its effects will be indicated and discussed.

2. NONLINEAR MODEL FOR REINFORCED CONCRETE SHELLS

The characteristic nonlinear behavior of the composite material, reinforced concrete, can be described by subdivision of the reinforced concrete cross section into different layers of steel and concrete.

The steel layers are represented by uniaxial components with an uniaxial bilinear material law, whereas some different biaxial constitutive laws are needed to realistically describe the nonlinear behavior of the concrete at different working state.

The behavior of concrete is nearly linear elastic before the tensile strength of concrete i.e. cracking is reached. After cracking, concrete will have still some limited bearing capacity due to the interaction between concrete and reinforcement. This post-cracking effect appears in reinforced concrete areas as Tension-Stiffening and in pure concrete areas without reinforcement as Tension-Softening. Modelling Tension-Stiffening and Tension-Softening leads to different anisotropic stress-strain relations.

For the description of all nonlinear concrete effects, eight different material laws for different zones are necessary. They can be identified by the principal stress relations as in Fig. 1 [2,7].

Zone 1 and 2 represent the undamaged concrete. In zone 1 the isotropic linear constitutive law with two initial elastic constants, Young's modulus E_0 and Poisson's ration v_0, can be used. In zone 2, the concrete can be described with nonlinear elastic stress-strain relations which are formulated e.g. by CEDOLIN and MULAS [1].

In zone 3 up to 6 the tensile strength of concrete is exceeded. In those cases cracks of concrete occur. But concrete will not lose its entire load bearing capacity immediately because of the effects of the Tension-Softening and Tension-Stiffening. For representing the behavior of reinforced-concrete shells, the popular smeared-crack model is accepted. The effects of Tension-Stiffening is taken into account as a function of the strain in the direction of the reinforcement. The aspect of the reduced shear capacity can be represented by a shear factor depending upon the reinforcement quantities. In zone 5 and 6, where the cross direction of a cracked cross-section concrete carries compressive stresses, the corresponding compressive strength has to be reduced as a function of the tensile stresses.

In zone 7 and 8, the concrete is damaged through crushing where the compressive strength of concrete is exceeded. In that case it could be assumed that concrete has lost its bearing capacity completely.

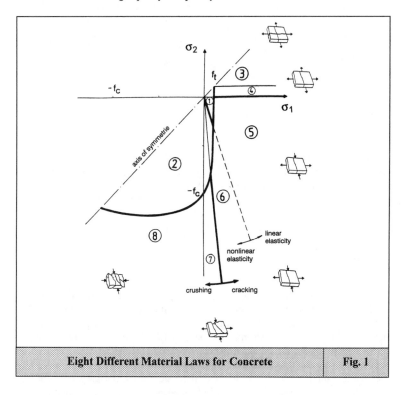

Eight Different Material Laws for Concrete | **Fig. 1**

3. NONLINEAR ANALYSIS AND DESIGN OF CROSS-SECTION

To reflect the actual structural behavior, the internal forces and moments and resistance should be calculated by nonlinear analysis method. Although the physical nonlinear behavior of the structure is taken into account in nonlinear analysis, the strong inconsistencies between nonlinear stress calculation and cross-section design can not be avoided. Two quite different constitutive laws are used for different purposes. In nonlinear analysis the tensile strength of concrete has to be considered in the constitutive law and the stress is calculated by means of mean material values. Afterwards to do a cross-section design, a different constitutive law based on fractile values divided by a partial safety factor, in which the tensile strength of concrete is neglected, will be used.

In addition nonlinear analysis requires an initial definition of reinforcement. Reinforcement based on a linear stress analysis can be taken as the initial reinforcement. Once cracks appear, the stiffness of the shell will be changed in comparison with the linear situation. This results in a redistribution of internal forces. The redistribution of internal forces is also influenced by the initial reinforcement. If the nonlinear internal forces are used to make a new design of the cross-section, the newly required reinforcement is certainly not the same as the first one. For the statical indeterminate cooling tower shell, the final reinforcement quantity can be determined through an iterative process. Usually the convergence can be reached after 2-3 iterations [2].

4. COMPARISON OF LINEAR/NONLINEAR BEHAVIOR OF RC COOLING TOWERS

In order to investigate the physical nonlinear behavior of RC cooling towers, a nonlinear analysis for an actual cooling tower with dimensions shown in Fig. 2 under one of important design load combination g + 1.75w is carried out, where g means dead load and w wind load. The nonlinear effects compared with those based on linear theory will be presented with respect to the overall behavior of the shell considering stress-, strain-, reinforcement-, and crack-distribution.

4.1 Distribution of cracks

Before discussing the nonlinear behavior of the shell, the situation of cracks on the surface of the shell shall be examined. Fig. 3 shows the distribution of the cracks on the outside and inside surface of the shell under the load level g + 1.75w. It may be observed that in a luv-meridian area from 0° to 30° along the circumference and from about 15 to 85 m along the height of the shell, a lot of horizontal cracks resulting from the principal membrane meridional tensile forces appear on both sides of the shell. At the same time many vertical cracks resulting from the ring bending moment occur on the upper flank-side of the outside surface of the shell. On the inside surface of the shell in luv-area, apart from horizontal cracks, there are many other vertical cracks until about 10° along the circumference. This means that in that area, the ring bending effects are activated due to the material physical nonlinearity.

With the development of the cracks, the structural behavior of the cooling tower shell will obviously be changed in comparison with the linear case.

4.2 Comparison of stress and internal forces

Because of the limited space, only the distribution of the stress and strain on the outside surface of the shell will be discussed.

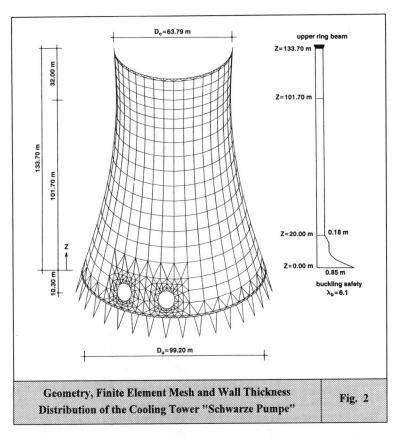

Geometry, Finite Element Mesh and Wall Thickness Distribution of the Cooling Tower "Schwarze Pumpe"

Fig. 2

The comparison of the first principal stress distribution σ_1 on the outside surface of the shell between the linear and nonlinear situations is shown in Fig. 4. For the nonlinear case, the stress is calculated with nonlinear determined reinforcement for $g + 1.75w$.

The characteristic behavior of so-called nonlinear stress redistribution can be observed clearly. Because of the cracking of concrete in the luv-meridian region, the stiffness of the shell in that area is reduced compared with the linear solution. Along the luv-meridian the linear maximal first principal tensile stress reaches 4.8 MN/m^2, which is far beyond the tensile strength of the existing concrete. For nonlinear calculations, the maximal first principal tensile stress in this area is limited by the tensile strength of concrete, 2.7 MN/m^2 (mean value) for this case. The changing stiffness results in a redistribution of internal stress and forces. The peaks

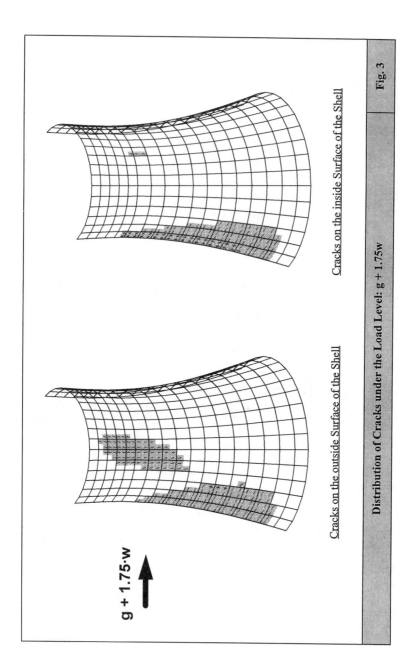

Cracks on the inside Surface of the Shell

Cracks on the outside Surface of the Shell

g + 1.75·w

Distribution of Cracks under the Load Level: g + 1.75w

Fig. 3

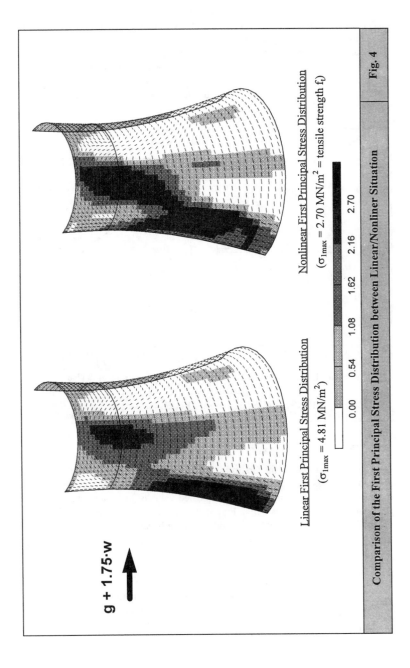

g + 1.75·w

Linear First Principal Stress Distribution

(σ_{1max} = 4.81 MN/m²)

Nonlinear First Principal Stress Distribution

(σ_{1max} = 2.70 MN/m² = tensile strength f_t)

| 0.00 | 0.54 | 1.08 | 1.62 | 2.16 | 2.70 |

Comparison of the First Principal Stress Distribution between Linear/Nonliner Situation

Fig. 4

of the tensile stress are decreased in the nonlinear case, but in addition the area of high tensile stress (below the tensile strength) will be broadened. The high tensile meridional forces up to a height of 80 m are reduced strongly and the internal energy transfers along the circumferential direction on the upper flank-side of the shell, where the ring bending effects are obviously activated with increased bending moment (on the inside surface of the shell, the concrete withstands strong compressive stresses along the circumferential direction in that area.). This result is shown in Fig. 5.

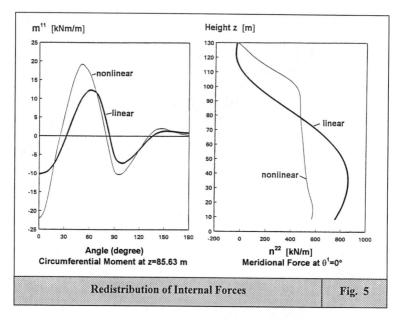

Redistribution of Internal Forces   Fig. 5

4.3 Comparison of strain

The distributions of the first principal strain ε_1 on the outside surface of the shell for the linear and nonlinear model are shown in Fig. 6. For the existing concrete, the first cracks appear when the smeared value of the strain reaches 0.08‰ (compare with the crack pattern in Fig. 3). The nonlinear crack capacity of the concrete allows strains much greater than those in the linear case. They will extend up to $\varepsilon_{1\,max} = 1.042$ ‰; this is about the half value of the yielding strain of the steel.

4.4 Comparison of reinforcement

To illustrate the influence of the nonlinear stress redistribution on the redistribution of the reinforcement, the calculation of the reinforcement is again based on

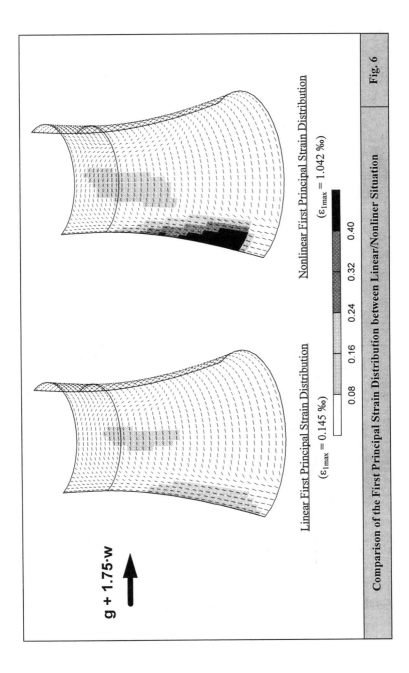

g + 1.75·w

Linear First Principal Strain Distribution
(ε_{1max} = 0.145 ‰)

Nonlinear First Principal Strain Distribution
(ε_{1max} = 1.042 ‰)

0.08 0.16 0.24 0.32 0.40

Comparison of the First Principal Strain Distribution between Linear/Nonliner Situation

Fig. 6

the same load combination: g + 1.75w. The limitation of the minimum reinforcement based on BTR [3], 0.3% of the cross-section, has been taken into account as well.

It should be pointed out that the final design of cooling towers must take into consideration temperature effects, which have been excluded from this comparing study.

In Fig. 7 the necessary reinforcement for both directions is plotted over the height of the shell. The amount of reinforcement a_s represents the sum of the area of the steel cross section for the interior and exterior steel layers.

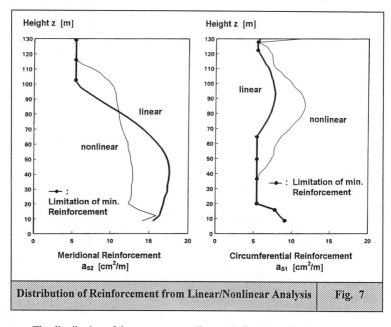

Distribution of Reinforcement from Linear/Nonlinear Analysis   Fig. 7

The distribution of the necessary nonlinear reinforcement for the assumed load combination is calculated with the help of the iterative process as mentioned earlier. As the first step the internal forces $n^{\alpha\beta}$ and $m^{\alpha\beta}$ at each point of the shell are calculated by means of the mean values of the material parameters under the consideration of the tensile strength of the concrete and based on the initial reinforcement resulted from linear analysis (just as shown in Fig. 7 with "linear"). The cross-section resistance is then determined by neglecting the tensile stress of the concrete; in addition the fractile values of the concrete compressive strength divided by a partial safety factor (1.75) and yielding stress of the steel layers are chosen to determine the reinforcement. The new obtained reinforcement now differs from the initial one (linear case) and the next calculation is necessary. It starts with this new reinforcement distribution. The iterative process will be finished when the final reinforcement meets the condition: $a_s^{(i+1)} \geq a_s^{(i)}$.

For the observed loading condition the nonlinear meridional reinforcement will be reduced in the lower part of the shell and increased in the upper part of the shell compared with the linear solution. In the circumferential direction the necessary reinforcement should be increased for almost all parts of the shell in the nonlinear case.

With respect to the design of the cooling tower shell, the ultimate load state must be analyzed considering temperature effects in addition. A further design requirement is the limit state of serviceability which is also not reviewed in this paper. The aim of this contribution has only been to demonstrate the effects of the nonlinear material behavior. Nevertheless all requirements for a successful cooling tower design [4,5,6] with respect to the limit state of the ultimate load and the limit state of serviceability, which also provides economic aspects, may be executed by the consideration of these nonlinear effects.

REFERENCES

[1] CEDOLIN,L./MULAS, M.G.
     Biaxial Stress-Strain Relation for Concrete, Journal of Engineering Mechanics
     Division, 1984
[2] GROTE, K.
     Anwendung geometrisch und physikalisch nichtlinearer Algorithmen auf
     Flächentragwerke aus Stahlbeton, Universität Kaiserslautern, Fachgebiet
     Baustatik, Bericht 1/1992
[3] VGB-Fachausschuß Bautechnik bei Kühltürmen Bautechnische Richtlinien,
     VGB-Verlag, Essen, 1990.
[4] WITTEK, U.
     Anwendung neuer Entwurfskonzepte für extrem windbeanspruchte Naturzug-
     kühltürme aus Stahlbeton, VGB-TB-602, Essen, 1991.
[5] WITTEK, U.
     Ein nichtlineares Entwurfskonzept für Naturzugkühltürme aus Stahlbeton,
     SFB 151 - Bericht Nr.23 Ruhr-Universität Bochum, 1992.
[6] WITTEK, U./ MEISWINKEL, R.
     Increasing the durability and safety of RC cooling towers using nonlinear
     design strategies, Computational Modeling of Concrete Structures, Proc.
     of Europ-C-1994, Innsbruck, Pineridge Press, UK 1994
[7] WITTEK, U./GROTE, K./MEISWINKEL, R.
     ROSCHE-Handbücher zu FE-Computerprogrammen für Rotationsschalen,
     Universität Kaiserslautern, Fachgebiet Baustatik, 1990

DYNAMIC FATIGUE OF HIGH-RISE
NATURAL DRAUGHT COOLING TOWERS

Wilfried B. Krätzig[1]

ABSTRACT

The paper aspires to draw the attention of engineers to crack-damage processes of cooling tower shells under temperature action and severe storms. Cracking of the shell shifts the response properties of the structure towards spectral ranges of higher wind excitations. These processes will be simulated numerically by physically/geometrically nonlinear computational techniques, simplified in a quasi-static regime. As the paper proves, cracking will weaken cooling tower shells considerably causing remarkable redistributions of stresses and additional deformations.

1. INTRODUCTION

In engineering design practice reinforced and prestressed concrete structures are generally analysed by means of elastic models. The overwhelming advantage of such a linear elastic analysis is the validity of the principle of superposition, due to which stresses or stress-resultants of single load cases can be computed separately and then be superposed to deliver respective values of load combinations.

Within such linear concepts repeated loadings cause identical results as one single load cycle. Obviously, linear analysis concepts are unable to model damage processes in structures, e.g. in cooling tower shells, in which macro-cracking of the shell leads to considerable redistributions of stresses and deformations. The present investigation demonstrates the action of dead weight, temperature effects and repeated high storms weakening cooling tower shells by a nonlinear evolution of crack-damage.

---

[1] Institute for Statics and Dynamics, Ruhr-University, 44780 Bochum, Germany

2. NONLINEAR MODELLING OF REINFORCED CONCRETE SHELLS

2.1 FROM STRUCTURAL TO MATERIAL POINT LEVEL

We start with a review over the nonlinear analysis concept and the material modelling of FE-discretized reinforced concrete shells. Nonlinear responses $V(P)$ are evaluated by incremental-iterative solution procedures applied to the (global) structural level and based on the tangent stiffness equation:

$$K_T \cdot \overset{+}{V} = P - F_i \rightarrow \overset{+}{V} = K_i^{-1} \cdot (P - F_i). \tag{1}$$

Herein, K_T abbreviates the tangential stiffness matrix, $\overset{+}{V}$ the increments of the global degrees of freedom and F_i the internal nodal forces of a certain equilibrium state. While the solution of the discrete boundary value problem is controlled on the structural level, nonlinear material properties are defined on the material point level. As explained in Fig. 1, the incidence-table of global and element DOFs leads from the structural level to an arbitrary finite element, the inverse discretization process then guides to the layered shell level with its strain tensor increments $\overset{+}{\alpha}_{(\alpha\beta)}$, $\overset{+}{\omega}_{(\alpha\beta)}$ and the strain distribution over the shell thickness h finally yields strain increments $\overset{+}{\gamma}_{ij}$ of arbitrary material points. From here, after evaluation of sundry incremental material laws, the iteration loop takes its described way back to the structural level.

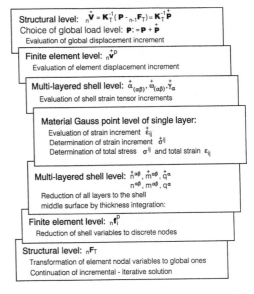

Figure 1. From structural material point level

2.2 INITIAL VALUE PROBLEMS ON MATERIAL POINT LEVEL

In material points, various constitutive laws of the reinforced concrete shell describing elasto-plastic actions, tension cracking, compression softening and nonlinear bond, are defined in rate-form of initial value problems

$$\dot{\sigma}^{ij} = C^{ijkl} \dot{\gamma}_{kl} + \overset{\circ}{\sigma}{}^{ij} \rightarrow \overset{+}{\sigma}{}^{ij} = C^{ijkl} \overset{+}{\gamma}_{kl} + \overset{\circ}{\sigma}{}^{ij}. \tag{2}$$

For simplicity, neighbourhoods Δh of finite thickness between selected material points (Fig. 2) are defined as material layers, across which all physical properties of the material points are assumed to vary linearly. By numerical integration over the shell thickness h according to

$$\overset{+}{n}{}^{\alpha\beta} = \int_h \mu\, \mu_\rho^\beta\, \overset{+}{\sigma}{}^{\alpha\rho}\, d\Theta^3 \quad = \overset{1}{E}{}^{\alpha\beta\lambda\mu}\, \overset{+}{\alpha}_{\lambda\mu} + \overset{2}{E}{}^{\alpha\beta\lambda\mu}\, \overset{+}{\omega}_{\lambda\mu}\ ,$$

$$\overset{+}{m}{}^{\alpha\beta} = \int_h \mu\, \mu_\rho^\beta\, \overset{+}{\sigma}{}^{\alpha\rho}\, \Theta^3\, d\Theta^3 \quad = \overset{2}{E}{}^{\alpha\beta\lambda\mu}\, \overset{+}{\alpha}_{\lambda\mu} + \overset{3}{E}{}^{\alpha\beta\lambda\mu}\, \overset{+}{\omega}_{\lambda\mu}\ , \tag{3}$$

incremental constitutive laws for the shell model [1] can be derived in the GAUSS-points. Herein μ_ρ^β denotes the shifter tensor from material points to the shell middle surface, and the following abbreviations have been introduced:

$$\overset{n}{E}{}^{\alpha\beta\lambda\mu} = \int_h C^{\alpha\beta\lambda\mu}\, (\Theta^3)^{n-1}\, d\Theta^3\ ,\quad n = 1, 2, 3\ . \tag{4}$$

To solve the above mentioned highly nonlinear initial value problem for a given strain increment $\Delta\gamma_{ij}$ following from $\overset{\bullet}{V}$, (2) has to be integrated:

$$\Delta\,\sigma^{ij} = \int_0^{\Delta\gamma_{kl}} C^{ijkl}\, \overset{+}{\gamma}_{kl} + \overset{o+ij}{\sigma}\ . \tag{5}$$

Internal solution algorithms have been developed for this purpose [2,3]; recently they are under rapid improvement to speed up the time-consuming solution processes.

To attain again the structural level according to Fig. 1, the strain increments $\overset{+}{\alpha}_{\alpha\beta}, \overset{+}{\omega}_{\alpha\beta}$ of (3,5) are connected with incremental displacement fields of geometrically nonlinear shell theories [4] for large deformations and medium rotations. The discretization process is based on the concept of high-precision NACS shell elements [5] as displacement field approximations. Finally the tangent stiffness equation (1) can be formulated with

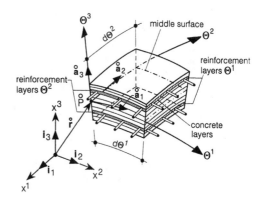

Figure 2. Layered reinforced concrete shell model

$$\mathbf{K}_T = \mathbf{K}_{ep} + \mathbf{K}_{\sigma\,ep} + \mathbf{K}_{u\,ep}\ ,\quad \mathbf{F}_i = \mathbf{F}_{i\,ep} + \mathbf{F}_{i\,\sigma\,ep}\ , \tag{6}$$

distinguishing elasto-plastic (ep) contributions of the (geometrically) linear, the initial stress (s) and the initial displacement (u) stiffness matrix as well as the internal nodal force vector.

2.3 NONLINEAR MATERIAL MODELLING OF REINFORCED CONCRETE

Reinforced concrete is a nonlinearly reacting composite. Its response is characterized by the following nonlinear phenomena:

- nonlinear stress-strain relation in compression,
- tension cracking at low stress level,
- elasto-plastic behaviour of the reinforcement,
- nonlinear bond between reinforcement and concrete.

Each of these phenomena can be responsible for stress redistributions, connected with essentially irreversible material processes, contributing to the damage evolution. Out of a large variety of modern modeling possibilities for reinforced concrete properties [6], our simulations are based on the following material models.

The nonlinear behaviour of concrete in compression is simulated due to the elasto-plastic-fracturing theory [7]. This model subdivides each stress increment resulting from a certain incremental strain into an elastic stress increase (el), a plastic stress decrease (pl) and a decrease due to microfracturing (fr):

$$\overset{+}{\sigma}{}^{ij} = \overset{+}{\sigma}{}^{ij}_{el} + \overset{+}{\sigma}{}^{ij}_{pl} + \overset{+}{\sigma}{}^{ij}_{fr} = (C^{ijkl}_{el} + C^{ijkl}_{pl} + C^{ijkl}_{fr}) \overset{+}{\gamma}_{kl}. \tag{7}$$

A series of required material parameters [7] as well as the failure envelope – see Fig. 3 – is fitted to well established material data.

The reinforced steel model covers elastoplastic behaviour in a multi-linear way for cyclic processes, under special attention of the BAUSCHINGER effect. To describe the nonlinearely fading bond, the tension-stiffening concept due to [8] is applied.

Concrete unter tension fails in a brittle manner leading to stiffness reductions in

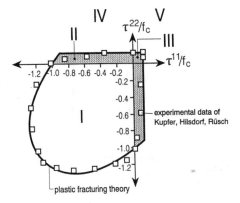

Figure 3. Applied plane failure envelope compared with experimental data

the material points affected. For the present analysis, cracking is modelled in a smeared manner neglecting localization phenomena. The analysis employs a principal stress criterion, by which cracks oriented at an arbitrary angle in a given frame of reference can be described. After exceedance of the tension strength, all stiffness coefficients of (7) orthogonal to an assumed crack are removed. Shear stiffnesses are partly retained due to aggregate interlocking. Closing and re-opening of cracks as well as the formation of secondary cracks can also be modelled [9].

3. WIND AND TEMPERATURE ACTION ON COOLING TOWER

3.1 COOLING TOWER K AT GERSTEINWERK POWER STATION

Fig. 4 portrays the cooling tower shell of the VEW power plant GERSTEINWERK, substation K, to which further investigations will be applied. Necessary material parameters can be found in [10]. The reinforced concrete shell of 123.90 m of height is composed of two different hyperboloidal middle surfaces

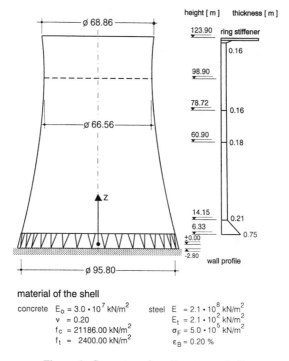

Figure 4. Geometry of cooling tower shell

$$r = r_o + \frac{a}{b} \sqrt{b^2 + (H_T - z)^2} \qquad\qquad (9)$$

with $r_o = 29.407$ m   respectively   2.579 m

         $a = 3.873$ m   30.701 m

         $b = 30.270$ m   85.230 m

         $H_T = 98.900$ m   98.900 m

for the lower respectively upper part.

The cooling tower shell is loaded due to BTR-regulations [11] by dead weight g, wind w and temperature action t [12].

3.2 INFLUENCE OF DIFFERENT SERVICE TEMPERATURES

Fig. 5 studies the influence of temperature actions on the collapse load level, related to wind by the load factor λ. It shows the following load superpositions:

g + λw: The cooling tower in out-of-service conditions (g only) is subjected to a gale up to collapse.

g + t + λw: The cooling tower in extreme winter service conditions (t: $\Delta t = 45$ K) is subjected to a gale up to collapse.

g + t − t + λw: The cooling tower has operated for a period of time in extreme winter conditions (g + t). After taking the tower out of service (g + t − t), it is subjected to a gale up to collapse.

g + t − 0.67t + λw: The cooling tower has operated in extreme winter conditions. After removal to 33% of permanent service temperature conditions, the tower is hit by a gale up to collapse.

The study demonstrates that the collapse level related to wind (from 2.25 to 2.30) does not differ considerably for varying service (temperature) conditions. The shapes of single response curves however deviate because of considerable pre-damage of the shell, especially in the case of extreme temperature conditions [13].

3.3 FATIGUE DUE TO CRACK-DAMAGE

Such extreme load combinations, e.g. (g + t + λw), cause considerable early crack-damage in the shell. This cracking can be simulated numerically in the GAUSS-points of all layers, as demonstrated for the inner and outer face of a tower shell on Fig. 6.

This figure shows the results of a continuous crack model, applied to (g + λw) for a different cooling tower with insufficient reinforcement, close before collapse ($\lambda = 1.83$). But even if the crack-damage is less severe, the change of dynamic properties of the structure is obvious.

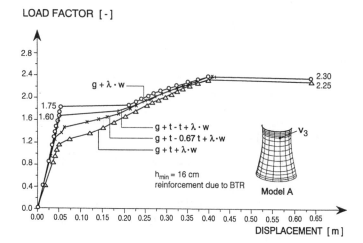

Figure 5. Influence of temperature actions

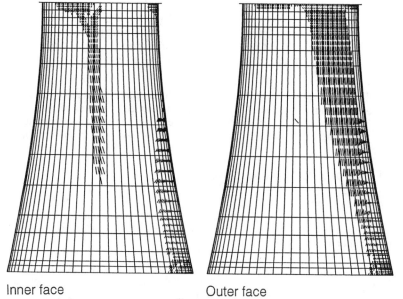

Inner face   Outer face

Figure 6.
Crack pattern of cooling tower shells at
VEAG power station BOXBERG III at $\lambda = 1.83$

4. REPEATED WIND ACTIONS

4.1 REPEATED MEDIUM STORMS

Because of the above mentioned linear analysis concepts the damage evolution in cooling tower shells, accessible only through physically nonlinear simulations, is largely unknown. We continue our studies of this behaviour with the cooling tower of Fig. 4 in extreme winter service conditions (g + t), subjected to 20 repeated storms up to $\lambda = 1.00$, storms which possess a return period in Germany of approximately 50 years.

As Fig. 7 demonstrates, wind actions of this intensity damage the structure only marginally. Compared to the response of a monotonic wind increase (g + t

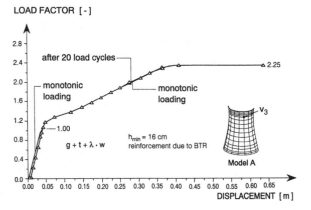

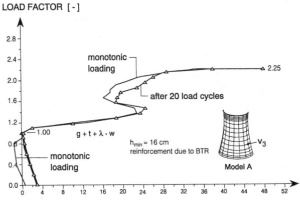

Figure 7. Several medium wind actions

$+ \lambda w$) up to collapse, the additional damage is initiated by temperature. At the first of 20 repeated storms the shell is largely cracked by temperature effects, but small damage contributions remain after temperature removal. Thus, additional wind damage for low intensities is of negligible order.

4.2 GALE CYCLES OF HIGH INTENSITY

This reported behaviour changes fundamentally for higher storm-intensities, as portrayed in Fig. 8 by example of 4 repeated storms $\lambda = 2.00$ correspon-

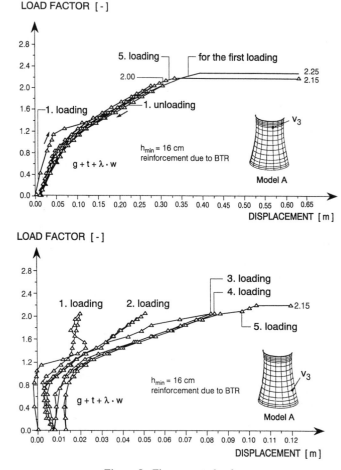

Figure 8. Five repeated gales

ding to a return period of around 5.000 years. In comparison to the monotonic response curve $(g + t + \lambda w)$ with collapse at $\lambda = 2.25$, each storm cycle damages the shell to a new damage level by a shift of the response to the right. Finally, the 5th load cycle is increased up to collapse level at $\lambda = 2.15 < 2.25$.

Fig. 9 ultimately simulates a complete cooling tower life of around 50 years with several severe storms from 50 to 500 years return period. Corrosion of steel, aging and spalling of concrete is considered, before a final gale leads to failure. The figure emphasizes nicely the continuous damage evolution, even leaving the damage increase of the dynamic magnification factor unconsidered.

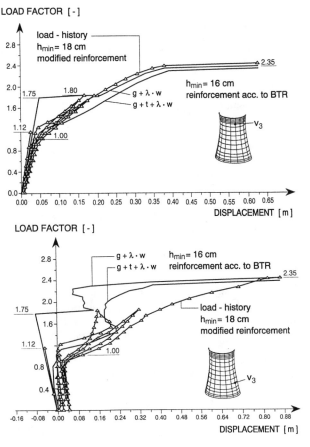

Figure 9. Damage of cooling tower during life time

5. CONCLUSIONS

The paper has demonstrated the ability of materially nonlinear simulations applied to damage processes for large reinforced concrete shells. The arising damage has been expressed in terms of inelastic tensile strains. Because of enforced brevity the more evident crack-simulations could not be recorded in more detail. Obviously, virgin cooling towers remain undamaged until extreme winter conditions and gales initiate a damage process. After this initiation, also events of medium intensity contribute to the further damage evolution [14].

REFERENCES

[1]   Basar, Y., Krätzig, W.B.: Mechanik der Flächentragwerke. Friedr. Vieweg & Sohn, Braunschweig 1985.

[2]   Bergan, P.G., Horrigmoe, G., Krakeland, B., Soreide, T.H.: Solution Techniques for Nonlinear Finite Element Problems. Int. Journ. Num. Meths. Engg., 11 (1978), 1677–1696.

[3]   Ortiz,M., Simo, J.C.: An Analysis of a New Class of Integration Algorithms for Elastoplastic Constitutive Relations. Int. Journ. Num. Meths. Engg., 23 (1986), 353–366.

[4]   Basar, Y., Krätzig, W.B.: A Consistent Shell Theory for Finite Deformations. Acta Mechanica 76 (1989), 73–87.

[5]   Harte, R., Eckstein, U.: Derivation of Geometrically Nonlinear Finite Shell Elements via Tensor Notation. Int. Journ. Num. Meths. Engg., 23 (1986), 367–484.

[6]   Hofstetter, G., Mang, H.A.: Computational Mechanics of Reinforced Concrete Structures. Vieweg Verlag, Braunschweig 1995.

[7]   Bazant, P.Z., Kim, S.-S.: Plastic Fracturing Theory for Concrete. ASCE, Journ. Engg. Mech. Div., 105 (1979), 407–428.

[8]   Warner, R.F.: Tension Stiffening in Reinforced Concrete Slabs. ASCE, Journ. Struct. Div. 104 (1978), 1885–1900.

[9]   Zahlten, W.: A Conribution to the Physically and Geometrically Nonlinear Computer Analysis of General Reinforced Concrete Shells. Techn. Report No. 90–2, Inst. Struct. Engg., Ruhr-University Bochum 1990.

[10]   Krätzig, W.B., Gruber, K., Zahlten, W.: Numerical Collapse Simulations to Check Safety and Reliability of Large Natural Draught Cooling Towers. Techn. Report No. 92-3, Inst. Struct. Engg., Ruhr-University Bochum 1992.

[11]   VGB–BTR Guideline: Structural Design of Cooling Towers. VGB–Technical Committee "Civil Engineering Technique of Cooling Towers – Bautechnik bei Kühltürmen", Essen 1990.

[12]   Krätzig, W.B., Meskouris, K.: Natural Draught Cooling Towers. An Increasing Need for Structural Research. IASS bulletin 34 (1993), 37–51.

[13]   Gruber, K.: Nonlinear Computer Simulations as Components of a Design Concept for Increasing Safety and Durability of Natural Draught Cooling Towers. Techn. Report No. 94–7, Inst. Struct. Engg., Ruhr-University Bochum 1994.

[14]   Basar, Y., Gruber, K., Krätzig, W.B., Meskouris, K., Zahlten, W.: Nichtlineare Analyseverfahren für Tragsicherheit, Schädigungsevolution und Restlebensdauer windbeanspruchter Naturzugkühltürme. Bauingenieur 67 (1992), 515–524.

Experiments for Ultimate Strength of Reinforced Concrete Cylindrical Panels under Lateral Loading and Comparison with FEM Analysis

Makoto Takayama [1], Kazuhiko Mashita [2], Shiro Kato [3],
Yasuhiko Hangai [4] and Haruo Kunieda [5]

Abstract

The present paper briefly describes the state of the art of reinforced concrete roofs in Japan. The emphasis is on a comparison of load carrying capacities between FEM analyses and experiments on micro concrete models recently performed for cylindrical reinforced concrete roofs.

Reinforced Concrete Shells in Japan

The engineering history of reinforced concrete shells in Japan can be written to originate from the Ehime Prefectural Hall, realized in 1953. During this period, great progress, both on theories and solutions for symmetric/antisymmetric loadings of shallow caps as well as on experimental methods, has been achieved mainly by Prof. Y. Tsuboi and his colleagues. This research has provided a fundamental basis for R/C shells, in both an academic and an engineering sense, which guided many shell constructions after 1960 as listed in Table 1.

From the late 1960's, R/C shell constructions were reduced in number year by year because of high labor costs. On the contrary, steel space and tension structure usage has increased, partly because, in Japan in this period, the cost of steel production began to decrease and prefabricated construction was realized, and partly because discretization numerical methods combined with digital computers were actively developed for structural analysis.

- -

[1] Professor, Dept. of Architecture, Kanazawa Institute of Technology
[2] Professor, Dept. of Architecture, Tokai University
[3] Professor, Dept. of Architecture and Civil Engineering,
    Toyohashi University of Technology
[4] Professor, Institute of Industrial Science, University of Tokyo,
    Roppongi 7-22-1, Minato-ku, Tokyo 105
[5] Professor, Disaster Prevention Institute, Kyoto University

Figure 1. Ehime Prefectural Hall (1953)

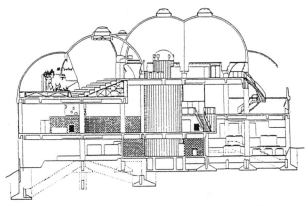

Figure 2. Kyusendo Forest Museum (1984)

In 1981 a new regulation for seismic design was officially introduced. This has required engineers to evaluate the ultimate strength of high rise buildings, and in some cases to investigate post critical behaviors. Advanced solution techniques, highly developed in this period, have also enabled engineers to analyse various shells with non-mathematical configurations. There has also been progress in long span prefabricated reinforced concrete (as Osaka City Gymnasium) and non-scaffolding concrete construction (as Memorial Hall).

At present, differing from the period of 1950 to 1960, a new era of reinforced concrete shells has started in Japan. Along with this current wave of construction, experimental research using concrete models of various configurations have been conducted at university laboratories to

Table 1. R/C shells in Japan

Name	Year	Type	Span (m)
Ehime Prefectural Hall	1953	Spherical shell	50
Gunma Music Center	1961	Folded Plate	60
Totsuka Country Club House	1961	Inverted Cylinder	36×77
Daisekiji Temple	1963	Double shell	68×59
Komazawa Olympic Park Gymnasium	1964	HP shells	68
Tokyo Roman Catholic Church	1964	HP shells	40×55
Ogaki Gymnasium	1980	HP shells	54
Kyusendo Forest Museum	1984	Combined domes	R=7.5
Memorial Church	1988	Combined arches	18
Kita-kyushu International Conference Center	1990	Prestressed cylindrical shells	26
Osaka City Gymnasium	1996	Precast concrete spherical shell	110

investigate ultimate strength and failure mechanisms. The present report describes some experimental results, with an emphasis on cylindrical roofs under lateral loading, and analyses the comparison of load carrying capacities between experiments and FEM analysis.

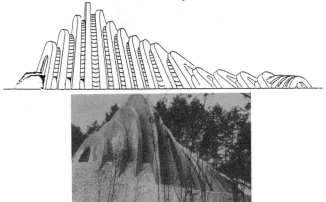

Figure 3. Memorial Church (1988)

Experiments for Ultimate Strength of Cylindrical Panels under Lateral Loading

The results of two series of experiments are compared with those based on FEM materially/geometrically nonlinear analysis. One is a series of panels under distributed loading, while the other series is under

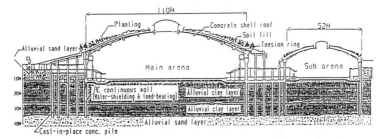

Figure 4. Osaka City Gymnasium (1996)

concentrated loading. The models in the experiments, which were carefully produced with minimum errors, are of micro concrete reinforced by steel wires. The material properties of the concrete and steel, the results, and the conventional FEM analytical scheme, and other data are briefly explained in the following paragraphs.

a. Shells under Distributed Loading

This discussion considers the ultimate strength of reinforced cylindrical panels under uniformly and partially distributed loading.

Eight micro concrete models, each designated T-i in Table 2, are experimentally tested to investigate the ultimate strength and the effects of both initial imperfections and loading patterns. The dimensions of the models, test apparatus, and loading points are shown in Figures 5 to 7 with the material properties given in Table 2. Each model is of the same configuration except for the initial imperfections, is reinforced in the same way by wires of 1.2 mm diameter with spacing of @40 mm at the middle surface, and is supported as a pin on the straight boundaries and as a roller on the circular edges. The details, including the FEM analysis adopted for comparison, are given in the references (Takayama 1991, 1992a, 1992b, 1995).

The models T-1 to T-3 are given an initial imperfection $w_0 = 0.0$mm, 4.0 mm and 8.0 mm, respectively, and are subjected to uniform loading. The mode of imperfection is taken as a sinusoidal wave in the arc direction, as shown in Figure 8, and the pattern is almost the same in the longitudinal direction, except for the ends. The five models T-4 to T-8 are tested under partial loading to see the loading effect with no imperfection.

Table 2. Test Specimens and ultimate strength

NAME	concrete				steel			ultimate strength		
	Ec	Pc	fc	ft	σy	σb	Es	P_u^{ex}	P_u^{an}	P_u^{ex}/P_u^{an}
T-1	1.80	0.25	21.2	2.1	363	428	2.14	74.3	71.8	1.04
T-2	2.00	0.14	25.4	2.8	363	428	2.14	29.9	19.4	1.54
T-3	2.00	0.19	32.6	2.4	363	428	2.14	10.4	7.18	1.45
T-4	1.96	0.17	26.3	2.8	393	432	2.07	86.6	96.4	0.90
T-5	1.94	0.16	33.7	3.0	393	432	2.07	6.96	10.6	0.66
T-6	2.06	0.18	33.1	3.1	393	432	2.07	16.2	19.4	0.83
T-7	2.06	0.17	27.9	2.8	393	432	2.07	52.8	50.0	1.06
T-8	1.91	0.19	24.3	2.9	452	472	2.18	9.81	10.3	0.95

t	:	shell thickness 8 mm for every case.
Ec	:	Young's modulus of concrete ($\times 10^4$ MPa)
Pc	:	Poisson's ratio of concrete
fc	:	Ultimate compressive strength of concrete (MPa)
ft	:	Ultimate tensile strength of concrete (MPa)
σy	:	yield strength of steel bars or wires (MPa)
σb	:	ultimate strength of steel bars or wires (MPa)
Es	:	Young's modulus of steel bars or wires ($\times 10^5$ MPa)
P_u^{ex}	:	experimental ultimate strength (kN) (applied total load)
P_u^{an}	:	FEM analytical ultimate strength (kN) (applied total load)

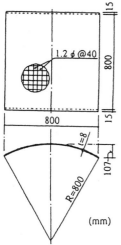

Figure 5. Shape and Dimensions of Model

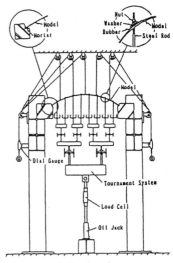

Figure 6. Testing Apparatus

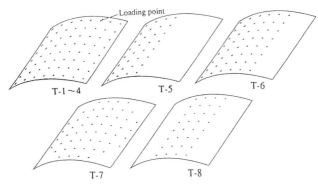

Figure 7. Loading Points and Mode

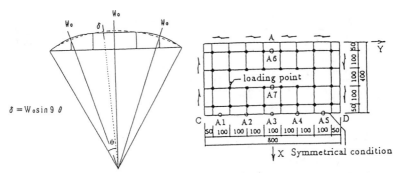

Figure 8. Mode of Initial Imperfection   Figure 9. Analytical Model

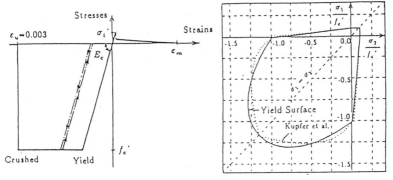

Figure 10. Stress-Strain Relation of Concrete   Figure 11. Yield Surface of Concrete

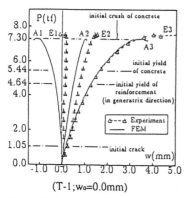

(T-1;w₀=0.0mm)

Figure 12. Load-Displacement Curves

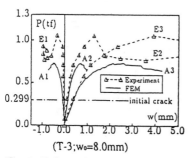

(T-3;w₀=8.0mm)

Figure 13. Load-Displacement Curves

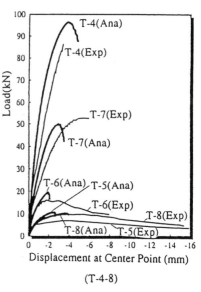

(T-4-8)

Figure 14. Load-Displacement Curves

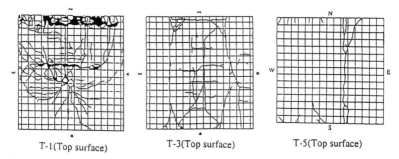

T-1(Top surface)   T-3(Top surface)   T-5(Top surface)

Figure 15. Crack Patterns

FEM geometrically and materially nonlinear analyses are conducted for comparison using 5 by 9 meshes as shown in Figure 9 , considering concrete cracking, tension stiffening, concrete crushing, and a cracked shear modulus (Cedolin 1977). Eight layers are used through the thickness with an additional two layers for smeared reinforcement. The stress strain relationship is assumed as perfect elasto-plastic under compression as shown in Figure 10, and is subjected to a flow rule just on the yield surface given in Figure 11 (Kupfer 1969). The details are given in the references (Takayama 1991, 1992b).

The load-displacement relationships and crack patterns are given in Figures 12 to 15 for several models, along with the ultimate strengths in Table 2. The displacements in the experiments are in good agreement with those from FEM analysis, in that the ratios of the experimental ultimate strength to analytical strength, P_u^{ex} / P_u^{an}, range around 1.0 with a narrow deviation. The mean value, E (P_u^{ex} / P_u^{an}), is 1.05 and the coefficient of variation, δ (P_u^{ex} / P_u^{an}), is 27 %.

b. Shells under Concentrated Loading

This discussion considers the ultimate strength of reinforced cylindrical panels under concentrated loading (Mashita,1993,1995a and 1995b).

Table 3. Test Specimens and ultimate strength

Type	t	Ec	Pc	fc	ft	ET	ST	P_u^{ex}	P_u^{an}	P_u^{ex}/P_u^{an}
Ma-1	8.7	2.36	0.24	56.3	3.0	BO	SP	2.45	2.35	1.04
Ma-2	9.1	1.91	0.26	59.0	2.5	BA	SP	2.16	2.35	0.92
Ma-3	7.9	1.91	0.26	59.0	2.5	BB	SP	1.96	1.67	1.18
Ma-4	10.3	2.02	0.19	55.7	2.9	BO	SR	2.06	1.96	1.05
Ma-5	10.3	1.96	0.19	59.6	3.1	BA	SR	1.86	1.18	1.58
Ma-6	9.3	1.96	0.19	59.6	3.1	BB	SR	0.69	0.34	2.00
Ma-7	11.5	2.02	0.19	55.7	2.9	BO	SC	0.88	1.35	0.82
Ma-8	8.9	2.23	0.18	53.1	3.4	BA	SC	0.59	1.49	1.20
Ma-9	9.1	2.23	0.18	53.1	3.4	BB	SC	0.29	0.20	1.50

t : shell thickness (mm)
ET (Edge type) : Depth of edge arch and beam of BO, BA, and BB are respectively 40×40, 40×20, and 20×40 mm.
ST (Support type) : SP, SR, and SC are respectively pin-support on four corners, roller-support on four corners, and two point-roller support, at each center of the straight edge beams. Ec, Pc, fc, ft, P_u^{ex}, and P_u^{an} are the same as those in Table 2.

An experimental study is conducted on nine small scale reinforced micro-concrete specimens, shown in Figure 16 and Tables 3 and 4. Three types of supporting conditions, designated as ST, are investigated. Type SP is a pin support at four corners, while SR is a roller support on four corners and SC is a two point-roller support, at each center of the straight edge beams.

Table 4. Material constants of steel bars or wires

Diameter	σy	σb	Es
0.85	1.86	162	250
1.2	2.35	364	468
D3	1.98	334	487

σy, σb, and Es are same as those in Table 2.

The design shell thickness (t_0) is 8 mm and the rise(L2) to chord width (L1) is 1/5. Each specimen is stiffened by four rectangular members, two edge beams (EB) on straight sides and two edge arches (EA) on circular sides. Their cross-section is represented by one of the three Types : BO, BA and BB. Each edge beam is reinforced by deformed bars D3 with diameter 3mm and wires with 0.85 mm diameter.

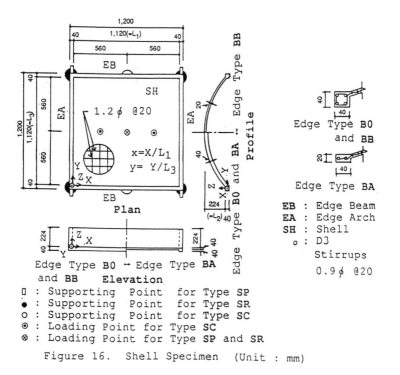

□ : Supporting Point for Type SP
● : Supporting Point for Type SR
○ : Supporting Point for Type SC
⊙ : Loading Point for Type SC
⊗ : Loading Point for Type SP and SR

Figure 16. Shell Specimen (Unit : mm)

FEM geometrically/materially nonlinear analyses are adopted for comparison using 10 by 10 meshes , considering concrete cracking, tension stiffening, concrete crushing, and a cracked shear modulus (Cedolin, 1977).

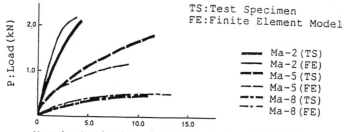

TS:Test Specimen
FE:Finite Element Model

Ma-2 (TS)
Ma-2 (FE)
Ma-5 (TS)
Ma-5 (FE)
Ma-8 (TS)
Ma-8 (FE)

Vertical Displacement(mm) at Loaded Point

Figure 17.   Load-Displacement Curves

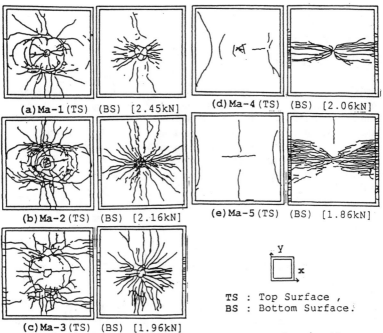

(a) Ma-1 (TS) (BS) [2.45kN]

(b) Ma-2 (TS) (BS) [2.16kN]

(c) Ma-3 (TS) (BS) [1.96kN]

(d) Ma-4 (TS) (BS) [2.06kN]

(e) Ma-5 (TS) (BS) [1.86kN]

TS : Top Surface ,
BS : Bottom Surface.

Figure 18.   Cracking Pattern on Test Specimens

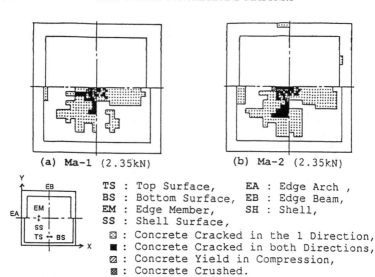

TS : Top Surface,   EA : Edge Arch ,
BS : Bottom Surface,   EB : Edge Beam,
EM : Edge Member,   SH : Shell,
SS : Shell Surface,
☒ : Concrete Cracked in the 1 Direction,
■ : Concrete Cracked in both Directions,
▨ : Concrete Yield in Compression,
▩ : Concrete Crushed.

Figure 19. Cracking Pattern on Finite Element Models

Ten layers are used through the thickness with an additional two layers for smeared reinforcement. The details are given in the references (Mashita,1993, 1995a and 1995b).

Cracking patterns at the surface and the bottom, and load displacement relationships are given for several types in Figures 17 to 19, while the comparison of ultimate strengths P_u , between experiments and FEM analyses are given in Table 3. The ratio of experiment to analysis, P_u^{ex} / P_u^{an}, is compured and shows good agreement in that the mean value, E (P_u^{ex} / P_u^{an}), is 1.19 and the coefficient of variation, δ (P_u^{ex} / P_u^{an}), is 22 %, after the case Ma-6 is discarded because of a somewhat large safe side discrepancy.

Conclusions

The present comparison of experimental ultimate strengths with those based on the FEM has proven that the FEM analytical method with commonly adopted assumptions may accurately predict the ultimate strength of cylindrical roofs under transverse loading, even when there are initial imperfections and partial and/or concentrated loading. The value of the average E (P_u^{ex} / P_u^{an}) minus δ (P_u^{ex} / P_u^{an}) for the experiment-to-analytical ratio of ultimate strength is 0.78 for distributed loading, and

0.97 for concentrated loading. This stochastic data might be effectively used for evaluating load carrying capacities of concrete roof shells.

References

1. Cedolin, L., 1977, "Finite Element Studies of Shear Critical R/C Beams", ASCE Journals, Vol. 103, No.EM3, pp. 395 - 410.
2. Kupfer, H. B., Hilsdorf, H. K. and Rush, H. 1969 , "Behavior of Concrete under Biaxial Stresses", Proc. of ACI, Vol. 66, No.8, Aug., pp. 656 - 666.
3. Mashita, K. 1993, "Ultimate Strength Analysis of R/C Roof-Type Cylindrical Shells with Boundary Members under Concentrated Loadings", Proc. of SEIKEN - IASS, Tokyo, JAPAN, Vol. 1, pp. 331 - 338.
4. Mashita, K., 1955a, "Ultimate Strength of Reinforced Concrete Cylindrical Shells Pin-Supported with Four Corners under Concentrated Loading", Proc. of IASS, Milan, Italy, Vol. 1, pp. 369 - 376.
5. Mashita, K., 1995b, "Study on Strength of Reinforced Concrete Roof-Type Cylindrical Shells under Concentrated Loading", Journal of Struct. Constr. Engng, AIJ, No. 474, Aug, pp. 137 - 145.
6. Takayama, M., 1991, "Comparison between Experimental and Numerical Analysis on Failure and Ultimate Strengths of Reinforced Concrete Shells with Initial Imperfections", Journal of Struct. Constr. Engng, AIJ, No.429, pp. 111 - 125.
7. Takayama. M., 1992a, "Experiments of the Effect on Initial Imperfection on Buckling Behavior of Reinforced Concrete Cylindrical Shells", Proc. of IASS-CSCE International Congress 1992 - Innovative Large Span Structures - Toronto, Canada, Vol. 2, pp.698-708.
8. Takayama, M., 1992b, "Failure Characteristics and Ultimate Strength of Reinforced Concrete Cylindrical Shells", Proc. of IASS-CSCE International Congress 1992 -Innovative Large Span Structures - Toronto, Canada, Vol. 2, pp. 698 - 708. 9.
9. Takayama, M. 1995, " Experiments of the Effect of Loading Condition on Buckling Behavior of Reinforced Concrete Cylindrical Shells", Proc. of IASS International Symp. 1995 - Spatial Structures : Heritage, Present and Fiture - Milano, Italy, pp. 425 - 432
10. Commitee of Shells and Spatial Structures, AIJ, Experimental Data of Reinforced Concrete Shells with Emphasis on Ultimate Strength Part1, 1993 March

Masonry Designers' Guide - A Comprehensive Design Tool

John H. Matthys, Ph.D., P.E., MASCE[1]

Abstract

The Masonry Designers' Guide (MDG) was produced to help those involved in the design, construction, and regulation of masonry structures apply the provisions of the Masonry Standards Joint Committee's (MSJC) Building Code Requirements for Masonry Structures and Specification for Masonry Structures. The MSJC documents have been adopted by the National (BOCA) and Standard (SBCCI) building codes. The MDG contents and its application are presented.

Introduction

Early masonry codes were totally empirical with such criteria as minimum wall thicknesses and maximum building heights. The empirical masonry code, ANSI A41.1, for years had been the basis for empirical design provisions for masonry. In the early 1960's masonry industry associations began the development of a technological database of masonry materials and assemblage performance through internally or externally sponsored research and testing programs. Results culminated in such design standards as the Brick Institute of America's (BIA) Recommended Practice For Engineered Brick Masonry in 1966 and the National Concrete Masonry Association's (NCMA) Specifications For Loadbearing Concrete Masonry in 1970. In 1970 the American Concrete Institute (ACI) Committee 531 published a report "Concrete Masonry Structures - Design and Construction" and in 1976 published Specifications For Concrete Masonry Construction. These documents served as the basis for ACI's Building Code Requirements For Concrete Masonry Structures published in 1979. Separate documents and different methodologies were used in clay masonry and concrete masonry design.

----

[1]Professor of Civil Engineering, Director of Construction Research Center, University of Texas at Arlington, Box 19347, Arlington, TX 76019-0347

Also little direction was given for addressing a common construction system -- combining clay and concrete masonry in a single element. For many years the use of masonry as a design/construction medium was confusing and often complicated for the design and construction professionals.

In the mid-1970's The Masonry Society (TMS) began development of a single structural masonry design standard that addressed both clay and concrete masonry. The TMS standard, completed in 1981, served as the source document for major changes to Chapter 24 of the 1985 Uniform Building Code (UBC). With the 1985 edition TMS has produced a commentary on the masonry chapter with each new version of the UBC Code.

The masonry industry associations recognized the need for a national design code covering all masonry materials. In 1978 the Masonry Structures Joint Committee (MSJC) was formed to develop a consensus standard for masonry design. In 1988 the Masonry Standards Joint Committee published a code containing design issues directed primarily to the designer and code enforcement officials and a specification with materials and construction issues principally directed to the contractor and inspector. Corresponding commentaries on each of the documents were produced. This represents the first time that a single rational design methodology for clay and concrete masonry was available for design professionals in the U.S.A. These standards are updated and revised as needed by the MSJC Committee. The 1992 edition made significant corrections and a few revisions to the 1988 version. At this writing the 1995 edition has completed public review. The revisions to the 1992 version include new material on seismic design, glass unit masonry, and masonry veneers.

Masonry Designers' Guide (1)

Based on the production and acceptance of the 1988 MSJC Code/Specifications, TMS, Council For Masonry Research, and ACI decided that a guide that specifically addressed the application of the MSJC documents with illustrative examples would be a significant benefit to those involved in the design, construction, and regulation of masonry structures. The results of these efforts is the Masonry Designers' Guide (MDG) - a comprehensive design tool with emphasis on the application of the MSJC 1992 Code/Specifications (2,3).

The 880+ page MDG is composed of four major parts divided into 16 chapters. Part I, General, is administrative and applies to all other parts. The Code

Reference Index and Specifications Reference Index tie the MDG discussions and design example problem procedures to the appropriate MSJC Code/Specifications section(s). The MDG Notations, Definitions, and Abbreviations chapter presents the MSJC Code/Specifications notations and definitions along with additions and abbreviations in the MDG.

MDG Part II, Material and Testing, primarily addresses the MSJC Specifications provisions as related to materials and testing. The MSJC Code dictates compliance with the MSJC Specifications. Chapter 3 on Materials covers clay or shale masonry units, concrete masonry units, stone masonry units, mortar, grout, masonry assemblages, reinforcement and connectors. Chapter 4 on Testing addresses MSJC Specifications requirements on testing frequency and quality assurance provisions as related to preconstruction and construction.

MDG Part III, Construction, addresses quality assurance, quality control, and hot and cold weather construction. Quality assurance includes the administrative policies and requirements related to quality control measures that will provide the owner's quality objectives. Chapter 5 addresses quality assurance items including organizational responsibilities, material control, inspection, testing and evaluating, noncomplying conditions, and records. Quality control is the systematic performance of construction, testing, and inspection operations of the contractor at the construction site to obtain compliance with the contract documents. Chapter 6 addresses quality control by examining the MSJC Specifications provisions for masonry construction preparation: storage and protection of materials, placement of materials (units, mortar, grout, reinforcement, connectors), and tolerances. The MSJC Specifications contain some requirements that are always mandatory and others that are optional. The latter become mandatory when required by the specifier. The MDG provides a compilation of these requirements in the form of a checklist for the user. The extent of the quality assurance and quality control program will vary with the type and size of project. Suggested applications of the MSJC Specifications QA/QC provisions to three typical types of masonry buildings (TMS Shopping Center, DPC Gymnasium, RCJ Hotel) are presented in the MDG. Chapter 7 addresses hot and cold weather masonry construction.

MDG Part IV, Design, basically covers the application of the MSJC Code provisions to the structural design of different types of masonry assemblages (beams, walls, columns, pilasters) for different types of masonry construction (multiwythe composite and noncomposite, single wythe, unreinforced and reinforced) based on the structural analysis of three typical types of masonry buildings (TMS Shopping Center, DPC Gymnasium, RCJ Hotel) for various load conditions. Note that these are the same buildings for which Quality Assurance/Quality Control recommendations are suggested in Part III, Construction. Chapter 8 on Design Philosophy and Methodology gives the background on material strength, design loads, masonry construction, and assemblage performance that have produced a structural design philosophy for masonry. The general design methodology for each structural component type (walls, columns, beams) as found in the Code including mathematical models is discussed and referenced to appropriate Code sections. Allowable stress design is the design method of the 1992 MSJC Code.

With the basic design philosophy and methodology established, appropriate application of the concepts found in the MSJC Code is accomplished by conducting structural analyses of typical masonry structures and presenting design aids and illustrative design examples. These structural application aspects are covered in MDG Chapters 9-16. A unique feature of the MDG is that application of the MSJC Code/Specifications provisions in all MDG chapters are based on the same three typical masonry structures:

TMS Shopping Center - a single story 16,000 square foot shopping center. North, east, west, and central fire walls are concrete masonry shear walls. The front south wall is primarily a glass curtain wall with one masonry shear wall element. Roof framing system consists of a one-way steel joist and steel beam system supported on steel columns and two wall construction options: unreinforced or reinforced concrete masonry. Seismic Zone 1

DPC Gymnasium - a single story 7,500 square foot gymnasium with four wall construction options (unreinforced noncomposite wall with pilasters, unreinforced composite wall, reinforced composite wall, single wythe reinforced hollow masonry wall). The roof framing system of gabled roof trusses and metal deck is supported by masonry bearing shear walls. Seismic Zone 2

RCJ Hotel - a four story hotel with nonloadbearing/loadbearing masonry shear walls, columns, and beams. The unreinforced wall system is used in Seismic Zone 2 and a reinforced wall system is used in Seismic Zone 4.

Plans, elevations, and sections are shown for each building. In addition an illustrative Design Problem Index for each building indicates for each of the 87 example problems the design issue under examination and its relation to the plans and elevations of the particular structure.

Chapter 9 deals with structural analysis aspects of gravity and lateral load distribution for the three basic structures. Both computer analysis and hand calculations have been included. Consideration of global gravity and lateral load on the building along with interwall and intrawall load distributions to and within the building components are covered.

Chapter 10 on Movements discusses causes and consequences of movements as related to masonry construction. Methods for determination of the magnitude of specific types of movements per the MSJC Code as well as ways to accommodate calculated movements (expansion joints, control joints) are discussed and illustrated.

Chapter 11 on Flexure addresses the structural design aspects of the MSJC Code for masonry elements where flexure may control. In particular the performance and Code provisions of unreinforced/reinforced walls and pilasters and reinforced beams are covered. The appropriate engineering equations for generating actual flexural stresses for both in-plane loading and out-of-plane loading and their relationship to Code requirements are covered.

Chapter 12 expands Chapter 11 into the flexural and axial load structural design aspects of columns, walls, and pilasters. The solution to the behavior of these elements is more complex than for flexure only. Thus iterative design procedures are presented along with graphical methods to simplify design calculations.

Chapter 13 on Shear presents the MSJC Code design requirements from the out-of-plane load viewpoint (masonry components) and the in-plane load aspect (shear walls). The mathematical models for both unreinforced and reinforced masonry are addressed and discussed in light of assemblage performance and Code allowables.

Chapter 14 on Reinforcement and Connectors addresses strength requirements, corrosion resistance and protection provisions, embedment criteria, and design aspects.

Chapter 15 presents the current empirical design provisions found in the Code, a design methodology well embedded in the masonry design profession and used for a significant percentage of structural masonry in the U.S.A. Specific criteria on restrictions, strength requirements, support provisions, and minimum wall thickness are discussed. Aspects of bonding wythes and anchoring intersecting walls, roofs, and floor diagrams are covered.

Chapter 16 addresses provisions for seismic design as related to masonry construction. Seismic resistant design of masonry buildings requires provisions for ductility not generally required for wind or other lateral loads. The Code's minimum requirements for the different seismic zones are intended to provide proper performance of masonry structures subjected to earthquake shaking.

Comprehensive Design Tool

The MDG is a comprehensive design tool - it provides a presentation of MSJC Code/Specifications requirements in a manner that is easy to understand and then illustrates the application by the appropriate industry professionals in the construction/design processes of masonry structures.

A significant amount of tabular/graphical material and illustrative design problems is included in the MDG for understanding and application purposes. Brief samples of such material available to the MDG user now follow. The Code and Specifications Reference Index directs users to specific pages of the MDG (text or illustrative example problems) where Code/Specifications provisions are addressed, explained, and illustrated. MDG Table 6.5.2 shows application of the MSJC Specifications in establishing Quality Assurance/Quality Control issues for three typical masonry structures. Product Specifications and Requirements for clay, concrete, and stone masonry units indicating coverage and options to the specifier is tabulated. See MDG Table 3.1.2. Section properties for clay masonry units/walls and concrete masonry units/walls provide information needed by the design professional for both structural design and construction consideration. See MDG Appendix Table 2. Movement of masonry units/assemblages due to various factors are required to be assessed by the Code. See MDG Table 10.2.2.

Nondimensional interaction diagrams for masonry beam columns reduce the manual computation in the design process for members with axial and bending loads. See

MDG Fig. 12.1-5 and Fig. 12.1-7. The MDG has similar nondimensional interaction diagrams for reinforced masonry walls.

The 87 illustrative example problems are directed to structural masonry analysis and design issues as found in typical masonry building structures. The format of each example is an introductory description of the structural issue, the associated structure, the load conditions, and material requirements. Typical calculations and discussion address the design issue in question. The Code/Specifications Reference keys the illustrative solution to the appropriate MSJC Code/Specifications sections. See MDG example 13.2-4.

The Masonry Designers' Guide -- written by TMS, funded by Council for Masonry Research, and published by ACI -- is a valuable reference to engineers, contractors, architects, inspectors, building code authorities, and educators in the design and construction of masonry structures according to the MSJC Code/Specifications.

With the approval of the 1995 edition of the MSJC Code/Specification, the Masonry Designers' Guide will be revised to reflect the corrections and revisions made to the 1992 MSJC Code/Specification. The 1995 MSJC Code/Specification can be obtained now from The Masonry Society (Phone 303/939-9700, Fax 303/541-9215). The second edition of the Masonry Designers' Guide, based on the 1995 MSJC documents, should also be available from The Masonry Society by the end of 1996 or spring of 1997.

References

1. Masonry Designers' Guide. Boulder: The Masonry Society, 1993.

2. Masonry Standards Joint Committee. Building Code Requirements for Masonry Structures (ACI 530/ASCE 5/TMS 402). Detroit: American Concrete Institute, 1992.

3. Masonry Standards Joint Committee. Specification For Masonry Structures (ACI 530.1/ASCE 6/TMS 602). Detroit: American Concrete Institute, 1992.

MDG Table 6.5.2 Quality Assurance/Quality Control Checklist For Masonry Buildings

R = Recommended for this project   -- = Not applicable to this project
N = Not recommended for this project   M = Mandatory per Specs. 530.1

Building	TMS Shopping Center		DPC Gymnasium				RCJ Hotel	
Wall Construction Type	A	B	A	B	C	D	A	B
Quality Assurance								
Certification								
Brick Units	--	--	R	R	R	R	R	R
CMU Units	R	R	R	R	R	--	R	--
Mortar Mix	R	R	R	R	R	R	R	R
Grout Mix	--	R	--	R	R	R	--	R
Reinforcing Steel	--	M	--	--	M	M	--	M
Joint Reinforcement	M	M	M	M	M	M	M	M
Anchor Bolts	M	M	M	M	M	M	M	M
Ties and Anchors	M	M	M	M	M	M	M	M
Metal Accessories	M	M	M	M	M	M	M	M
Procedures								
Hot Weather Construction	R	R	R	R	R	R	R	R
Cold Weather Construction	R	R	R	R	R	R	R	R
Cleaning Method and Materials	R	R	R	R	R	R	R	R
Material Samples								
Brick Units	--	--	R	R	R	R	R	R
CMU Units	R	R	R	R	R	--	R	--
Colored Mortar	R	R	--	--	--	--	--	--
Sample Panel	R	R	R	R	R	R	R	--
Joint Reinforcement	R	R	R	R	R	R	R	R
Anchor Bolts	R	R	N	N	N	R	N	N

MDG Table 3.1.2   Product Specifications and Requirements - Concrete Masonry Units

ITEM	C 55	C 73	C 90	C 129	C 744
Classification					
Grade MW		E			
Grade N			E		
Grade S			E		
Grade SW		D			
Type I, Moisture Control	E		E	E	
Type II, Nonmoisture	E		E	E	
Weight Class			E	E	
Physical Requirements					
Absorption, Water	E	E	E		
Compressive Strength	E	E	E	E	
Dimensions	E	E	E	E	E
Distortion					E
Facing Requirements					U
Craze Resistance					E
Chemical Resistance					E
Adhesion					E
Abrasion					E
Surface Burn					E
Color and Change					E
Soil and Clean					E
Moisture Content	I		I	I	
Product Standard					E
Visual	E	E	E	E	E

E   Specification Entry
D   Default Specification
I   Type I units, singly
U   Base Unit Specification Applies

MDG Table 10.2.2   Estimated Moisture Movement, %

No.	Material	Reversible Moisture Movement			Irreversible Moisture Movement		
		Mean	Standard Deviation	Characteristic Value	Mean	Standard Deviation	Characteristic Value
1	Clay Brick	0.0200	ND	ND	+ 0.020	0.020	+ 0.051
	Concrete						
2	Gravel	0.0400	0.0120	0.060	- 0.060	0.015	- 0.085
3	Crushed Stone	0.0700	0.0210	0.105	- 0.060	0.015	- 0.085
4	Limestone	0.0250	0.0030	0.030	- 0.035	0.003	- 0.040
5	Light Weight	0.0450	0.0090	0.060	- 0.060	0.018	- 0.090
	Concrete Masonry						
6	Sand & Gravel	0.0080	0.0010	0.010	- 0.023	0.003	- 0.028
7	Expanded Shale	0.0100	0.0030	0.015	- 0.031	0.009	- 0.046
	Concrete Masonry Units						
8	Sand & Gravel	0.0090	0.0010	0.011	- 0.027	0.004	- 0.034
9	Light Weight	0.0140	0.0040	0.020	- 0.042	0.120	- 0.062
10	Sand Lime	0.0300	0.0120	0.050	- 0.025	0.009	- 0.040
11	Mortar Shrinkage	0.0400	0.0122	0.060	- 0.180	0.042	- 0.249
12	in Masonry	0.0280	0.0126	0.049	- 0.126	0.050	- 0.209
	Stone						
13	Granite	0.0040	0.0020	0.0072	ND	ND	ND
14	Limestone	0.0025	0.0012	0.0045	ND	ND	ND
15	Marble	0.0015	0.0012	0.0035	ND	ND	ND
16	Sandstone	0.0700	ND	ND	ND	ND	ND

MDG Appendix Table 2   Section Properties - Hollow Clay Unit Walls - ASTM C 652 Units

Property per ft of wall		Mortar Bedding		Fully Grouted
		Face-Shell	Full	
4" Walls		Actual Unit Thickness = 3 1/2"		
Area	A, in.^{2}	18.0	22.0	42.0
Moment of Inertia	I, in.^{4}	34.0	35.7	42.9
Section Modulus	S, in.^{3}	19.4	20.4	24.5
Kern Eccentricity	e_k, in.	1.07	0.93	0.58
Radius of Gyration	r, in.	1.37	1.27	1.01

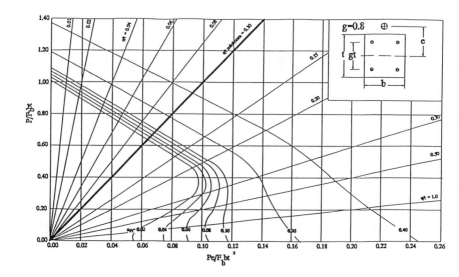

**MDG Fig. 12.1-5   Non-Dimensional Column Interaction
Diagram - Compression Controls**

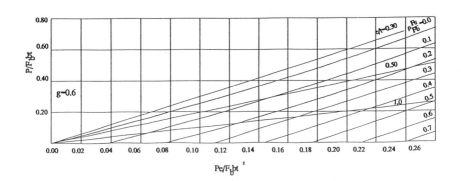

MDG Fig. 12.1-7   Column Design Aid - Tension Controls

MDG Example 13.2-4   **RCJ Hotel - Design of Unreinforced Masonry Shear Wall for In-Plane Lateral Loads**

Design the shear wall on Grid Line C between Grid Lines 1 and 2 using Wall Construction Option A (Unreinforced) and Building Construction Option II. Use hollow concrete masonry units. Seismic Zone 2.

Calculations and Discussion	Code Reference

Consider 8 in. wall and design 1st floor wall section for different load combinations. For unreinforced wall two critical parameters need to be checked:

1) No tension is developed under minimum dead load and maximum lateral load, i.e., load case of $0.9\,D + E$.   5.3.1

2) Compressive stress at the other end of the wall complies with the Code allowables using the unity equation, i.e., load case $D + L + (E + W)$.

The above two cases are considered below. Other cases may be checked.

Loads

Lateral loads are calculated using hand calculations in MDG Example 9.2-3. Gravity loads are given in MDG 9.1.3. Maximum shear and moment develop at the base section since wall acts as a free-standing cantilever.

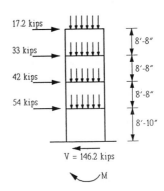

Update of Building Code Requirements for Masonry, 1992 to 1995 Editions

J. Gregg Borchelt, P.E., MASCE[1]

Abstract

The third edition of *Building Code Requirements for Masonry Structures (ACI 530/ASCE 5/TMS 402)*, its mandated *Specification for Masonry Structures (ACI 530.1/ASCE 6/TMS 602)*, and their Commentaries were approved in 1995. This paper provides a review of the changes made from the 1992 editions. The changes result from actions of the Masonry Standards Joint Committee, based on the use of these documents. The major changes of the Code and it Commentary include a revision to and inclusion of provisions for masonry structures in seismic areas and new provisions for glass unit masonry and anchored masonry veneer. Other changes clarify design intent and emphasize quality assurance. Changes have also been made to the Specification and its Commentary. The inclusion of requirements for construction of glass unit masonry is the primary change. The Specification has been brought into conformance with contractual requirements used by the American Institute of Architects. The Specification has been rewritten in the format prepared by the Construction Specifications Institute. An index has been added to aid the user in locating requirements.

Introduction

The design and construction of masonry was significantly changed when the first edition of *Building Code Requirements for Masonry Structures (ACI 530/ASCE 5)* and *Specifications for Masonry Structures (ACI 530.1/ASCE 6)* and their commentaries were published in 1988. This represented an effort of more than ten years by a joint committee of the American Concrete Institute (ACI) and the American Society of Civil Engineers (ASCE). These documents combined design and construction requirements for the most common structural masonry

----

[1]Director Engineering & Research, Brick Institute of America, 11490 Commerce Park Drive, Reston, VA 22091

Update of Building Code Requirements for Masonry,
1992 to 1995 Edition

materials into a single, unified location. The Code, Specifications and their commentaries received status as consensus standards under the public review processes of ACI and ASCE. The use of these design and construction requirements was then recognized by reference in the 1989 edition of the Standard Building Code and in the 1990 edition of the National Building Code. Reference to the later editions of the Code and Specification has been included in subsequent editions of these model building codes.

The Masonry Society (TMS) joined ACI and ASCE as a sponsor of the committee charged with oversight of these documents in 1989. This committee is now known as the Masonry Standards Joint Committee (MSJC). The MSJC is responsible for the preparing changes and updates to the documents. Revised editions of the Code, the Specifications, and their associated Commentaries were approved in 1992. The changes from 1988 to 1992 editions primarily corrected omissions, clarified content, and refined the documents.

The MSJC has continued to work on changes to the Code and Specification in order to properly cover all types of masonry materials and construction. Format changes were requested to bring the contents in agreement with other design and construction industry documents. Clarification was needed in several instances. Research on and recorded performance of masonry and masonry structures continue to advance the knowledge about masonry. The deliberations of the MSJC resulted in changes from the 1992 to the 1995 editions which are much more comprehensive than those made earlier. There are three major changes to the Code. The requirements for seismic design have been moved from an appendix to a chapter in the Code. Additionally, a modified strength design process replaced an allowable stress method. A chapter on design of glass unit masonry was added. An index which references Code sections and Specification articles on a comprehensive list of subjects was added. A chapter on design of veneer, specifically anchored veneer, was added. The Specification was reformatted to the three part format of the Construction Specifications Institute. Provisions for construction of glass unit masonry were added. As with the 1988 and the 1992 editions, these changes were submitted to public review. Several editorial changes were made as a result of public review comments.

The 1995 edition of the Code, Specification, and their Commentaries have been accepted by ACI, ASCE and TMS through their consensus processes. The 1996 edition of the National Building Code has adopted the MSJC Code by reference.

In this paper the appropriate Code Section or Specification Article is given in parentheses to aid the reader in locating the changes.

Update of Building Code Requirements for Masonry,
1992 to 1995 Editions

Changes to the Building Code Requirements for Masonry Structures

Throughout the Code editorial changes were made to provide clarity, to result in consistency within the Code, or to provide correct reference to other documents.

Included in these are the insertion of metric dimensions in the text. Standards cited in the Code were updated to the most recent edition available at the time of public review.

An implicit assumption for masonry construction was made clear by a change to the Scope (1.1.1). Masonry elements were identified as those consisting of masonry units bedded in mortar. This fact is the basis for the allowable stresses found in the Code. There are new, dry-stacked masonry elements and stone cladding systems which are used, but do not transfer stress or undergo deformations in the same manner as units bedded in mortar. These latter methods of masonry construction are not covered.

The MSJC Code was prepared with the assumption that the construction would be monitored by a quality assurance program. This was emphasized by appropriate additions (1.2.3, 3.1.1).

Definitions were added as necessary to explain new provisions and to clarify industry terms which were not given in the prior editions (2.2). The latter type include:

> *Cavity wall* - A multiwythe noncomposite masonry wall with a continuous air space within the wall (with or without insulation), which is tied together with metal ties.

> *Dimensions, specified* - Dimensions specified for the manufacture or construction of a unit, joint, or element.

> *Wall, masonry bonded hollow* - A multiwythe wall built with masonry units arranged to provide an air space between the wythes and with the wythes bonded together with masonry units.

The allowable axial force on reinforced columns and pilasters received an important change. The contribution of laterally-tied longitudinal reinforcing steel was included (7.3.2.1). This contribution to the allowable compressive force is $0.65\, A_{st}\, F_s$ as was used in previous masonry design standards. This required a change in the maximum compressive stress in flexure of reinforced masonry. The calculated stress due to the axial load must be less than that permitted on the section as an unreinforced element. (7.3.2.2)

Update of Building Code Requirements for Masonry,
1992 to 1995 Editions

The allowable shear stress in reinforced masonry was modified to:

$$f_v = V/bd$$

The notation j, the ratio of distance between centroid of flexural compressive forces and centroid of tensile forces to depth d, was deleted. (7.5.2.1)

Epoxy coating of reinforcement was added as a means of providing corrosion resistance. An increase of 50% in both development length (8.5.2) and in lap length (8.5.7.1) for epoxy-coated bars was mandated.

The terminology used in ASCE 7, *Minimum Design Loads for Buildings and Other Structures* for seismic classification and wind loads was changed. Since the MSJC Code references ASCE 7 this required corresponding changes in the terminology used to limit the use of the empirical design of masonry (9.1.2.1, 9.1.2.2). Empirical design was prohibited from use of the design of the lateral force-resisting systems for structures in Seismic Performance Categories B or C. (9.1.2.1)

There are several additional changes to the empirical requirements. Since empirical design is unique to masonry and is based on successful performance rather than allowable stress provisions it is important to carefully establish the criteria of empirical design. Thus the thickness of multi-wythe walls for lateral support requirements was specifically stated (9.5.1).

The requirements for foundation walls have been revised (9.6.3). Drainage requirements are included. Lateral bracing, both vertical and horizontal, is necessary. Soil drainage and grading conditions must be met. Running bond is required and the mortar must be Type M or S. These restrictions permitted an increase in the maximum depth of unbalanced backfill. Reinforced basement walls are no longer included in empirically designed foundation walls.

The most significant changes to existing Code requirements are in the seismic design requirements. In the earlier editions these were included in an Appendix. Thus they had to be specifically identified by the local or state code body before the seismic provisions became part of the governing building code. Since the seismic requirements are now a chapter in the Code they are now enforced when the MSJC Code is adopted.

The Appendix contained prescriptive requirements which were based on Seismic Zones and working stress seismic loads. ASCE 7 changed from working stress design loads to strength level (factored) earthquake loads. Further, structures were assigned to Seismic Performance Categories which are different from Seismic Zones. Thus, the design approach used in the new Chapter 10 differ greatly from that of Appendix A.

Update of Building Code Requirements for Masonry,
1992 to 1995 Editions

used in the new Chapter 10 differ greatly from that of Appendix A.

The design method used with strength level loads is a modification of the
allowable stress design procedures found in Chapters 6 and 7. The allowable
stress values are multiplied by a factor of 2.5 to approximate a nominal strength.
This nominal strength is then reduced by a strength reduction factor to achieve a
design strength. The strength reduction factors are:

$\phi = 0.8$ for axial load and flexure, except for flexural tension in
            unreinforced masonry
$\phi = 0.4$ for flexural tension in unreinforced masonry
$\phi = 0.6$ for shear
$\phi = 0.6$ for shear and tension on anchor bolts embedded in masonry

The design strength of masonry structures and masonry elements must be equal to
or greater than the required strength from ASCE 7 (10.2.2).

Another new aspect is the imposition of calculated story drift of 0.007 times the
story height rather than using an induced moment from 2.25 times the drift.

The prescriptive requirements in the new Chapter 10 are the similar to these of the
Appendix:

- requirements are triggered by increasing seismic force
- limits to the design method applicable
- limits to materials and bond patterns
- more attention to connections
- increased ductility from increasing amounts of prescriptive reinforcement
  and lateral ties.

The provisions differentiate between elements which are a part of the lateral
force-resisting system and those which are not. The amount of prescriptive
reinforcement in the lateral force-resisting system did not change dramatically.
However, the change from Seismic Zones to Seismic Performance Categories also
changes the application so it is difficult to compare directly. Minimum
dimensions for walls and columns have been deleted. New requirements for
elements which are not part of the vertical and lateral load resisting system
provide isolation and increase ductility. Partial grouting of lateral load-resisting
walls is prohibited in Seismic Performance Categories D and E (10.6.3).

The empirical design methods, permitted in Seismic Performance Category A, are
exempted from this pseudo-strength requirement (10.2.1). Further, glass unit
masonry and veneer are not covered by the seismic requirements (10.1.1).

Update of Building Code Requirements for Masonry,
1992 to 1995 Editions

Chapter 11 contains new requirements for glass unit masonry. These are
empirical requirements for nonload-bearing elements in interior or exterior walls
(11.1.1). The glass units are hollow or solid. They are classified as standard or
thin units, with specified thicknesses based on whether the units are hollow or
solid (11.2).

The empirical design procedure correlates design wind pressure and maximum
panel area. Thin-unit exterior glass masonry is limited to a design wind pressure
of 958 Pa (20 psf). Maximum dimensions between structural supports, based on
unit thickness, are given. Exterior, interior, and curved panels are addressed
(11.3).

Support provisions require that in-plane loads not be transferred to the glass unit
masonry. Elements supporting glass unit masonry have their deflection limited to
span length divided by 600 (11.4). Expansion joints are required at the top and
sides. They must accommodate displacements of the supporting structure (11.5).

Mortar is limited to Type S or N mortar (11.6). The requirements for glass unit
masonry are based on the presence of reinforcement in the panels. Specific size
and spacing of bed joint reinforcement is given (11.7).

The new Chapter 12 covers the requirements for design and detailing of anchored
masonry veneer and its anchors. Key definitions are:

Veneer, masonry - A masonry wythe which provides the exterior finish of a
wall system and transfers out-of-plane load directly to a backing, but is not
considered to add load resisting capacity to the wall system.

Veneer, anchored - Masonry veneer secured to and supported laterally by the
backing through anchors and supported vertically by the foundation or other
structural elements.

Backing - The wall or surface to which the veneer is secured. The backing
shall be concrete, masonry, steel framing, or wood framing.

Anchored veneer must meet several general design requirements and is either
designed rationally or detailed by prescriptive requirements. As implied by the
definition of masonry veneer, the veneer wythe is not subjected to the flexural
tensile stress limitations of Chapter 6 (12.1.1). Materials and construction must
comply with the MSJC Specification except for Articles 1.4 on compressive
strength, 3.4D on wall ties and 3.6A on field prism testing (12.1.2). Chapter 12
does not now include adhered veneer nor stone veneer (12.1.3). The MSJC plans
to add requirements for these types of veneer in the future.

Update of Building Code Requirements for Masonry,
1992 to 1995 Editions

General requirements for design and detailing (12.2) are:

- water penetration resistance of the backing of exterior veneer
- presence of flashing and weep holes
- accommodating differential movement
- exclusion of indicating the specified compressive strength of masonry on the drawings.

If the designer chooses the rational design alternate (12.3) the following conditions must be met:

- loads are distributed using principles of mechanics
- out-of-plane deflection of the backing is limited to maintain veneer stability
- any masonry element, other than the veneer wythe, is designed in accordance with other applicable chapters of this Code
- the veneer wythe is not subject to the provisions of Chapter 6
- the scope, general design requirements, reinforcement provisions for other than running bond, and seismic requirements of Chapter 12 apply.

The remainder of Chapter 12 is comprised of prescriptive detailing requirements for anchored veneer. These are based on the successful performance of anchored veneer provisions in model building codes. The use of these prescriptive requirements is limited to areas where the wind velocity pressure, as defined in ASCE 7, does not exceed 1190 Pa (25 lb/ft^2) (12.4.1).

Vertical support of anchored veneer must be provided by noncombustible structural supports except that veneer less than 5.5 m (18 ft) high may be supported by preservative-treated wood foundations. There are limits to the height of the veneer above the foundation with wood frame backing. Anchored masonry veneer with cold-formed steel frame backing must be supported at each floor above given maximum heights. Support deflection requirements of span length divided by 600 or 7.6 mm (0.3 in.) apply (12.5).

Masonry units used in anchored veneer must be at least 67 mm (2-5/8 in.) in actual thickness (12.6).

Anchors are to be made of corrugated sheet metal, sheet metal, wire, or joint reinforcement, and may be adjustable. Size and placement requirements are given. Anchor spacing depends on the type of anchor. Maximum spacing and anchors around openings are required (12.7).

Each backing type has specific requirements for the type of anchors permitted, means of connecting the anchor, and for the dimension between the inside of the veneer wythe and backing (12.8-12.10).

Update of Building Code Requirements for Masonry,
1992 to 1995 Editions

With the recognition that masonry units laid in other than running bond have
lower flexural bond strength than running bond masonry, there is a requirement
for joint reinforcement in the veneer wythe when laid in other than running bond
(12.11).

Requirements for anchored veneer in Seismic Performance Categories are
cumulative as the exposure increases. The veneer wythe is first isolated from
seismic forces resisted by the structures (12.12.1). The weight of the veneer
wythe is supported at each floor level and the frequency of anchors is increased.
Horizontal joint reinforcement is required (12.12.2). Finally, vertical expansion
joints, which provide articulation, and mechanical attachment of the anchors to
the joint reinforcement is required (12.12.3).

Changes to Building Code Commentary

The purpose of the Commentary is to provide discussion and background to the
Code provisions. Thus many of the changes to the Code discussed in the
preceding section of this paper have corresponding changes in the Code
Commentary. Complete chapters were added to the Code Commentary for the
seismic design requirements (10), glass unit masonry (11), and veneer (12). Such
changes are not discussed in this paper. The following are changes to the Code
Commentary which were made without changes to the Code.

Masonry columns and pilasters are subject to a limit of axial load to prevent
premature stability failure caused by eccentrically applied axial load. The means
of determining the eccentricity to use in this calculation is clarified as the actual
eccentricity of the applied compressive load (6.3.1).

Changes to the Specification for Masonry Structures

As with the Code there are several editorial changes made for consistency. The
most comprehensive change to the Specification was, in fact, an editorial change.
The format was changed from four individual parts based on materials to the three
part format of the Construction Specifications Institute. Thus, the Specification is
now comprised of:

Part 1 - General
Part 2 - Products
Part 3 - Execution.

This global change resulted in a relocation of items and a new numbering system
for Specification Articles. While this is a radical change in format, the
Specification is now in a format which is much more widely accepted. No
changes in content resulted from this reformatting.

Update of Building Code Requirements for Masonry,
1992 to 1995 Editions

New definitions include (1.2):

Dimensions, specified - Dimensions specified for the manufacture or construction of a unit, joint, or element.

Glass unit masonry - Nonload-bearing masonry composed of glass units bonded by mortar.

Grout pour - The total height of masonry to be grouted prior to erection of additional masonry. A grout pour consists of one or more grout lifts.

Mean daily temperature - The average daily temperature of temperature extremes predicted by a local weather bureau for the next 24 hours.

Wall, masonry bonded hollow - A multiwythe wall built with masonry units arranged to provide an air space between the wythes and the wythes bonded together with masonry units.

Standards referenced in the Specification have been updated to the most recent published version available at the time of public review (1.3). New standards may reflect changes to this Specification and will be given when those changes are discussed.

More specific references to quality assurance requirements are also added to the Specification (1.6A). These match changes made to the Code and give guidance to the duties of the independent testing laboratory.

Since glass unit masonry design was added to the Code there are logically added Specification requirements for glass unit masonry. Glass unit masonry cannot be laid during cold weather construction periods (1.8C10). Mortar materials are limited to Types S or N (2.1B). The glass units are described (2.3D). Panel anchors are specified (2.4C6). Mortar mixing requirements for glass unit masonry are stated (2.6A4). Bed and head joint placement with thinner joints is called for (3.3B5). Panel anchor placement and spacing is given (3.4F).

Epoxy coating reinforcing bars are now permitted (2.4A). Stainless steel was added as a material for joint reinforcement, anchors and ties (2.4D). Since these were added, coatings for corrosion protection now apply to carbon steel only. Further, the term moist environment was changed to a mean relative humidity exceeding 75% (2.4E).

A limit of 0.2 percent chloride ion in mortar admixtures replaced an outright ban (2.6A3).

Grout slump is to be measured in accordance with ASTM C 143 (2.6B2).

Update of Building Code Requirements for Masonry,
1992 to 1995 Editions

Source quality tests for stone masonry are referenced (2.8A).

Wetting clay or shale masonry units with an initial rate of absorption less than 0.00031 grams per min. per mm^2 (0.2 grams per min. per in^2) is prohibited (3.2C2).

Cross webs of hollow units are to be fully mortared when necessary to confine grout or loose-fill insulation (3.3B3b).

The contractor is instructed to notify the Architect/Engineer when bearing of a masonry wythe on its support is less than two-thirds of the wythe thickness (3.3B6c).

Installation wall ties was modified for better load transfer (3.4D5) and ties are now required within 305 mm (12 in.) of openings and at a maximum spacing of 0.9 m (3 ft.) (3.4D6).

Changes to Specification Commentary

The changes to the Specification Commentary are associated with the changes to the Specifications noted above.

Closing

This Code, Specification, and their Commentaries are in use and as such receive scrutiny by the users. Suggested improvements and requests for clarification are received by the sponsoring associations and the committee members. Many of these suggestions will result in changes to the provisions. The MSJC is working on provisions for limit states design, prestressed masonry, and other veneer systems. There are currently eleven task groups under eight subcommittees which are working to revise portions of the documents. These are dynamic documents and future editions will incorporate the work of the MSJC.

Notation

A_{st} = total area of laterally tied longitudinal reinforcing steel in a reinforced masonry column or pilaster, mm^2 (in.^{2})

b = width of section, mm (in.)

d = distance from extreme compression fiber to centroid of tension reinforcement, mm (in.)

f_v = calculated shear stress in masonry, Mpa (psi)

Update of Building Code Requirements for Masonry,
1992 to 1995 Editions

F_s = allowable tensile or compressive stress in masonry, Mpa (psi)

j = ratio of distance between centroid of tensile forces to depth, d

V = design shear force, N(16)

ϕ = strength reduction factor

References

Building Code Requirements for Masonry Structures (ACI 530/ASCE 5/TMS 402, American Society of Civil Engineers, Washington, D.C., American Concrete Institute, Detroit, MI and The Masonry Society, Boulder, CO, 1988, 1992, 1995

Specifications for Masonry Structures (ACI 530.1/ASCE 6/TMS 402), American Society of Civil Engineers, Washington, D.C., American Concrete Institute, Detroit, MI and The Masonry Society, Boulder, CO, 1988, 1992, 1995

Proposed Limit States Design Provisions
for Masonry

Mark B. Hogan, P.E.
Vice President of Engineering
National Concrete Masonry Association

Introduction

Limit States Design is the latest development in the structural design of masonry. This design method is based on evaluating the structural performance of masonry subjected to specific design load conditions. Existing design methods for masonry (the empirical design method and the allowable stress design method) are effective method of design, however these methods do not predict structural performance as accurately as the limit states design method. The limit states design method is intended to more accurately predict the response of masonry structures to specified loading conditions.

Structural properties of masonry are based on expected values (mean value of a measured property) by the limit states design method. The expected value is then modified by a strength reduction factor which takes into account material variability and other differences between the mean measured value and the actual strength of the masonry in the construction. Limit states design criteria is supported by laboratory research and by the analysis of this research to establish structural properties used in predicting response of the structure. Analogous to strength design of reinforced concrete and load and resistance factor design of other materials, limit states design is considered to be the most predictable method of assessing the structural performance of masonry. Factored design loads are used by this design method in accordance with national standards[1] to establish values which have a low probability of exceedance during the life of the structure.

The method is being proposed by the Limits States Design Subcommittee of the Masonry Standards Joint Committee as an alternative to the existing methods of designing masonry. The national masonry design standard currently includes an allowable stress design method (for both reinforced masonry and plain masonry) and an empirical design method. Many structural engineers prefer the limit states design approach for masonry, however most masonry structures are not complex and will continue to be designed by existing methods. The Limit States Design Subcommittee has drafted the standard for consideration by the Masonry Standards Joint Committee. Once accepted by the main committee and subjected to public review, comment and balloting the proposed Limits States Design Standard (LSDS) will be published as an alternative design method for reference by building codes.

Limit States

The draft LSDS identifies three primary limit states, a cracking limit state, a yield limit state, and a strength limit state. The LSDS establishes criteria for the design of masonry elements based on these limit states. For example, the design of plain masonry is not permitted to exceed the cracking limit state. As another example, reinforced masonry designed to resist frequently occurring wind loads is not permitted to exceed the yield limit state, while reinforced masonry designed to resist infrequently occurring seismic loads is not permitted to exceed the strength limit state.

Cracking Limit State

The expected moment strength at the cracking limit state occurs when the stress in the masonry equals the modulus of rupture of the masonry.

Yield Limit State

The expected moment strength at the yield limit state shall be based on the following:

(a) Strain in the extreme tensile reinforcement equals the expected yield strain, ϵ_{ye}.

(b) The expected flexural strength shall be calculated by assuming a masonry stress of $0.85 f_{me}$ to be uniformly distributed over a compression zone bounded by the edges of the cross section and a straight line parallel to the neutral axis at distance, a, from the extreme compression fiber. The distance a shall be taken as $0.85c$. Tensile force shall be taken to be $A_s f_{ye}$ in the extreme tensile reinforcement, and shall be taken to be $A_s E_{se} \epsilon_s$ for tensile reinforcement nearer to the neutral axis. The force in the compression zone shall be equated to the sum of the total tensile forces in all of the reinforcement outside of the compression zone and the factored axial loads on the section. The tensile strength of masonry shall be neglected.

Strength Limit State

The expected moment strength at the strength limit state shall be based on the following:

The expected flexural strength shall be calculated by assuming a masonry stress of $0.85 f_{me}$ to be uniformly distributed over a compression zone bounded by the edges of the cross section and a straight line parallel to the neutral axis at distance, a, from the extreme compression fiber. The distance shall a be taken as $0.85c$. Tensile force shall be taken to be $A_s f_{ye}$ in all tensile reinforcement. The force in the compression zone shall be equated to the sum of the total tensile forces in all of the reinforcement outside of the compression zone and the factored axial loads on the section. The tensile strength of masonry shall be neglected.

Design Assumptions

Limit states design of masonry is based on the requirement that the required strength due to factored loads equals or exceeds the design strength. The design strength is equal to the nominal strength of the member times the applicable strength reduction factor, ϕ. For axial load this requirement is expressed as follows:

$$P_u \leq P_n \, \phi$$

For flexure the requirement is expressed as follows:

$$M_u \leq M_n \, \phi$$

In these expressions the required strength (P_u and M_u) is based on design loads which have been factored such that the probability of exceedance during the life of the structure is minimized. The nominal strength (P_n and M_n) is based on the strength of masonry members which has been documented through research on the properties of masonry. The strength reduction factor is a multiplier which reduces the nominal strength based on the inherent variability of materials and construction methods. Strength reduction factors proposed for the LSDS are shown in the following table:

Strength Reduction Factors for Strength Design of Masonry	
Description	ϕ
Reinforced masonry subjected to flexure or flexure with axial load; in which flexural reinforcement is placed in a single line parallel to the edge under compression:	0.9
Reinforced masonry subjected to flexure or flexure with axial load; in which flexural reinforcement is distributed over the depth of the member:	1.0

Shear in Reinforced Masonry:
     Shear Walls
     Other Elements
	0.85
	 0.65

Embedment and Lap Splices of Reinforcement:	0.8
Bearing:	0.9

Anchor Bolts Embedded in Masonry:
     Tension or Shear Governed by Masonry
     Tension or Shear Governed by the Bolt
	0.5
	 0.9

Plain Masonry:
     Axial Compressive Strength
     Flexural Strength
     Shear Strength
	0.6
	 0.6
	 0.8

The LSDS specifies an expected modulus of elasticity, E_{me}, in accordance with the following:

$$E_{me} = 550 \, f_{me}$$

The expected shear modulus of masonry:

$$G_{me} = 0.4 \, E_{me}$$

The value of expected modulus of elasticity of steel reinforcement, E_{se}, is taken as 29,000,000 psi (200,000 MPa).

The expected value for yield strength of reinforcement is based on mill tests or is based on the values given in the following table:

Values of Expected Yield Strength of Reinforcement	
Grade of Steel	f_{ye} **psi**
40	54,000
60	68,000
75	82,000

The length of embedment (development length) for reinforcement is determined in accordance with the following equation.

$$\ell_d = \frac{1}{\phi} \left[\frac{0.15 \, d_b^2 \, f_{ye}}{K\sqrt{f_{me}}} \right] \leq 52 \, d_b$$

where:

d_b = diameter of reinforcement, in. (mm)
f_{me} = expected compressive strength of masonry used in the design, psi (MPa)
f_{ye} = expected yield strength of reinforcement, psi (MPa)
K = the lesser of the masonry cover, clear spacing between adjacent reinforcement, or 3 times d_b, in.
ℓ = span length, in. (mm)
ϕ = strength reduction factor

The shear strength of masonry determined in accordance with the LSDS considers both the masonry and the reinforcement to be effective in resisting shear. The expected shear strength, V_e, is determined by the following:

$$V_e = V_{me} + V_{se}$$

for $\dfrac{M}{Vd} < 0.25$, $V_{e(max)} = 6\sqrt{f_{me}}\, A_n$

for $\dfrac{M}{Vd} \geq 1.0$, $V_{e(max)} = 4\sqrt{f_{me}}\, A_n$

The expected shear strength, provided by masonry, V_{me}, for sections not in a plastic hinge zone is determined by:

$$V_{me} = \left[4.0 - 1.75\left(\frac{M}{Vd}\right)\right] A_n \sqrt{f_{me}} + 0.25 P$$

The expected shear strength provided by masonry, V_{me}, for sections in a plastic hinge zone is:

$$V_{me} = 1.0\, A_n \sqrt{f_{me}} + 0.25\, P$$

The expected shear strength provided by reinforcement, V_{se}, is determined by:

$$V_{se} = \frac{A_v}{s}\, f_{ye}\, d_v$$

A_n	=	net cross-sectional area of masonry, in.^{2}
A_v	=	cross-sectional area of shear reinforcement, in.^{2}
d_v		
f_{me}	=	expected compressive strength of masonry used in the design, psi (MPa)
f_{ye}	=	expected yield strength of reinforcement, psi (MPa)
P	=	axial load acting on a section, lb (N)
s	=	spacing of reinforcement, in.
V_e	=	expected shear strength, lb (N)
V_{me}	=	expected shear strength provided by masonry, lb (N)
V_{se}	=	expected shear strength of shear reinforcement, lb (N)

The above length of embedment, ℓ_d is also the required minimum length of lap for lap splicing of reinforcement.

Research Substantiation

Research to substantiate the provisions of the proposed LSDS has been conducted by a number if institutes and was coordinated through the Technical Coordinating Committee for Masonry Research (TCCMAR)[4]. The TCCMAR program was a multi year effort involving 28 research projects whose collective objective was to document the structural performance of masonry and to substantiate a limit states design approach for masonry. The program began with the documentation of masonry components and fulminated in the testing of a full scale masonry structure. An overview of the TCCMAR Research categories is shown in the following table:

TCCMAR RESEARCH CATEGORIES
1   Materials
2   Analytical and Computer Modeling
3   Wall Tests
4   Wall Intersection Tests
5   Floor and Roof Diaphragm Test
6   Construction Practices
7   Small Scale Model Tests
8   Design Methods
9   Full Scale Tests
10 Design Recommendations and Limit States   Design Development

By design, each phase of the program builds upon and takes advantage of knowledge grained in the previous phase. A cross section of researchers, consultants, industry representatives and design professionals reviewed each research task so that revisions reflected the broadest cross section of expertise.

Materials—Testing the behavior of both clay masonry prisms and concrete masonry prisms was conducted to determine if sufficient similarities between these materials allows both to be included in a single design methodology. Load vs deformation data for masonry prisms was used to determine the distribution of stress within members subjected to flexure. The research resulted in a recommended strength design procedure for flexure. It was also verified that the procedure is applicable to both clay and concrete masonry.

Shear Wall Panel Test—Thirty shear wall panel tests with various levels of horizontal and vertical reinforcement were conducted. The 6 foot by 6 foot shear panels were subjected to various levels of simultaneous compression and lateral load. These conditions simulated the actual state of stress that load bearing shear walls experience. These tests documented failure modes and load vs deflection relationships. The test results verified design recommendations for masonry shear walls which includes an evaluation of the effectiveness of both horizontal and vertical reinforcement.

Three Story Shear Wall Tests—These tests verified that the performance observed in testing 6 foot by 6 foot shear panel could be extrapolated to predict the performance of three story shear walls with load bearing floors built into the wall at each level.

Coupled Shear Wall Tests—Coupled shear walls connected by floor diaphragms were tested to investigate the extent to which load and moment is transferred between each element. Precast plank running parallel to the wall (non-bearing direction) was tested in one series and plank perpendicular to the wall (bearing direction) was tested in another series.

Out of Plane Wall Tests—This task is an extension of previous research on tall reinforced walls loaded perpendicular to the surface and included both static and dynamic load tests. The purpose of these tests is to verify that the distribution of stress and flexural design model developed under material investigations predicted performance of these full scale walls. Load vs deflection data under earthquake simulated dynamic loading was also documented.

Intersections—Story height walls consisting of a stem wall connected to a flange wall were constructed and tested by loading parallel to the stem wall through the floor diaphragm. The integrity of the intersection was investigated as well as the strength and load deflection properties of the wall.

Floor to wall intersections were also tested in which precast plank bearing wall connections were evaluated. These tests consisted of loading the plank laterally to document the capacity of the connection to transfer shear from the diaphragm into the shear wall.

Floor Diaphragms—Tests on several floor diaphragm systems were performed to investigate the capacity of horizontal diaphragms to distribute lateral loads to shear walls. Floors were connected to concrete masonry stem walls which simulated the shear wall. Surprisingly the capacity of the intersection and stem wall exceeded the capacity of plank diaphragm in these tests.

Reinforcement Bond, Splices and Anchorage—Several methods of testing reinforcing bond to grout were performed. Bars were pulled out of grouted specimens in one series. In another test, series bars were lap spliced within a grouted specimen and pulled apart. Horizontal bars anchored around vertical reinforcement were also tested. Beam specimens with bars lap spliced were tested in flexure to verify the splice behavior in a prototype specimen.

Scale Model Tests—A three story, quarter scale building of reinforced concrete masonry was tested on the shaking table under simulated earthquake loads. Building response was documented including lateral displacements at floor levels, cracking patterns and strength of members.

Full Scale Building Test

The experimental phrase of the TCCMAR program culminated with the testing of a full scale five story concrete masonry structure. The full scale building consists of a precast plank load bearing wall structure. The lateral loads simulated an earthquake loading condition. Response and performance of the building was predicted by computer models and limit states design criteria. The building verified our ability to predict performance based on the testing of materials, components and sub-assemblages as well as analytical models. The analytical phase of the TCCMAR program and technology transfer tasks resulted in the proposed LSDS. There has have been a wealth of information generated which has been incorporated into the design provisions. The program has generated increased interest in the masonry shear wall system. Wide spread acceptance and use of the system by structural engineers is occurring as the documented performance and design criteria is put into use.

References

1. *Minimum Design Loads for Building and Other Structures*, ASCE 7. American Society of Civil Engineers, New York, NY, 1995

2. *Draft Masonry Limit States Design Standard*. Masonry Standards Joint Committee, November 1995.

3. *Commentary for the Draft Masonry Limit States Design Standard*, Masonry Standards Joint Committee, November, 1995.

4. TCCMAR Research Program (Technical Coordinating Committee for Masonry Research):
    • 1.2(a)-1   Hamid A.A., Assis, G.F., Harris, H.G., *Material Models for Grouted Block Masonry*, August 1988. 67 pgs.

    • 1.2(a)-2   Assis, G.F., Hamid A.A., Harris, H.G., *Material Models for Grouted Block Masonry*, August 1989. 134 pgs.

    • 1.2(b)-1   Young, J.M., Brown, R.H., *Compressive Stress Distribution of Grouted Hollow Clay Masonry Under Strain Gradient*, May 1988. 170 pgs.

• 2.1-3   Nakaki, D. & Hart, G., *Uplifting Response of Structures Subjected to Earthquake Motions*, August 1987. 200 pgs.

• 2.1-6   Hart, G.C. Jaw, J.W., Low, Y.K., *SCM Model for University of Colorado Flexural Walls*, December 1989. 31 pgs.

• 2.1-12 Hart, G., Englekirk, R., Jaw, J.W., Huang, S.C., Drag, D.J., *Seismic Performance Study, 2-Story Masonry Wall-Frame Building Designed by Tentative Limit States Design Standard*, February, 1992. 146 pgs.

• *2.2-3 Ewing, R.D., *Finite Element Analysis of Reinforced Masonry Building Components Designed by a Tentative Masonry Limit States Design Standard*, March 1992.

• 2.3-3   Kariotis, J., Rahman, M., El-Mustapha, A., *Investigation of Current Seismic Design Provisions for Reinforced Masonry Shear Walls*, January 1990. 48 pgs.

• 2.3-4   Kariotis, J., Rahman, A., Waqfi, O., Ewing, R., *Version 1.03 LPM/I - A Computer Program for the Nonlinear, Dynamic Analysis of Lumped Parameter Models*, February 1992. 252 pgs.

• 2.3-7   Kariotis, J., Waqfi, O., *Recommended Procedure for Calculation of the Balanced Reinforcement Ratio*, February 1992. 74 pgs.

• 2.4(a)-1 Yeomans, F.S., Porter, M.L., *Dynamic Analysis of Diaphragms*, Dec. 1993 (revised Oct. 1995)

• 3.1(b)-1 Seible, F., and LaRovere, H., *Summary of Pseudo Dynamic Testing*, February 1987. 46 pgs.

• 3.1(c)-1 Merryman, K., Leiva, G., Antrobus, N., Klingner, R., *In-Plane Seismic Resistance of Two-Story Concrete Masonry Coupled Shear Walls*, September 1989. 176 pgs.

• 3.1(c)-2 Leiva, G., Klingner, R., *In-Plane Seismic Resistance of Two-Story Concrete Masonry Shear Walls with Openings*, August 1991. 304 pgs.

• 3.2(a)-1 Hamid, A., Abboud, B., Farah, M., Hatem, K., Harris, H., *Response of Reinforced Block Masonry Walls to Out-of-Plane Static Loads*, September 1989. 120 pgs.

• 3.2(b)-1 Agbabian, M., Adham, S., Masri, S., Avanessian, V. Traina, *Out-of-Plane Dynamic Testing of Concrete Masonry Walls*, Volumes 1 and 2, July

1989. 220 pgs.

• 4.1-2   He, L., Priestley, M.J.N., *Seismic Behavior of Flanged Masonry Shear Walls - Final Report*, November 1992. 280 pgs.

• *4.2-1 Hegemier, G. Murakami, H., *On the Behavior of Floor-to-Wall Intersections in Concrete Masonry Construction: Part I: Experimental*

• *4.2-2 Hegemier, G., Murakami, H., *On the Behavior of Floor-to-Wall Intersections in Concrete Masonry Construction: Part II: Theoretical*

• 5.1-1   Porter, M., Sabri, A., *Plank Diaphragm Characteristics*, July 1990. 226 pgs.

• 6.2-1   Scrivener, J., *Bond of Reinforcement in Grouted Hollow-Unit Masonry: A State-of-the-Art*, June 1986. 53 pgs.

• 6.2-2   Soric, Z. and Tulin, L., *Bond Splices in Reinforced Masonry*, August 1987. 296 pgs.

• 7.1-1   Paulson, T., Abrams, D., *Measured Inelastic Response of Reinforced Masonry Building Structures to Earthquake Motions*, October 1990. 294 pgs.

• *8.1-1 Hart, G., *A Limit State Design Method for Reinforced Masonry*, June, 1988.

• 9.1-2   Kariotis, J.C., Waqfi, O.M., *Trial Designs Made in Accordance with Tentative Limit States Design Standards for Reinforced Masonry Buildings*, February 1992. 220 pgs.

• 9.4-1   Seible, F., Hegemier, G.A., Priestley, M.J.N, Kinglsey, G.R., Kurkchubasche, A., Igarashi, A., Weeks, J.S., *The U.S. - TCCMAR Five-Story Full-Scale Masonry Research Building Test, Part I - Executive Summary*, January 1994. 71 pgs.

Proposed Prestressed Masonry
Design Provisions

Matthew J. Scolforo, P.E.[1], MASCE

Abstract

Provisions for prestressed masonry are a logical extension of the rapid advancement of rational design methods and construction requirements for structural masonry. The arduous task of conducting an extensive review of the current knowledge base for prestressed masonry was the first obstacle hurdled. Following this, the Prestressed Masonry Subcommittee was formed in 1992 within the Masonry Standards Joint Committee with the charge to develop code and specification provisions for prestressed masonry. The subcommittee has produced draft code, specification and commentary provisions for prestressed masonry that are under balloting within the committee. This paper discusses these draft provisions, explaining the subcommittee's rationale and the approval requirements. Also discussed is the interesting process that developed to bring these provisions to fruition. It is a process that benefitted greatly from the experiences and comments of researchers and practicing engineers in the international masonry community.

Introduction

Prestressed masonry is a viable alternative system to unreinforced and reinforced masonry. Prestressing is used to reduce or eliminate tensile stresses in masonry due to external loads by using controlled precompression. The precompression is generated by steel prestressing tendons, either bars or strands, which are contained in openings in the masonry and which may be grouted. The prestressing tendons can be pre-tensioned, in which case they are stressed against external abutments prior to placing the masonry, or post-tensioned, in which case they are stressed against the masonry after the masonry has been placed and has reached a predetermined strength. Most construction applications to date have involved post-

---

[1]Staff Engineer, Brick Institute of America, 11490 Commerce Park Drive, Reston, VA 22091

tensioned, ungrouted masonry for its ease of construction and overall economy. Although not very common, pre-tensioning has been used to construct prefabricated masonry panels. A more detailed review of prestressed masonry systems and applications can be found elsewhere (Schultz and Scolforo, 1991).

Although applications of prestressed masonry date to the turn of the century, the study and use of prestressed masonry in the United States has increased most significantly over the last 10 years. Extensive research has been conducted on the behavior of prestressed masonry systems in the United States and in many other countries. Summaries of prestressed masonry research and proposed design criteria are presented in the literature (Schultz and Scolforo, 1992; VSL, 1990; Curtin et al., 1988; Phipps and Montague, 1976). Provisions for prestressed masonry have been included in the British masonry code since 1985 (Phipps, 1992; BS 5628, 1985) and are under development for inclusion in future editions of the Canadian and Australian masonry codes. Currently, there are at least five proprietary systems being marketed in the United States. This points to the need for national consensus code and specification requirements to ensure public safety.

In 1992, the Masonry Standards Joint Committee (MSJC) formed the Prestressed Masonry Subcommittee to develop code and specification requirements for prestressed masonry. The MSJC has jurisdiction over the *Building Code Requirements for Masonry Structures* (ACI 530/ASCE 5/TMS 402) and the *Specification for Masonry Structures* (ACI 530.1/ASCE 6/TMS 602). In this paper, these standards are termed the MSJC Code and Specification, respectively (ACI 530, 1995). The Prestressed Masonry Subcommittee is chaired by Dr. Hans Ganz of VSL International.

Work of the Prestressed Masonry Subcommittee has included development of six drafts of provisions to date. Persons knowledgeable about prestressed masonry systems are currently somewhat few in number. Consequently, drafts were reviewed by subcommittee members as well as over fifteen persons throughout the world with expertise in the design and application of prestressed masonry systems. Comments and suggestions on the proposed criteria have come from Australia, Canada, England, France, Sweden and Switzerland. The provisions are currently under MSJC Main Committee balloting and it is likely that they will be included in the 1998 edition of the MSJC Code and Specification.

This paper provides an overview of the proposed design and specification requirements currently under review by the MSJC (Draft, 1995). The contents of the proposed provisions and the subcommittee's rationale behind the provisions are given. Because the current MSJC Code is based on an allowable stress design methodology, the Prestressed Masonry Subcommittee chose to develop design requirements that are primarily based on allowable stresses. In this way, design requirements could be integrated into the existing code requirements as a chapter of the MSJC Code. This, in fact, is what has been proposed. Specification of materials and methods of construction were blended into the structure of the current MSJC Specification where appropriate. The first subject addressed by the Prestressed Masonry Subcommittee was the flexural and axial compression behavior of prestressed masonry members.

Axial Compression and Flexure Requirements

The requirements for prestressed masonry members subjected to axial compression and flexure are separated into those for members with laterally unrestrained prestressing tendons and those for members with laterally restrained prestressing tendons. This separation was necessary because the flexural behavior of a prestressed masonry member significantly depends upon the lateral restraint of the prestressing tendon. Lateral restraint of a prestressing tendon is typically provided by grouting the cell or void containing the tendon before or after transfer of prestressing force to the masonry. Alternatively, lateral restraint may be provided by building the masonry into contact with the tendon or the tendon's protective sheathing at periodic intervals along the length of the prestressing tendon.

The proposed method of design of prestressed masonry members with prestressing tendons that are not laterally restrained is an allowable stress design procedure. The method of design of prestressed masonry members with laterally restrained prestressing tendons is a combination of allowable stress design and strength design. A moment strength check is included for members with laterally restrained prestressing tendons to ensure adequate strength and ductility of the member in flexure under strength level loading.

Laterally unrestrained prestressing tendons - Since masonry members with unrestrained prestressing tendons are equivalent to masonry members subjected to applied axial loads, the design approach for unreinforced masonry in MSJC Code Chapter 6, Unreinforced Masonry, has been adopted for convenience and consistency. Buckling of masonry members under prestressing force must be avoided for members with laterally unrestrained prestressing tendons. The prestressing force is to be added to the design axial load for all stress and load computations and in the computation of the eccentricity of the axial resultant.

Allowable compressive stresses for prestressed masonry with unrestrained prestressing tendons address two distinct loading stages; stresses immediately after transfer of prestressing force to the masonry member and stresses after all prestress losses and gains have taken place. The magnitude of allowable axial compressive stress and bending compressive stress after all prestress losses and gains occur are consistent with those for unreinforced masonry in MSJC Code Chapter 6. Immediately after transfer of prestressing, allowable compressive stresses and applied axial load are increased by 20 percent. This means that the factors of safety at transfer of prestress may be lower than those after prestress losses and gains occur. The first reason for this is that the effective precompression stress at the time of transfer of prestressing almost certainly decreases over time and masonry compressive strength most likely increases over time. Second, loads at the time of transfer of prestressing, namely prestress force and dead loads, are known more precisely than loads throughout the remainder of service life.

Cracking of prestressed masonry with laterally restrained or unrestrained prestressing tendons under permanent loads is to be avoided. The prestressing force and the dead weight of the member are permanent loads. It is not desirable to permit cracking under permanent loading conditions due to the potential for significant water penetration, which may precipitate corrosion of the prestressing tendons and accessories and damage to interior finishes. Masonry provides a flexural tensile resistance to cracking, as reflected by the allowable flexural tensile stress values stated in MSJC Code Chapter 6 for unreinforced masonry. However, tensile stress under prestressing force and dead loads alone is not permitted in the proposed criteria as a conservative measure. The subcommittee deemed this a reasonable restriction that is reflective of current practice for prestressed masonry members.

Laterally restrained prestressing tendons - Prestressed masonry members with laterally restrained prestressing tendons require a modified design approach from the criteria in MSJC Code Chapter 6 for unreinforced masonry. If the prestressing tendon is laterally restrained, the member cannot buckle under its own prestressing force. Any tendency to buckle will induce a lateral deformation which is resisted by an equal and opposite restraining force provided by the prestressing tendon. Such members are susceptible to buckling under axial loads other than prestressing and this condition must be checked in accordance with the axial load limit in MSJC Code Chapter 6. For the rare case of eccentrically prestressed masonry members, the prestressing force must be considered in the computation of the eccentricity of the axial resultant for use in the MSJC Code equation for the buckling force. In addition to a buckling load check, the proposed criteria do not permit flexural tension due to the prestressing force and dead loads alone for the reasons noted earlier. These are the extent of the provisions for such members under service level loads.

There are proposed restrictions on compression and flexure which are based on a moment strength computation and apply to strength level loading. Computation of the moment strength of prestressed masonry members with laterally restrained prestressing tendons is similar to the method for prestressed concrete (ACI 318, 1989). The proposed equation for the unbonded prestressing tendon stress at the moment strength condition, f_{ps}, is as follows:

$$f_{ps} = f_{se} + 100,000 \left(\frac{d}{l} \right) \left[1 - 0.7 \left(\frac{f_{pu} A_{ps}}{b d f'_m} \right) \right]$$

where f_{se} is the effective prestress, d is the effective depth, l is the tendon length, f_{pu} is the ultimate tendon strength, A_{ps} is the area of the tendon, and b is the effective width. This equation was derived analytically and is consistent with the equation in the British code for prestressed masonry (Phipps, 1992). Recent tests of prestressed masonry members have supported the use of this equation (Graham and Page, 1995).

The ratio of equivalent compression zone to effective depth is not permitted to exceed 0.425. This limitation will require significant yielding of the prestressing tendons prior to masonry compression failure. In such a situation, the moment strength is determined by the strength of the prestressing tendon, so a strength reduction factor equal to 0.8 has been proposed.

Shear Requirements

The enhancement of shear resistance provided by prestressing is represented by a Coulomb friction expression, as is done for unreinforced masonry in MSJC Code Chapter 6. The proposed allowable shear stress is based on an exact shear stress, which assumes a parabolic shear stress distribution. Also proposed are allowable shear stress equations to limit the principal tensile stress and the principal compressive stress. These new equations bear some explanation.

In order to avoid masonry crushing, a limitation on the principal compressive stress is imposed. This mode of failure has been observed in experimental tests of masonry shear walls that were simultaneously loaded in shear and axial compression. A safety factor of 4.0 has been applied, which is consistent with the safety factor for compression in MSJC Code Chapters 6 and 7. For fully grouted masonry, the strength is nearly isotropic. Hence, for any load direction the principal compressive stress may not exceed a single allowable compressive stress equal to one-quarter of the specified compressive strength of masonry. For ungrouted or partially grouted masonry, the strength may drop below f'_m for certain load directions due to the anisotropic strength of masonry. This fact is recognized by limiting the principal compressive stress to 15 percent of f'_m, again assuming a safety factor of 4.0. In this case, the compressive strength of ungrouted or partially grouted masonry along a principal compression axis oriented at approximately 45 degrees is taken as six-tenths of the compressive strength of masonry loaded perpendicular to bed joints.

Prestressing Tendon Requirements

The proposed allowable prestressing tendon stresses are based on criteria established for prestressed concrete (ACI 318, 1989). Allowable prestressing tendon stresses are for jacking forces and for the state of stress in the prestressing tendon immediately after the prestressing has been applied, or transferred, to the masonry member. When computing the prestressing tendon stress immediately after transfer of prestress, consideration must be given to all sources of short term prestress losses. These sources include anchorage seating loss, elastic shortening of masonry, and friction losses.

The state of stress in a prestressed masonry member must be checked for all stages of loading. For each loading condition, the effective level of prestress should be used in the computation of stresses and member strength. Sources of change of prestress that must be considered are anchorage seating loss, elastic shortening of masonry, creep of masonry, shrinkage of concrete masonry, relaxation of prestressing tendons, friction loss, and moisture expansion of clay masonry. The loss of effective prestress due to stress relaxation of the prestressing tendon is dependent upon the level of prestress, which changes with time-dependent phenomenon such as creep, shrinkage, and moisture expansion of the masonry. Appropriate formula for predicting prestress loss due to relaxation as a function of these phenomenon have been presented in the literature (Shrive, 1988; Lenczner, 1987, 1985).

Friction losses are minimal or nonexistent for most post-tensioned masonry applications because prestressing tendons are usually straight and contained in cavities. For jacking systems which incur anchorage losses, manufacturer's information should be used to compute prestress losses. Changes in prestress due to thermal fluctuations may be neglected if masonry is prestressed with high-strength prestressing steels because the magnitude of the prestress change will be negligible. If mild strength bars are used, thermal changes can significantly influence the level of prestress and this should be considered when computing effective prestress.

Seismic Requirements

MSJC Code Chapter 10, Seismic Design Requirements, includes strength design provisions for masonry members subjected to strength level seismic loads, as stipulated in ASCE 7. These provisions will apply to prestressed masonry members with one exception. The moment strength of prestressed masonry members with laterally restrained prestressing tendons is computed in accordance with the proposed strength design method as discussed earlier rather than the strength design method in MSJC Code Chapter 10.

The proposed provisions do not require a mandatory quantity of non-prestressed reinforcement nor bonded prestressing tendons for all masonry members. However, the minimum reinforcement quantities stated in Chapter 5 for masonry columns and Chapter 10 for Seismic Performance Category C, D, and E buildings will be required for prestressed masonry members. The cross-sectional area of bonded prestressing tendons may contribute to these minimum reinforcement requirements.

Prestressing Tendons, Anchorages and Couplers

The proposed specification will govern the materials and methods of construction that will be permitted for prestressed masonry members. Much of these criteria are based on what is standard practice for prestressing of concrete members.

Prestressing tendons will be required to comply with ASTM A 421, A 416, A 615, A 706, or A 722 with a minimum yield strength of 60 ksi (414 MPa). The buildability aspects of prestressed masonry favor the use of rods or rigid strands in ungrouted construction with mechanical anchorage. Mild strength steel bars are being used in prestressed masonry installations in the United States (Schultz and Scolforo, 1991).

Due to the extensive variety of hardware available for anchorage and coupling of prestressing tendons, the proposed requirements are somewhat general for these. The tendon anchorages, couplers and end blocks must be designed to withstand the prestressing operation and effectively transfer prestress force to the masonry member without distress to the masonry or the prestressing accessories. Because the actual stresses around post-tensioning anchorages are quite complicated, a refined analysis is recommended whenever possible. Appropriate formulae are available in the literature (Guide, 1990) to size prestressing tendon anchorages when experimental data or more refined analysis are not available. Additional guidance on design and details for post-tensioning anchorage zones is given in the references (Sanders et al., 1987). Typical anchorage and coupling devices for prestressed masonry are shown in Figure 1.

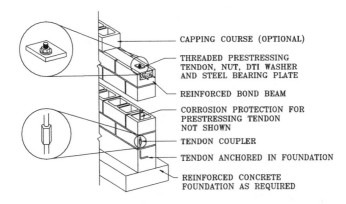

CAPPING COURSE (OPTIONAL)

THREADED PRESTRESSING TENDON, NUT, DTI WASHER AND STEEL BEARING PLATE

REINFORCED BOND BEAM

CORROSION PROTECTION FOR PRESTRESSING TENDON NOT SHOWN

TENDON COUPLER

TENDON ANCHORED IN FOUNDATION

REINFORCED CONCRETE FOUNDATION AS REQUIRED

Figure 1. Typical anchorage and coupling devices for prestressed masonry

Corrosion Protection Systems

Corrosion protection of the prestressing tendon and accessories is required by the proposed code requirements for masonry walls that will be subject to a moist and corrosive environment. The methods of corrosion protection required in the Post Tensioning Institute's *Specifications for Unbonded Single Strand Tendons* (Specifications, 1990) are recommended to provide a clear and significant corrosion protection system for single tendon systems. Masonry unit, mortar and grout cover is not permitted by themselves to be sufficient corrosion protection for prestressing tendons due to the variable permeability of masonry and the sensitivity of prestressing tendons to corrosion. Corrosion protection by galvanized coating is not permitted for tendons with yield strengths in excess of 100 ksi (690 MPa) because of the potential for hydrogen embrittlement. Poor experiences in England with galvanized high strength tendons in prestressed masonry members have demonstrated the need for this proposed restriction. For high strength tendons, protection against corrosion is commonly provided by a number of measures. Typically, a proprietary system is used which includes sheathing the prestressing tendon with a waterproof plastic tape or duct.

Protection of anchorage devices typically include filling the opening of bearing pads with grease, grouting the recess in bearing pads, and providing drainage of cavities housing prestressing tendons with base flashing and weep holes.

Discussion of the various corrosion protection systems used for prestressed masonry is available in the literature (Garrity, 1995). One example of a corrosion protection system for the prestressing tendon is shown in Figure 2.

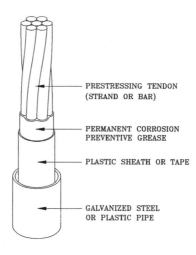

Figure 2. Typical corrosion protection system for the prestressing tendon

Construction Issues

Tolerances for prestressing tendon placement and methods of application and measurement of prestressing force are stated in the proposed specification. The designer and contractor are directed to the Post Tensioning Institute's *Field Procedures Manual for Unbonded Single Strand Tendons* (Field, 1994) or similar literature before conducting the work. Critical aspects of the prestressing operation which require inspection include handling and storage of the prestressing tendons and anchorages, installation of the anchorage hardware into the foundation and capping members, integrity and continuity of the corrosion protection system for the prestressing tendons and anchorages, and the prestressing tendon stressing and grouting procedures.

Often times the masonry member will be prestressed prior to 28 days after construction. The appropriate compressive strength of the masonry at the time of prestressing must be used to determine allowable prestressing levels. This strength will likely be a fraction of the 28 day compressive strength of the masonry. It cannot be taken greater than the specified compressive strength of masonry, even if the tested strength is greater. Assessment of the compressive strength at prestress transfer is by testing of masonry prisms or by a record of strength gain over time of masonry prisms constructed of similar masonry units, mortar, and grout when subjected to similar curing conditions.

For high strength tendons, the prestress is applied with hydraulic jacks and is checked against tendon elongation curves for accuracy. For mild strength bars, calibrated DTI spring washers and torque wrenches may be used. DTI spring washers are often used for prestressed masonry systems in England and the United States for their speed, simplicity and economy.

Appendix A - References

ACI 318, Building Code Requirements for Reinforced Concrete, 1989 Edition, American Concrete Institute, Detroit, MI.

ACI 530/ASCE 5/TMS 402, Building Code Requirements for Masonry Structures, and ACI 530.1/ASCE 6/TMS 602, Specification for Masonry Structures, American Society of Civil Engineers, New York, NY, 1995.

BS 5628, Code of Practice for the Use of Masonry, Part 2: Reinforced and Prestressed Masonry, British Standards Institution, London, England, 1985.

"Chapter 2 - Post-Tensioning Systems," Post-Tensioning Manual, 5th Edition, Post-Tensioning Institute, Phoenix, AZ, 1990, pp. 51-206.

Curtin, W.G., Shaw, G., and Beck, J.K., Design of Reinforced and Prestressed Masonry, Thomas Telford Ltd., London, England, 1988, 244 pp.

"Draft - Prestressed Masonry Design and Specification Provisions," Masonry Standard Joint Committee, July 1995.

Field Procedures Manual for Unbonded Single Strand Tendons, 2nd Edition, Post-Tensioning Institute, Phoenix, AZ, 1994, 62 pp.

Garrity, S.W., "Corrosion Protection of Prestressing Tendons for Masonry," Proceedings, Seventh Canadian Masonry Symposium, McMaster University, Hamilton, Ontario, June 1995, pp. 736-750.

Graham, K.J. and Page, A.W., "The Flexural Design of Post-Tensioned Hollow Clay Masonry," Proceedings, Seventh Canadian Masonry Symposium, McMaster University, Hamilton, Ontario, June 1995, pp. 763-774.

"Guide Specifications for Post-Tensioning Materials," Post-Tensioning Manual, 5th Edition, Post-Tensioning Institute, Phoenix, AZ, 1990, pp. 208-216.

Lenczner, D., "Creep and Loss of Prestress in Stack-Bonded Brick Masonry Prisms, Pilot Study - Stage II," Department of Civil Engineering, University of Illinois,

Urbana-Champaign, IL, August 1987, 29 pp.

Lenczner, D., "Creep and Stress Relaxation in Stack-Bonded Brick Masonry Prisms, A Pilot Study," Department of Civil Engineering, Clemson University, Clemson, SC, May 1985, 28 pp.

Phipps, M.E. and Montague, T.I., "The Design of Prestressed Concrete Blockwork Diaphragm Walls," Aggregate Concrete Block Association, England, 1976, 18 pp.

Phipps, M.E., "The Codification of Prestressed Masonry Design," Proceedings, Sixth Canadian Masonry Symposium, Saskatoon, Saskatchewan, Canada, June 1992, pp. 561-572.

Sanders, D.H., Breen, J.E., and Duncan, R.R. III, "Strength and Behavior of Closely Spaced Post-Tensioned Monostrand Anchorages," Post-Tensioning Institute, Phoenix, AZ, 1987, 49 pp.

Schultz, A.E. and Scolforo, M.J., "An Overview of Prestressed Masonry," The Masonry Society Journal, V. 10, No. 1, August 1991, pp. 6-21.

Schultz, A.E. and Scolforo, M.J., "Engineering Design Provisions for Prestressed Masonry, Part 1: Masonry Stresses and Part 2: Steel Stresses and Other Considerations," The Masonry Society Journal, V. 10, No. 2, February 1992, pp. 29-64.

Shrive, N.G., "Effects of Time Dependent Movements in Composite and Post-Tensioned Masonry," Masonry International, V. 2, No. 1, Spring 1988, pp. 1-34.

"Specifications for Unbonded Single Strand Tendons," Post-Tensioning Manual, 5th Edition, Post-Tensioning Institute, Phoenix, AZ, 1990, pp. 217-229.

VSL Report Series, "Post-Tensioned Masonry Structures," VSL International Ltd., Berne, Switzerland, 1990, 35 pp.

Structural Behaviour of High Strength Concrete Columns

Robert Park[1], F.ASCE

Abstract

The flexural strength and ductility of high strength concrete columns is discussed, including the compressive stress block parameters which take into account the special characteristics of high strength concrete and the use of very high strength reinforcing steel as a means of providing adequate confinement for ductile behaviour. The results of laboratory tests on high strength concrete columns subjected to simulated seismic loading are reported which examined the strength and ductility when high strength transverse reinforcement and mixed grade longitudinal reinforcement are used.

Introduction

The use of high strength concrete, together with high strength steel reinforcement, appears to be an attractive proposition for heavily loaded columns of building structures. Figs 1 and 2 show typical stress-strain curves for normal and very high strength concrete and steel. Some aspects of the use of high strength concrete and steel in columns is considered in this paper.

Flexural Strength of High Strength Concrete Columns

The shape of stress-strain curve for high strength concrete (see Fig. 1) implies that the actual shape of the concrete compression stress distribution in a high strength concrete member at the flexural strength is near triangular, whereas for normal strength concrete the shape is more curved (see Fig. 3). Also the falling branch of the stress-strain curve is more steep for high strength concrete implying more brittle behaviour of the concrete at failure.

---

[1]Professor of Civil Engineering, University of Canterbury, Private Bag 4800, Christchurch, New Zealand

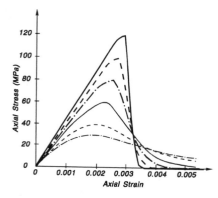

Fig. 1   Stress-Strain Curves for
          Unconfined Concrete Cylinders
          in Uniaxial Compression

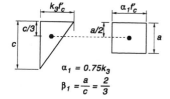

Fig. 2   Typical Stress-Strain
          Curves for Normal and
          Very High Strength Steel

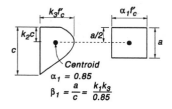

(a)   Normal strength concrete   (b)   High Strength Concrete

Fig. 3   Equivalent Concrete Compressive Stress Blocks

The parameters of the equivalent rectangular concrete compressive stress block, as recommended in the building code of the American Concrete Institute (ACI 318 1989), were derived originally using data which had very few experimental results for concrete cylinder strengths f'_c greater than 55 MPa. In more recent years there has been controversy as whether those stress block parameters apply to higher strength concrete. It is of note that the recent structural concrete codes of several countries (for example, Canada, New Zealand and Norway) have introduced parameters for stress blocks which take into account the special characteristics of high strength concrete.

In the recent New Zealand concrete design standard (NZS 3101 1995) it is recommended that the currently used parameters for the equivalent rectangular concrete compressive stress block are applicable up to $f'_c = 55$ MPa. For $f'_c > 55$ MPa it is assumed that $k_3 = 1.0$ (see Fig. 3b) since the actual shape of the stress block is slightly curved with an average stress of about 0.5 f'_c. For the full range of f_c' NZS 3101 (1995) recommends that the equivalent rectangular concrete compressive stress block be taken to have a mean stress of $\alpha_1 f_c'$ uniformly distributed over a depth a, where $\beta_1 = a/c$, where

$$
\begin{aligned}
\alpha_1 = \quad & 0.85 & \text{for } f_c' \leq \;\; 55 \text{ MPa} \\
\text{or} \quad & 0.85 - 0.004 \, (f_c' - 55) & \text{for } f_c' > \;\; 55 \text{ MPa} \\
& \text{but not less than 0.75, which is reached when } f_c' = 80 \text{ MPa}
\end{aligned} \tag{1}
$$

$$
\begin{aligned}
\beta_1 = \quad & 0.85 & \text{for } f_c' \leq \;\; 30 \text{ MPa} \\
\text{or} \quad & 0.85 - 0.008 \, (f_c' - 30) & \text{for } f_c' > \;\; 30 \text{ MPa} \\
& \text{but not less than 0.65, which is reached when } f_c' = 55 \text{ MPa}
\end{aligned} \tag{2}
$$

It is also to be noted that the above parameters α_1 and β_1 defining the equivalent rectangular concrete compressive stress block are for unconfined concrete. The extreme fibre concrete compressive strain is taken as 0.003.

Example of the Effect of Concrete Strength On the Size of Column Section Required for Flexure and Axial Load

For reinforced concrete columns carrying heavy compressive loads the use of high strength concrete can lead to a dramatic reduction in the amount of longitudinal reinforcement, and/or the size of column cross section. For example, consider a 400 mm square section of a column constructed from high strength concrete with $f_c' = 70$ MPa. Let the section contain eight 24 mm diameter Grade 430 longitudinal reinforcing bars distributed around the perimeter, giving $p_t = 2.26\%$ where p_t is the ratio of the area of longitudinal steel to the gross area of the column section. The cover to the longitudinal reinforcement is 31 mm. When the column carries an axial load of $0.53 f_c' A_g = 5{,}940$ kN, where $A_g =$ gross area of column, the nominal flexural strength is 523 kNm using the concrete stress block parameters given by Eqs 1 and 2.

Consider the compressive strength of the concrete to be reduced to $f_c' = 30$ MPa and the column still to contain eight 24 mm diameter Grade 430 longitudinal bars distributed around the perimeter with 36 mm cover to the longitudinal reinforcement. The column cross section would need to be increased to 530 mm square to have about the same flexural strength of 523 kNm as the high strength concrete column when carrying the same axial load of 5,940 kN. That is, in this example the column with $f_c' = 70$ MPa has 57% of the concrete area of the equivalent column with $f_c' = 30$ MPa and the same area of longitudinal steel, meaning a substantial saving of concrete volume.

The Role of Transverse Reinforcement

High strength concrete is more brittle than ordinary strength concrete (see Fig. 1). A high strength concrete column, if unconfined, on reaching its flexural strength at an extreme fibre compressive strain of about 0.003 will fail suddenly and violently with very little ductility. Transverse reinforcement in the form of hoops or spirals can be used to improve the ductility of columns by confining the compressed concrete core. The other important roles of transverse reinforcement are to prevent premature buckling of the compressed longitudinal reinforcement and to provide shear resistance. Transverse reinforcement has a particularly important role in seismic design.

For columns designed for ductile performance the required quantity of transverse reinforcement to confine the concrete core is based on considerations of the passive confinement from that reinforcement arising from the lateral expansion of the compressed concrete due to internal microcracking when f_c' is approached. The design equations for the quantity of confining reinforcement for normal strength concrete columns generally assume that the transverse reinforcement reaches the yield strength as a result of the lateral expansion of the concrete. However, high strength concrete exhibits less internal microcracking than normal strength concrete for a given imposed axial compressive strain near f_c' (ACI Committee 363 1984, Li et al 1994). The lower lateral expansion of high strength concrete means that the stress in the transverse reinforcement at the peak load of a high strength concrete column may be less than the yield strength of the transverse steel, if high strength transverse reinforcement is used. NZS 3101 (1995) permits high strength transverse reinforcement to be used providing that the value of the yield strength used in strength design does not exceed 800 MPa for calculations of concrete confinement and lateral restraint against buckling of longitudinal bars. A limitation of 500 MPa on that stress is imposed for shear calculations to restrict the width of diagonal tension cracks at service loads.

High strength concrete columns require large quantities of transverse reinforcement to achieve ductile behaviour. One obvious way of reducing the congestion of transverse reinforcement would be to use high strength steel. A significant amount of research in Japan, New Zealand and North America has been conducted investigating the confinement of high strength concrete by high strength transverse reinforcement. For example, Fig. 4 compares experimental stress-strain curves for unconfined and confined high strength concrete obtained from concentric load tests on either prisms with 240 mm square cross sections reinforced transversely by helices, or cylinders with 240 mm diameter circular sections reinforced transversely by spirals (Li et al 1994). The remarkable improvement in the compressive strength and ductility of due to the confinement is evident.

The Provision of Transverse Reinforcement for Confinement

Design codes specify equations for the quantities of transverse reinforcement required to confine the compressed concrete of ductile columns

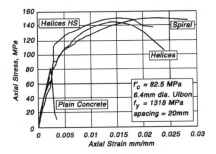

Fig. 4 Measured Stress-Strain Relationships for High Strength Concrete in Compression Loaded at Slow Strain Rate (Spiral and Helices) and High Strain Rate (Helices HS) (Li et al 1994)

in seismic design. The NZS 3101 (1995) equations for the required quantities of transverse reinforcement to confine the compressed concrete include as variables the ratio of column gross area to core area A_g/A_c, the longitudinal steel ratio p_t, the axial load ratio $N/f_c'A_g$, where N = axial compressive load, the yield strengths of the longitudinal and transverse reinforcement f_y and f_{yt}, and the concrete compressive cylinder strength f_c'. In particular the required amount of confining steel increase linearly with axial load ratio. The equations were derived by Watson et al (1994) from the results of theoretical cyclic moment-curvature analyses incorporating cyclic stress-strain models for confined concrete and reinforcing steel. The equations, take into account the efficiency of different arrangements of transverse reinforcement and are aimed at achieving specified levels of curvature ductility. At low axial load levels separate requirements for transverse reinforcement to prevent premature buckling of compressed longitudinal reinforcement become more critical. The NZS 3101 (1995) equations for transverse confining reinforcement have been shown by analyses conducted by Li et al (1994) to be applicable to high strength concrete columns with f_c' at least up to 70 MPa, providing the stress in the transverse steel is assumed not to exceed 800 MPa. NZS 3101 (1995) limits f_c' for ductile structures to 70 MPa.

As an example, consider a 400 mm square column section with f_c' = 70 MPa and with an axial load ratio $N/f_c'A_g$ = 0.53. The confining steel requirements of NZS 3101 (1995) can be satisfied by using sets of two 16 mm diameter overlapping hoops of Grade 430 steel with vertical spacing of 65 mm between centres of hoop sets, using an arrangement similar to that shown in Fig. 8. However, this is very dense transverse reinforcement. If Grade 800 steel is used the transverse reinforcement could be reduced to sets of two 12 mm diameter overlapping hoops with vertical spacing of 82 mm between centres of hoop sets. It can be concluded that confinement of high strength concrete columns is much more conveniently provided by high strength reinforcement.

If individual rectangular hoops are used, the ends of each hoop require anchorage by either at least a 135° bend around a longitudinal bar plus an extension of six or eight transverse bar diameters into the concrete core or by lap welding. However, if the bars are available in long lengths they could be bent in the factory into spirals or helices of the required size. Japanese manufactured ultra high strength Ulbon reinforcement, with a yield strength of at least 1,300 MPa, delivered bent into square helices, as shown in Fig. 5, can be rapidly made into reinforcement cages. The eight longitudinal bar column cage shown in Fig. 6 was made first as two four bar column cages separately (with one bar in each corner of each helix shown in Fig. 7) and then the smaller cage was inserted inside the larger cage and rotated into its final position (Li et al 1994). The minimum inside diameter of the bend for Ulbon steel is five times the diameter of the transverse bar.

Fig. 5   Reinforcement Prebent to Shape of Square Helices

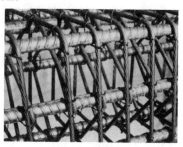

Fig. 6   Two Square Helices Forming Transverse Reinforcement

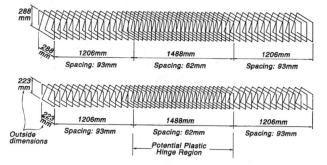

Fig. 7 Bent Shape of Square Helices Used for a Test Column

<u>Simulated Seismic Load Tests on High Strength Concrete Columns</u>

Five reinforced concrete columns were constructed, each with a 350 mm square cross section in the critical regions (see Fig. 8) (Li et al 1994). The

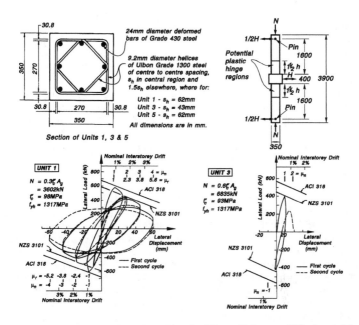

Fig. 8 Details of Column Sections, Test Set Up and Measured Horizontal Load Versus Horizontal Displacement Hysteresis Loops for Two Column Units Tested by Li et al(1994).

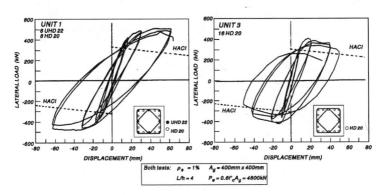

Fig. 9 Measured Horizontal Load Versus Horizontal Displacement Hysteresis Loops and Details For Two Column Units Tested by Satyarno et al (1993)

column units were tested under constant axial compressive load and reversible static horizontal loading was applied through a column stub as shown in Fig. 8. The axial compressive load level was either $0.3f_c'A_g$ or $0.6f_c'A_g$. At the stage of testing the column units, the compressive cylinder strengths of the concrete were f_c' = 98 MPa for Units 1 and 2, and f_c' = 93 MPa for Units 3, 4 and 5. The transverse reinforcement was from either New Zealand manufactured steel with a measured yield strength of 453 MPa or Japanese manufactured Ulbon steel with a measured yield strength of 1,317 MPa at the 0.2% offset strain. The deformed longitudinal reinforcement was from New Zealand manufactured steel with a measured yield strength of 446 MPa. Typical stress-strain curves for the grades of steel used are shown in Fig. 2. The quantities of Ulbon transverse confining reinforcement provided in the potential plastic hinge regions of Units 1 and 3 were 99% and 62%, respectively, of the quantities required for ductile columns by NZS 3101 (1995).

Fig. 8 shows the horizontal load versus horizontal displacement hysteresis loops measured for Units 1 and 3. Also shown in Fig. 8 by sloping straight lines (which include the PΔ effect) are the nominal theoretical flexure strengths of the columns calculated using the concrete equivalent rectangular compressive stress block of ACI 318 (1989) and NZS 3101 (1995) using the measured material strengths and a strength reduction factor ϕ of unity. It is evident that for the column with an axial load level of $0.6f_c'A_g$ the ACI approach resulted in a too high (unsafe) estimate of the column flexural strength. For all column units, spalling of the cover concrete occurred at the maximum lateral load carrying capacity and was sudden and violent.

Units 1 with axial load level of $0.3f_c'A_g$ exhibited some, but limited, ductility. Unit 3 with axial load level of $0.6f_c'A_g$ displayed no ductility once the flexural strength was reached at the onset of spalling of the cover concrete. The inability of these high strength concrete column units to exhibit ductile behaviour can be attributed to two reasons. First, a considerable loss in moment capacity occurred when the cover concrete spalled. The A_c/A_g ratio for the column sections was 0.69 on average and hence loss of concrete cover left a much smaller core section. Second, the heavily loaded Unit 3 contained only 62% of the quantity of confining reinforcement required by NZS 3101 (1995).

High Strength Concrete Columns With Mixed Grades of Longitudinal Reinforcement

The flexural ductility of reinforced concrete columns when undergoing plastic hinge rotation can be improved by using mixed grades of normal strength and ultra high strength longitudinal reinforcement. Yielding of such columns when loaded into the inelastic range commences in the normal strength steel and then at higher curvatures in the high strength steel, thus delaying the degradation of the flexural strength of the column. The effectiveness of using mixed grades of reinforcement was first examined by Watanabe and Ohsumi

(1991). Further analytical and experimental studies have been conducted at the University of Canterbury (Satyarno et al 1993 and Sato et al 1993). Typical stress-strain relations for the reinforcing steel used are shown in Fig. 2. The very high strength reinforcement used was manufactured in Japan.

The columns tested by Satyarno et al (1993) had 400 mm square sections and the compressive strength of the concrete was 50 MPa. For the longitudinal reinforcement, 20 mm diameter deformed bars with yield strength of 497 MPa and 22 mm diameter ultra high strength deformed bars with yield strength of 1,080 MPa were used. The ratios of the area of the total longitudinal reinforcement to the column sectional area, p_t were in the range 3.14 to 3.51%. The transverse reinforcement was from 7.4 mm diameter Ulbon steel with a yield strength of 1,250 MPa. The hoop sets were at 80 mm centres and provided a volumetric ratio of confining reinforcement of p_s = 1.56%. The axial load applied was $0.6f'_cA_g$. The details of two of the column sections can be seen in Fig. 9.

The column units were tested subjected to axial and horizontal loading as cantilever columns. Fig. 9 show the measured horizontal load versus horizontal displacement hysteresis loops of a column unit containing mixed normal strength and ultra high strength longitudinal reinforcement, and a column unit with all normal strength longitudinal reinforcement. As can be seen from Fig. 9, a positive stiffness of the column is maintained at higher inelastic displacements when mixed ultra high strength and normal strength longitudinal steel is used. For all columns with ultra high strength longitudinal reinforcement serious buckling of those bars eventually occurred and led to the final failure.

CONCLUSIONS

The conventional equivalent rectangular concrete compressive stress block parameters used in flexural strength calculations of normal strength concrete columns need modification when high strength concrete is used. High strength concrete columns supporting high compressive loads require very large quantities of confining reinforcement to achieve ductile flexural behaviour. One method for reducing congestion of the transverse reinforcement is to use high strength confining steel. Such steel could be conveniently supplied from the factory prebent into continuous spirals or helices of the required size and shape to enable the rapid fabrication of reinforcing cages.

High strength concrete columns tested at the University of Canterbury, using concrete with f'_c of almost 100 MPa, demonstrated that the yield strength of very high strength transverse reinforcement may not be reached at the stage when the column attains peak flexural strength. Other high strength concrete columns tested at the University of Canterbury investigated the advantages of using mixed grades of longitudinal reinforcement in order to improve column ductility.

Acknowledgements

The contributions of Dr H Tanaka, Dr Li Bing, Mr I Satyarno and Mr Y Sato are gratefully acknowledged. Financial support for this research project was received from the Cement and Concrete Association of New Zealand, the New Zealand Concrete Society, Firth Certified Concrete, Pacific Steel Ltd, Koshuha-Netsuran Co. of Japan, Taisai Corporation of Japan, WG Grace (NZ) Ltd and the New Zealand Ministry of External Relations and Trade.

References

ACI 318. (1989). "Building Code Requirements for Reinforced Concrete", American Concrete Institute, Detroit.

ACI Committee 363. (1984). "State of the Art Report on High-Strength Concrete", ACI Journal, Proc. Vol. 81, No. 4, pp 364-411.

Li, B, Park, R and Tanaka, H. (1994). "Strength and Ductility of Reinforced Concrete Members and Frames Constructed Using High Strength Concrete", Research Report 94-5, Department of Civil Engineering, University of Canterbury, New Zealand.

NZS 3101. (1995). "Concrete Design Standard", Wellington, New Zealand.

Sato, Y, Tanaka, H and Park, R. (1993). "Reinforced Concrete Columns With Mixed Grade Longitudinal Reinforcement", Research Report 93-7, Department of Civil Engineering, University of Canterbury, New Zealand.

Satyarno, I, Tanaka, H and Park, R. (1993). "Concrete Columns Incorporating Mixed Ultra High and Normal Strength Longitudinal Reinforcement", Research Report 93-1, Department of Civil Engineering, University of Canterbury, New Zealand.

Watanabe, F and Ohsumi, K. (1991). "Improvement of Flexural Behaviour of Reinforced Concrete Sections by Combined Use of Different Grade Longitudinal Bars", Proceedings of International Conference on Evaluation and Rehabilitation of Concrete Structures and Innovations in Design, American Concrete Institute, Hong Kong, pp 927-940.

Watson, S, Zahn, F A and Park, R. (1994). "Confining Reinforcement for Concrete Columns", Journal of Structural Engineering, ASCE, Vol. 120, No. 6, pp 1798-1824.

Ten year performance of a high performance concrete
used to build two experimental columns

Éric Dallaire[1]
Michel Lessard[2]
Pierre-Claude Aïtcin[3]

Abstract

Two experimental columns were built in 1984 in a 26-story high-rise building in Montréal. The two columns were built next to each other : one was part of the building structure and the other was a dummy used for long term compressive strength and strain measurements. Four series of specimens were cored from the dummy column since the construction while the active column was cored for the first time 10 years after the construction.

After 10 years of service life, the compressive strength of this silica fume high performance concrete does not show any regression. The concrete is still testing 95 MPa and its elastic modulus is equal to 45 GPa. No significant differences have been found between the values obtained from the active and the dummy columns. No alteration of the very dense microstructure of this high performance concrete could be observed.

Introduction

High performance concrete was first used in the construction of columns of some high-rise buildings in the early seventies. However, it took several years to see them used in other applications like offshore structures, bridges and highways [HOFF, 1993]. Therefore, high performance concrete is a quite new material and there is definitely a lack of information on its long term behaviour, especially in field condition, and in spite of the fact that there has been a lot of publication on HPC the last few years.

[1-2]Research engineer, Département de génie civil, Université de Sherbrooke, Sherbrooke (Québec) Canada J1K 2R1
[3]Professor, Département de génie civil, Université de Sherbrooke, Sherbrooke (Québec) Canada J1K 2R1

Laboratory experimentation, which is most of the time done in almost ideal condition, do not always represent real life conditions. Real life rimes with ambient temperature changes, delivery time variation from half an hour to 2 hours, difficult placing conditions due to steel congestion or thin members casting. Moreover, different results are often noted when comparing the behaviour of small laboratory specimens and elements found in real structure, especially when studying the creep and shrinkage of the concrete. Therefore, it is important to have some field data through monitoring real structure made of high performance concrete.

Historical background of the project

During the summer of 1984, two high-performance concrete columns were cast in the basement of a 26 floor building in downtown Montreal [AÏTCIN et al., 1987]. One of those columns is part of the supporting columns for the building, while the other, which is slightly shorter and located right next to the other, supports no load (Fig. 1). This latter column acts as a control (dummy) column since the construction of the high-rise. It served for the removal of several cores in order to follow the evolution of concrete properties over time [AÏTCIN et al., 1990]. The active column was cored for the first time in 1994, 10 years after it was constructed.

Figure 1 - The two experimental columns

Both columns had a cross section of 1100 mm x 1100 mm and were instrumented in such a way as to allow the monitoring of concrete temperature during hydration, as well as the measurements of deformations due to shrinkage and the deformations of the active column due to service loading. Several samples were casted during the course of the concrete placement in order to carry out several characterisation tests in the laboratory. Such tests included the measurement of compressive strength, flexural strength, modulus of elasticity in compression, and chloride-ion permeability [AÏTCIN et al., 1987, 1990].

At the time of construction, these experimental columns demonstrated that it was possible to make a good quality high-performance concrete in a central mixing plant using commonly available materials, despite warm concreting conditions.

Characteristics and early-age properties of concrete

Fresh concrete characteristics

The composition and properties of the fresh, high-performance concrete used for the construction of the two concrete columns are shown in Table 1 and 2. Since the time needed to transport the concrete from the mixing plant to the construction site was 45 minutes, a given dose of superplasticizer was added to the concrete on-site in order to attain the desired slump values. The concrete was placed in the formwork in two layers using a 1.5-m^3 bucket and was then internally consolidated with vibrators.

Table 1 - Composition of the concrete used to build the two experimental columns

	kg/m^3
Water (including the water of the superplasticizer)	135
Cement	500
Silica Fume	30
Coarse aggregate	1,100
Fine aggregate	700
Superplasticizer (plant)	14 L
Superplasticizer (job site)	±7 L
Retarding agent	1.8 L
Water/Cement	0.25

Table 2 - Properties of fresh concrete measured on the job site

Slump, mm	240
Air content, %	3.2
Temperature, ° C	22

Temperature progress of the concrete during setting

Each of the columns were instrumented with 12 thermocouples in order to establish the temperature profile of the concrete during the first 72 hours. The 12 thermocouples were positioned in three different locations in order to monitor the temperature gradients between the surface and the centre of the columns (Fig. 2) [AÏTCIN et al., 1987, 1990]. Figure 2 also presents the mean temperature profile of the concrete in different locations. The maximum gradient between the centre of the

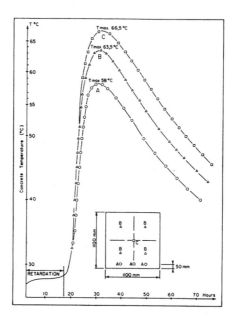

Figure 2 - Mean temperature profiles and schematics of thermocouple locations

column and the exterior surface was attained approximately 30 hours after the placement of concrete. The maximum gradients for either column did not exceed 17°C/m [AÏTCIN et al., 1987, 1990]. This value is inferior to the 25°C/m generally admissible as the maximum value to prevent the risk of cracking due to thermal gradients [BAMFORTH, 1980]. Finally, it is important to emphasize the late setting of the concrete which did not begin until approximately 18 hours after placement. This late setting is a consequence of the high dosage of superplasticizer and set retarder which was used because of the 45-minute transport time in non rush-hour traffic.

Mechanical properties measured during the first year

The samples taken during the placement of the two columns allowed the characterisation of certain mechanical properties of the concrete over the course of the first year. These properties consisted of compressive strength, modulus of elasticity under compression, and flexural strength. The results obtained during the first year are presented in Table 3 [LAPLANTE et al., 1986]. It must be emphasized that all of the specimens taken during the sampling were stored in a lime-saturated water until the time of testing.

The strength measured after 2 days is relatively low due to the high dosage of set retarder in the mixture. A significant strength gain between 28 days and one year of curing was observed. Similarly, a non-negligible gain in stiffness (modulus of elasticity) was obtained between 28 days and one year. On the other hand, no increase in flexural strength was noted over the same elapsed period [LAPLANTE et al., 1986].

Table 3 - Mechanical properties measured during first year
(100 x 200 mm standard specimens)

	Compressive strength (MPa)	Flexural strength (MPa)	Elastic modulus (GPa)
2 d	23.7	-	-
7 d	68.6	-	-
28 d	83.6	11.0	40
91 d	97.0	11.0	43
1 year	100.4	10.7	46

Concrete deformation

In addition to thermocouples, each column was instrumented with four vibrating cords in order to measure the axial and diametrical deformations of the concrete. Two of the vibrating wires were installed at the interior of the columns, while two other were positioned on the exterior surface of the columns. This made it possible to establish deformation gradients between the surface and inner concretes as well as the difference in deformation between the two columns which can lead to the evaluation of the differences in response related to the presence of service load of the active column.

Table 4 summarizes the results of shrinkage measurements. The deformation measurements taken from the control column before the removal of the formwork, in other words in the first four days after placement, demonstrated that the shrinkage attributed to cement hydration during this period was on the order of 250 µm/m [LAPLANTE et al., 1986]. Subsequently, a drying shrinkage of 330 µm/m was measured at the surface of the control column over a period of four years, while during this same period, the shrinkage measured at the centre of the column was practically nil, approximately 30 µm/m [LAPLANTE et al., 1986]. The resulting shrinkage deformation gradient carried a concentration of tensile stresses at the surface. Nevertheless, microstructural observations made with a scanning electron microscope on concrete samples taken from the outer surface of the columns after four years indicated the presence of limited amounts of cracking and deterioration. Damage attributed to the shrinkage deformation gradient can then be considered to be marginal [AÏTCIN et al., 1990, LAPLANTE et al., 1986].

Table 4 - Shrinkage strain in the dummy column

Period	Measured strain (µm/m)	
	Inside the column	At the surface of the column
0 to 4 d	250	not measured
4 d to 4 years	30	330

The deformations of the active column caused by service loading enabled the calculation of creep due to sustained axial loading which was approximately 300 µm/m after four years of service [LAPLANTE et al., 1986].

Development of compressive strength and modulus of elasticity over the years

The development of the compressive strength of the concrete as well as the behaviour of the columns was regularly monitored following their construction. The characterisation of the properties was first done on samples taken at the time of placement. After this period, the properties were monitored over 10 years from cores taken from the columns.

The compressive strength values after two, four, and seven years were determined on cores taken from the control (dummy) column only, while the 10 year measurements were determined using cores from both instrumented columns. Similarly, the elastic modulus was measured using the cores obtained from both columns after 10 years. The results of these measurements are presented in Table 5.

Table 5 - Compressive strength and elastic modulus over 10 years

	Water-cured specimens 100 x 200 mm				Cores 95 x 190 mm			
	7 d	28 d	91 d	1 year	2 year	4 year	7 year	10 year
Compressive strength (MPa)	70.8	85.4	89.7	97.6	83.0*	83.9*	87.5*	95.1†
Elastic modulus (GPa)	-	40	43	46	-	-	-	45

* cores from dummy column only
† cores from active and dummy columns

The results presented in Table 5 could be interpreted as if the concrete had some strength loss between the first and second year. It should be noted, however, that strength between 7 days to 1 year was determined by testing 100 x 200 mm standard cylinders that were cured under optimal conditions in lime-saturated water until the time of testing. The curing water penetrated the surface of the specimens and was able to help hydrate the cement grains in this peripheral zone, thus contributing to the increase in compressive strength. Such additional hydration of the outer concrete can lead to certain confinement of the specimen. This phenomenon of additional hydration should not take place in the cast columns. Therefore, it is completely normal that the strength measurement of cores taken from the columns is slightly less than that measured for the standard water-cured cylinders. Furthermore, it is possible that the consolidation of the concrete columns is different from that of the concrete used when casting the specimens. Therefore, it can be hypothesized that the strength difference noted between the first and second years can be attributed to these two effects and not to a loss of strength for the high-performance concrete. This hypothesis is furthermore supported by the fact that between the second and the tenth years, a slight but constant strength gain was measured for each of the cores. Therefore, it seems that rather than losing strength over long-term, the high-performance concrete of the columns exhibited some strength gain.

Finally, chloride-ion permeability measured on samples taken after four years had a mean value of 300 coulombs. This clearly demonstrates that the concrete in the columns had a very low permeability, such as that of latex-modified concrete [AÏTCIN et al., 1990, WHITING, 1984].

1994 coring campaign

In 1994, 10 years after the placement of the columns, samples were taken during a new coring campaign in order to measure compressive strength and modulus of elasticity of the concrete columns. Complete stress-strain curves of samples tested in compression were generated for three tested specimens. The mean compressive strengths measured after 10 years are presented in Table 5. The individual values, the mean and standard deviation of each measurement are presented in Tables 6 and 7.

Table 6 - Individual compressive strength values for the 10-year old cores, MPa

Sample number	Active column	Dummy column
1	99.3	90.9
2	97.2	103.4
3	-	84.8
Average	98.3	93.0
σ	1.5	9.5

Table 7 - Individual elastic modulus values for the 10-year old cores, GPa

Sample number	Active column	Dummy column
1	45	44
2	46	46
Average	45.5	45

Table 6 shows that the results obtained from the concrete of the loaded column are more consistent, and that the mean strength is slightly higher than that of the control column. The large difference in strength of the control column is due to the larger number of cores taken during previous coring campaigns which resulted in a very dispersed coring. This was not the case for the loaded column which was cored for the first time after 10 years and where the cores were able to be taken very closely to one another.

The complete stress-strain curves of the concrete from the columns were determined by three tests: one coming from the control column and the other two from the loaded column. These tests were carried out on a stiff testing machine having a 5000 kN capacity. This testing machine was operated on a deformation-control mode in order to enable the recording of the complete stress-strain relationship, up to the maximum stress and into the post-peak portion of the curve.

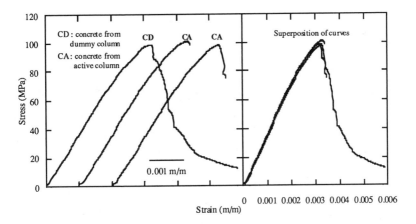

Figure 3 - Compressive stress-strain relation of the ten year old concrete
from the experimental columns

The superposition of the three curves in Figure 3 show that the stress-strain
relationship is practically identical in the three cases. The maximum deformation
corresponding to the failure of concrete was approximately 0.0036 m/m (Table 8).
This ultimate strain is slightly higher than the 0.003 m/m value used in the Canadian
Code [CSA A23.3-M84, 1984].

Table 8 - Results from complete stress-strain curves in compression

Sample number	Active column		Dummy column	
	σ max	ε peak	σ max	ε peak
1	100.1	0.0039	97.3	0.0034
2	97.3	0.0035		

Conclusion

The results obtained over 10 years for a high-performance concrete containing
silica fume used for casting columns in a high rise in Montreal demonstrate that there is
no decrease in compressive strength over time. Rather, a slight strength gain is
obtained for tested cores. Ten years after its placement, the silica fume concrete had an
approximate core strength of 100 MPa and a modulus of elasticity of 45 GPa, with a

deformation value of 0.0036 m/m at failure. This high compressive strength value also results in a very low chloride-ion permeability value which was 300 coulombs after four years.

During the first four days after casting, the concrete located at the interior of the columns developed a shrinkage value of 250 μm/m. This shrinkage can be solely chemical in nature since the interior temperature of the columns was already stabilised. However, the skin concrete exhibited an additional shrinkage of 330 μm/m after four years due to a slow desiccation. Unfortunately, it is impossible to evaluate to what depth the shrinkage manifested itself, yet observations with a scanning electronic microscope demonstrated that thermal gradients generated at the surface and shrinkage gradients due to differential drying did not cause any damage to the microstructure of the high-performance concrete.

References

AÏTCIN, P.-C., LAPLANTE, P., BÉDARD, C. (1987) *Development and Experimental Use of a 90 MPa Field Concrete*, American Concrete Institute SP-87, H.G. Russell Editor, Detroit, Michigan, pp. 51-70.

AÏTCIN, P.-C., SARKAR, S.L., LAPLANTE, P. (1990) *Long-Term Characteristics of a Very High Strength Concrete*, Concrete International, Vol. 12, No. 1, pp. 40-44.

BAMFORTH, P.B. (1980) *In Situ Measurement of the Effect of Partial Portland Cement Replacement Using Either Fly Ash or Ground Granulated Blast-Furnace Slag on the Performance of Mass Concrete*, Proceedings, Instn. Civ. Engrs, part 2, september, pp. 777-800.

CANADIAN STANDARDS ASSOCIATION (1984) *Design of Concrete Structures for Buildings with Explanatory Notes*, CSA A23.3-M84, Rexdale, Canada, 485 p.

LAPLANTE, P., AÏTCIN, P.-C. (1986) *Field Study of Creep and Shrinkage of Strength Concrete*, Proceedings, 4th RILEM International Symposium on Creep and Shrinkage of Concrete, Evanston, Ill. Aug., pp. 777-786.

WHITING, D. (1984) *In Situ Measurements of the Permeability of Concrete to Chloride Ions*, In Situ / Nondestructive Testing of Concrete, American Concrete Institute SP-82, , Detroit, pp. 501-524.

Applications of High-Performance Concrete in Columns and Piers

S. K. Ghosh[1]

Abstract

This paper discusses applications of high-performance concrete in columns of high-rise buildings and in bridge piers. In multistory building applications, the performance characteristics of interest have been high strength and/or high stiffness. In bridge applications, the performance characteristics of importance are high strength and enhanced durability.

Introduction

The tallest concrete building in the world in 1959 was 371 ft (179 m) tall. By 1989, the height of the tallest concrete building had risen to 969 ft (295 m). In the 1990s, even taller, record-setting concrete buildings are being built outside the United States. Many factors contributed to this dramatic increase in height. The most important amongst them were: (1) the development of efficient structural systems, and (2) the availability of high-strength concrete, which has invariably been used in the lower-story columns of these buildings. The second aspect is discussed in this paper.

The application of high-strength concrete in highly seismic regions has lagged behind its application in regions of low seismicity. One of the primary reasons has been a concern with the inelastic deformability of high-strength concrete structural members under reversed cyclic loading of the type induced by earthquakes. This paper discusses the current state of application of high-strength concrete in buildings in a number of U. S. cities that are subject to high seismic risks. It examines code requirements concerning the confinement of concrete columns that are part of the lateral force resisting system of a structure in a region of high seismicity. The adequacy of these requirements, when high-strength concrete is used in the columns, is commented on.

The application of high-performance concrete in bridges has been largely confined to the superstructure. Its usage in bridge piers is explored and discussed in this paper.

---

[1] Director, Engineering Services, Codes and Standards, Portland Cement Association, Skokie, Illinois 60077-1083

High-Performance Concrete in Multistory Buildings

Figure 1 shows a series of nine concrete buildings, each of which, with the exception of Two Prudential Plaza, was the tallest concrete building in the world at the time of its completion. It is clear that the growth in the height of concrete buildings has gone hand-in-hand with the availability of higher- and higher-strength concrete.

It is readily apparent that all the record-setting concrete buildings illustrated in Fig. 1 are in the United States. In fact, an amazing seven out of the nine buildings are located in Chicago, a city that in many ways has pioneered the evolution of high-strength concrete technology. However, in recent years, there has been an impressive spread in the availability of ultra-high-strength concrete (with specified compressive strength in excess of 10,000 psi or 69 MPa). Figure 2 shows that 12,000 psi (83 MPa) or higher-strength concrete has been used in the last three or four years in Atlanta, Cleveland, Minneapolis, New York, and Seattle.

With changing economic times, the title of the tallest concrete building in the world passed overseas for the first time in 1992 to Hong Kong when the 1227 ft (374 m) high Central Plaza was completed. Table 1 shows that two buildings taller than the Central Plaza are currently under construction: the twin Petronas Towers in Kuala Lumpur, Malaysia, and the Jin Mao Building in Shanghai.

Table 1: The Ten Tallest Buildings in the World

Building	City	Year	Stories	Height * m	Height * ft	Material	Use
1 Petronas Tower 1	Kuala Lumpur	UC95	98	450	1476	Mixed	Multiple
2 Petronas Tower 2	Kuala Lumpur	UC95	98	450	1476	Mixed	Multiple
3 Sears Tower	Chicago	1974	110	443	1454	Steel	Office
4 Jin Mao Building	Shanghai	UC98	88	418	1371	Mixed	Multiple
5 World Trade Center North	New York	1972	110	417	1368	Steel	Office
6 World Trade Center South	New York	1973	110	415	1362	Steel	Office
7 Empire State Building	New York	1931	102	381	1250	Steel	Office
8 Central Plaza	Hong Kong	1992	78	374	1227	Concrete	Office
9 Bank of China Tower	Hong Kong	1988	72	368	1209	Mixed	Office
10 Sky Central Plaza	Guangzhou	UC95	68	364	1193	Concrete	Office

* Height is measured from sidewalk level of main entrance to structural top of building (Television, radio antennas, and flag poles are not included).

Source: Council on Tall Buildings and Urban Habitat
            Lehigh University, Bethlehem, Pennsylvania

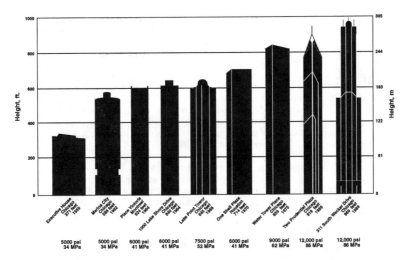

FIG. 1. High-Strength Concrete in High-Rise Construction

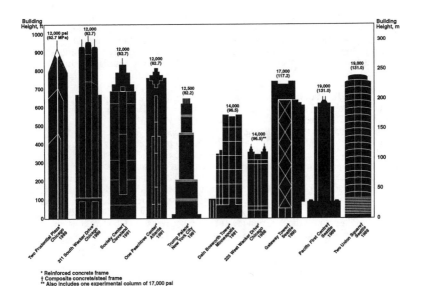

* Reinforced concrete frame
† Composite concrete/steel frame
** Also includes one experimental column of 17,000 psi

FIG. 2. Ultra-High-Strength Concrete Shapes New Skylines

The 78-story Central Plaza office tower rises above a three-level basement. The tower itself is composed of three sections: the base, forming the main entrance and public circulation areas; the tower with 57 office floors, a skylobby, and five mechanical plant floors; and the tower top, which includes six mechanical plant floors and a 335 ft (102 m) tall tower mast. Above the tower base, wind resistance is provided by the external facade frames acting as a tube. The core has a shearwall arrangement, and just above the tower base it carries approximately 10% of the total wind shear. Edge transfer beams of the tower base structure allow alternate columns to be dropped from the facade, thereby opening up the public area at ground level. The increased column spacing, together with the elimination of spandrel beams in the tower base, results in the external frame no longer being able to carry the wind loads acting on the building. Over the height of the tower base, the core transfers all of the wind shears to the foundations. A thick slab at the underside of the transfer beam transfers the total wind shear from the external frame to the inner core below. In the lower levels, concrete with a 28-day cube strength of 8700 psi (60 MPa) was used, with pulverized fuel ash as partial cement replacement to avoid the need for cooling. This was the first time that government approval was given for the use of such high-strength concrete in private sector work in Hong Kong.

The twin Petronas Towers of Kuala Lumpur City Centre, now under construction in Malaysia, will soar past Hong Kong's Central Plaza to 1476 ft (450 m) to the tip of the spires. Each tower will contain 98 stories including mezzanines and mechanical penthouse floors. Each tower has five levels below grade and is surrounded by a six-level podium base above grade level. The structural system for each tower consists of a large concrete core that has walls of varying thickness, and a cylindrical concrete tube with sixteen cast-in-place concrete circular columns. Eight outrigger walls link the core to the cylindrical tube at the four corners of the core at the 38th floor mechanical double story, to improve the efficiency of the overall system. Concrete strengths for the concrete core and cylindrical tube columns vary from 12 ksi to 6 ksi (80 MPa to 40 MPa). Because of the mass required to minimize overturning and the perception of motion, concrete stresses are generally low. High concrete strength is required, however, for high modulus of elasticity of up to 7000 ksi (48,300 MPa) for increased stiffness.

The Jin Mao Building is a multi-use development including office, hotel, retail, parking, and service spaces, which is currently under construction in Shanghai. The development consists of an 88-story, 1371 ft (418 m) tall tower with an adjacent low-rise building. The tower, one of the most slender in the world, has an aspect ratio of 8:1 when considering the full building height (7:1 to the last fully occupied floor). The only vertical elements of the system are an octagon-shaped reinforced concrete core wall, eight exterior composite mega-columns, and eight exterior steel mega-columns. The composite mega-columns (composed of reinforced concrete and structural steel) are linked to the core by eight structural steel outrigger trusses in two-story high spaces within the mid-height of the building and one multiple-story high space at the top of the building. The concrete strength for the core and mega-columns varies from C60 (7500 psi or 52 MPa) at the foundation to C40 (5000 psi or 34 MPa) at the top of the structure.

Advantages of High-Strength Concrete

The three biggest advantages of high-strength concrete that makes its use attractive in high-rise building are that it provides more
- strength / unit cost
- strength / unit weight
- stiffness / unit cost

than most other building materials including normal weight concrete.

Commercially available 14,000 psi (97 MPa) concrete costs significantly less in dollars per cu. yd. than 3 1/2 times the price of 4000 psi (28 MPa) concrete. In fact, the unit price goes up relatively little as concrete strength increases from 4000 to 10,000 psi (28 to 69 MPa). Thus, high-strength concrete gives the user more strength per dollar.

The unit weight of concrete goes up only insignificantly as concrete strength increases from moderate to very high levels. Thus, more strength per unit weight is obtained, which can be a significant advantage for construction in high seismic regions, since earthquake induced forces are directly proportional to mass.

The modulus of elasticity of concrete remains largely proportional to the square root of the compressive strength of concrete at the age of loading even for high-strength concrete. The user thus obtains a higher stiffness per unit weight and unit cost. Indeed, it is quite common for a structural engineer to consider and specify high-strength, high-performance concrete for its stiffness, rather than for its strength.

It is important to mention that the specific creep (ultimate creep strain per unit of sustained stress) of concrete decreases significantly as the concrete strength increases. This is indeed a fortunate coincidence without which the application of high-strength concrete in highrise buildings would have been seriously hampered. Because of the lower specific creep, high-strength concrete columns with their high stress levels suffer no more total shortening than normal-strength concrete columns with their lower strength levels. Otherwise, the problem of differential shortening of vertical elements within highrise buildings would have been aggravated by the use of high-strength concrete in columns.

High-Strength Concrete in Regions of High Seismicity

The highest concrete strength ever used in a building has been 19,000 psi (131 MPa) in the composite columns of Seattle's 62-story, 759-ft (231 m) high Two Union Square. Seattle is in Uniform Building Code Seismic Zone 3 (areas of little or no seismicity are in Zone 0; areas of the highest seismicity are in Zone 4). The strength was obtained by use of: what may be a record low water to cementitious materials ratio of 0.22 (this is the single most important factor in increasing strength and reducing shrinkage and creep); the strongest of available cements; a superplasticizer which reduces the need for water and provides the necessary workability; a very high cement content; a very strong, small (3/8 in. or 10 mm), round glacial aggregate available locally; silica fume (increasing strength by about 25%); a design strength obtained at 56 rather than the usual 28 days; and an extraordinarily thorough quality assurance program. The 19,000 psi (131 MPa) strength was the byproduct of the design requirement for an extremely high modulus of elasticity of 7.2 million psi (49,650 MPa). The stiffness was desired in order to meet the occupant-comfort criterion for the completed building. The same concrete strength was later used by the same structural engineering firm in the composite columns of the shorter 44-story Pacific First Centre.

Concrete with 9500 psi (66 MPa) compressive strength has been used at 600 California in San Francisco, and at 1300 Clay in Oakland, both composite buildings. Concrete with 8000 psi (55 MPa) compressive strength has been used in several all-concrete Bay Area buildings, including the 19-story Fillmore Building.

The spread in the use of high-strength concrete in Southern California has been hampered by the City of Los Angeles Code provision restricting the strength of concrete to a maximum of 6000 psi (41 MPa). Even then, concrete strength in excess of 6000 psi (41 MPa) has been used in the Great American Plaza office-hotel-garage complex in San Diego, a 14-story residential building at 5th and Ash in San Diego, and in the 22-story Pacific Regent (senior citizen housing) at LaJolla.

Confinement of High-Strength Columns

ACI 318 *Building Code Requirements for Reinforced Concrete* (1992) as well as the *Uniform Building Code* (1994) contains specific requirements concerning the confinement of concrete columns that are part of the lateral force resisting system of a structure in a region of high seismicity.

Column (factored axial compressive force on member $> A_g f'_c / 10$) flexural strength is determined such that the sum of the design flexural strengths of the columns framing into a joint (calculated for the factored axial forces, consistent with the direction of the lateral forces considered, resulting in the lowest flexural strengths) exceeds the sum of the design flexural strengths of the girders framing into the same joint by a factor of at least 1.2. This requirement is intended to result in frames where the flexural yielding of columns is restricted. Shear design is based on required shear strengths that correspond to the development of a moment at each column end that is equal to either (a) the maximum probable flexural strength of the section associated with the range of factored axial loads on the column, or (b) the column end moment corresponding to the development of probable flexural strengths at the ends of beams framing in. The required shear strength may never be less than the factored shear force determined from analysis of the structure. The configuration and spacing of the transverse reinforcement within the regions of potential plastic hinging at the two ends are established to confine the concrete core and to restrain the longitudinal compression bars from buckling (see Figure 3). Portions of the column outside of the regions of potential plastic hinging are also governed by specific transverse reinforcement requirements.

The following comments can be made based on available research:

Strength — The ACI rectangular stress block yields accurate strength prediction in flexure and axial compression as long as $f'_c \leq 8$ ksi (55 MPa). For $f'_c > 8$ ksi (55 MPa), in the absence of substantive enhancement in the current confinement requirements, accurate strength prediction for members subject to combined bending and axial compression would appear to require an adjustment of the ACI rectangular stress block. The latest New Zealand code (1995) has adopted a specific adjustment; the latest Canadian Standards Association standard (1994)

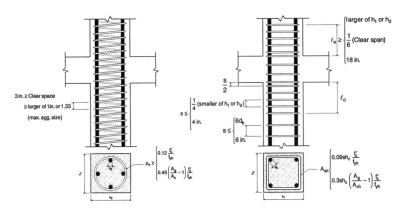

Confinement Requirements for Spirally Reinforced Columns   *Confinement Requirements for Tied Columns*

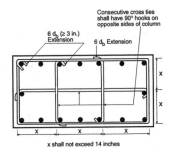

Transverse Reinforcement in Columns

FIG. 3. Code Requirements for Transverse Reinforcement in Reinforced Concrete Columns

has adopted a different adjustment. A slight variation of the Canadian adjustment has been proposed for adoption in ACI 318. Other modifications of the ACI rectangular stress block have also been proposed. These different options need to be evaluated, and a proper assessment made. As an alternative to the use of a modified stress block, accurate strength prediction should also be obtainable through sophisticated analysis using the confined core only, ignoring the cover concrete.

Inelastic Deformability — The degree of confinement of concrete required in a compression member depends on the axial load level and the performance criteria (e.g., what percentage of the axial load strength needs to be sustained up to what drift level). At low axial load levels, prevention of buckling of the longitudinal steel, rather than confinement of the compression concrete, becomes the primary function of the transverse reinforcement. The ACI 318 confinement requirements are independent of the axial load level. The ACI requirements are generally satisfactory for normal-strength concrete compression members ($f_c' \leq 8$ ksi or 55 MPa) at axial load levels up to 0.4 P_o. The same requirements cease to be satisfactory for high-strength concrete compression members ($f_c' > 8$ ksi or 55 MPa) at axial load levels lower than 0.4 P_o (but not lower than 0.2 P_o, probably not lower than 0.3 P_o).

Indications are that the higher degree of confinement required in high-strength concrete compression members subject to high axial load levels should be provided partly in the form of higher yield strength transverse reinforcement. In tests carried out by Mugurama (1990) at the University of Kyoto, high-yield strength transverse reinforcement ($f_y > 116$ ksi or 800 MPa) has been shown to be advantageous to axial load levels exceeding 0.4 P_o. It has been pointed out (ACI-ASCE Committee 441), however, that the amount of transverse reinforcement provided for Mugurama's test columns was as high as 230% of that required by the seismic provisions of ACI 318. For relatively small columns, Mugurama used a 12 longitudinal bar arrangement with an extremely congested scheme of transverse reinforcement.

One reason for using higher strength steel for transverse reinforcement in high-strength concrete columns is to allow larger spacing of ties. However, caution must be exercised in using this approach. Use of high-strength material for transverse reinforcement may result in relatively large tie spacing, while satisfying the seismic provision of ACI 318. This large spacing may lead to early buckling of the longitudinal bars. Further research is needed to establish the maximum spacing of transverse reinforcement permitted when very high-strength material is utilized for such reinforcement in high-strength concrete columns.

High-Performance Concrete in Bridge Piers

Jobse and Moustafa (1984) conducted a study to determine the potential for high-strength concrete in highway bridges. The study analytically and experimentally examined the advantages of using high-strength concrete in hollow prestressed square and circular piers. A wall thickness of 6 in. (150 mm) was used in all cases. A 15 ft (4.6 m) outside diameter hollow circular pier had a prestressing steel area of 20.2 in.[2] (13,000 mm[2]). A 10 ft (3 m) square hollow pier had a prestressing steel area of 18.4 in.[2] (11,900 mm[2]) concentrated in the corners.

Interaction diagrams were developed for each of these configurations for a variety of slenderness ratios. The benefit of increasing concrete strength was found to be more than directly proportional to the strength increase.

According to Jobse and Moustafa (1984): "High-strength concrete permits the adaptation of thin wall members to pier construction for major bridge spans. The strength of these sections permits their use when tall piers are required. The increased load carrying capability permits longer spans or fewer piers. The lighter section permits construction with minimum disruption to surrounding terrain."

A state-of-the-art survey conducted by Poston et al. (1986) indicated that multiple-column bents accounted for 83 percent of the piers used in U. S. bridges and single-column bents accounted for 17 percent. The use of solid cross sections was much more prevalent than hollow cross sections. However, the study found that the use of hollow cross-sections increased dramatically with height. The use of high-strength concrete in lieu of normal-strength concrete reduces the column cross section and stiffness. The magnification of moment due to this increase in slenderness of compression members not braced against sidesway must be considered in their design.

Chen and Van Lund (1988) conducted analytical design studies of reinforced high-strength and normal-strength concrete bridge piers and columns subjected to static and seismic loading. Flexible and rigid foundation systems were incorporated into the analysis. It was observed that the use of high-strength concrete in bridge piers and columns supported by rigid foundation systems can potentially reduce column shears and moments due to seismic loading. Savings of 33 percent in the material cost of the piers may be possible when high-strength concrete is used in lieu of normal-strength concrete. Additional economy is achieved because smaller bearings and foundations can be used.

For piers and columns that are supported by flexible foundation systems, high-strength concrete will not reduce column shears and moments due to seismic loading as much as with rigid foundation systems. However, cost savings will still be possible because less material is required in the columns.

Because foundations represent a significant portion of the cost of a bridge, greater savings can be realized by using high-strength concrete in the piers and columns of bridges supported by rigid foundations rather than in those supported by flexible foundations.

In order to quantify the potential benefits of using high-strength concrete in reinforced concrete pier columns, a complete column analysis was performed by Zia, Schemmel and Tallman (1989) for a specific set of cross sections and concrete compressive strengths. Rectangular and circular short column sections, both solid and hollow, were considered. Biaxial bending was examined in the case of the rectangular column sections. Twenty-eight day compressive strengths ranged from 6000 psi (41 MPa) to 12,000 psi (83 MPa). The primary goal of this investigation was to determine the range of application of various column cross-sections and concrete compressive strengths. Those combinations of cross sections and concrete strength which demonstrated a potential to provide superior benefits were further examined for their cost-effectiveness.

The use of high-strength concrete in the column members increased their axial load capacity almost in proportion to the increase in the compressive strength. The flexural capacity in the tension control region is unaffected by the increase in concrete strength. In the compression control region, the increase in flexural capacity is related to the level of axial load. The analysis of the piers indicated that the cost to carry a specific level of axial load decreases as the concrete strength is increased. In all cases, the most cost-effective designs were those which used 12,000 psi (83 MPa) concrete.

Zia et al. (1989) have pointed out that the improved durability of high-strength concrete provides additional benefits. Since high-strength concrete is less permeable than normal-strength concrete, deterioration of the concrete and corrosion of the reinforcement is reduced. These are important advantages for bridge piers, as this type of element is likely to have direct contact with seawater or other corrosive agents.

In spite of the above advantages, the use of high-performance concrete in bridge piers is only beginning. The new bridge over the Elron (La Technique 1994) on national road 165 between Brest and Quimper (Finistère) in Brittany, France, with a total length of 2625 ft (800 m), has five spans. The largest, in the center, is 1312 ft (400 m) long. Flanking the central span, two towers having a total height of 384 ft (117 m), of which 272 ft (83 m) is above the deck, support sets of cable stays. The towers are made of HPC 80 high-performance concrete (compressive strength = 80 MPa or 11,600 psi) except for their top quarter, made of HPC 60 concrete (compressive strength = 60 MPa or 8700 psi). The Normandy Bridge over the river Seine at its very mouth (La Technique 1994) has the longest cable-stayed span in the world, 2800 ft (856 m) long. The towers of the Normandy Bridge are 665 ft (203 m) high. They have an inverted Y shape. The legs are 398 ft (121 m) high, and are connected at two levels: at level of foundations by a prestressed tie-beam and at deck level by a very strong transverse beam into which the deck is embedded. Above the connection level of the legs, the tower consists of a head 267 ft (81 m) high where 23 pairs of stays are anchored. Concrete for the legs is high-performance concrete B60 (compressive strength = 60 MPa or 8700 psi).

Conclusions

The usage of high-strength, high-performance concrete in the columns of multi-story buildings and in bridge piers (or towers of cable-stayed bridges) has strong economic advantages associated with it. Such usage is already widespread in tall buildings, but not so prevalent in bridges.

References

ACI-ASCE Committee 441 (to be published). *High-Strength Concrete Columns: State of the Art.* American Concrete Institute, Detroit, Mich.

ACI Committee 318 (1992). *Building Code Requirements for Reinforced Concrete and Commentary, ACI 318-89 (Revised 1992) and ACI 318R-89 (Revised 1992).* American Concrete Institute, Detroit, Mich.

Chen, R. L., and Van Lund, J. A. (1988). "Seismic Design of High-Strength Concrete Bridge Piers and Columns." *Transportation Research Record 1180*, Transportation Research Board, Washington, D.C., 49-58.

Concrete Design Standard, NZS 3101: 1995, Part 1 and *Commentary on the Concrete Design Standard, NZS 3101: 1995, Part 2*, Standards Association of New Zealand, Wellington, N. Z.

CSA A23.3-94 Design of Concrete Structures (1994). Canadian Standards Association, Rexdale, Ontario, Canada.

Jobse, H. J., and Moustafa, S. E. (1984). "Applications of High Strength Concrete for Highway Bridges." *J. Prestressed Concrete Institute*, 29(3), 44-73.

La Technique Francaise de Béton Précontraint (1994). XIIᵉ Congrès de la Fédération Internationale de la Précontrainte, Washington, D.C., 263-270, 595-602.

Mugurama, H., and Watanabe, F. (1990). "Ductility Improvement of High-Strength Concrete Columns with Lateral Confinement." *Special Publication 121*, American Concrete Institute, Detroit, MI, 47-60.

Poston, R. W., Diaz, M., and Breen, J. E. (1986). "Design Trends for Concrete Bridge Piers." *J. American Concrete Institute*, 83(1), 14-20.

Uniform Building Code (1994) Vol. 2. International Conference of Building Officials. Whittier, Calif.

Zia, P., Schemmel, J. J., and Tallman, T. E. (1989). *Structural Applications of High Strength Concrete*. Report No. FHWA/NC/89-006, Department of Civil Engineering, North Carolina State University, Raleigh, N.C.

Lateral Load Behavior of
High Strength Fiber Reinforced Concrete Columns

Ramesh Nagarajah[1], ASCE Student Member
David H. Sanders[2], ASCE Assoc. Member

Abstract

The paper presents the results of an experimental investigation on the flexural, shear and ductility capacity of high strength concrete columns with and without steel fibers. As part of the investigation, three 2/3 scale test columns were subjected to constant axial load and cyclic lateral loads. The primary variables considered in this study were the spacing of lateral steel and the crimped fiber volume fraction (i.e. spacings of 76 mm (3 in.) and 102 mm (4 in.), and crimped fiber volume fractions of 0%, 0.75% and 1.25%). The fibers had an aspect ratio of 70. The ductility ratio and maximum compressive strain of confined high strength concrete for different percentages of fibers and lateral volumetric ratio were compared. It was observed from the experimental studies, that the addition of steel fibers increased the ductility ratio. Also, steel fibers maintained the strength and strain at peak load of the confined reinforced concrete columns even when the percentage of lateral steel was reduced. Columns with high confinement and no steel fibers, and columns with lower confinement and steel fibers performed well. All of the columns failed in flexure and shear did not control.

Introduction

Many studies have demonstrated the economy of using high strength concrete (HSC) columns in high rise buildings, however the use of HSC has not been limited to high rise buildings [8]. Studies indicate that HSC also enhances the economy of structures, in addition to reducing column size

---

[1]Design Engineer, Willdan Associates, Sacramento, California.
[2]Assistant Professor, Dept. of Civil Engineering, University of Nevada Reno.

and producing a more durable material [12]. In recent years, there has been some concern regarding the use of high strength concrete columns in seismic areas. The application of HSC in high seismic regions has lagged behind its application in low seismic regions. One of the primary reasons has been the concern over the ductility of HSC columns. When short randomly distributed fibers are added to reinforced concrete, the fibers improve the integrity of the material in addition to the tensile property of the concrete [7,9]. In addition, HSC columns with axial loads above 20%P_o have had problems reaching the moment capacity predicted based on the rectangular stress block [3]. It has been observed, that the addition of fibers improves the strength and strain at peak load of confined reinforced concrete columns. Several studies have investigated the impact of steel fibers on beam–column joints [10,15]. Each of these studies and others have found fibers to increase energy absorption capacity and decreased the amount of damage in the joint for a given loading. The studies have also indicated that the amount of transverse reinforcement can be reduced if steel fibers are added. Studies on reinforced concrete beams with steel fibers [11,14] found an increase in shear strength by including steel fibers. Therefore, the addition of steel fibers should be able to improve the ductility and shear strength of HSC columns. However, information on the ductility and shear strength of HSC fiber reinforced columns has been limited.

Test Specimens

Three columns were tested. Each had the same size and shape as shown in Figures 1 and 2. Their cross section was 305 mm by 305 mm (12 by 12 in.). Height from the top of the footing to the lateral load point was 1.83 m (6 ft) . The footing had a cross section of 2.13 m by 2.13 m (7 by 7 ft) and a height of 610 mm (2 ft). The specimens were designed and reinforced so that hinging would occur in the column rather than in the footing.

A combination of fly ash, silica fume slurry, superplasticizer, water retardant and steel fibers were used to obtain a concrete mix, that achieved the desired high strength for the columns. Nevada Type II cement was used with Class F type fly ash. The amount of fly ash was 15% by weight of the cement, while the amount of silica fume slurry was 10% of the total weight of the mix. Grace Daratard 17 and WRDA19 were each added at 50 oz per cubic yard of concrete. The mix proportions for all the specimens were the same except the percentage of fibers. One column was cast without steel fibers (A1), while the other two columns (A2 & A3) were cast with 0.75% and 1.25% fiber respectively. The steel fibers were crimped and had a length of 51 mm (2 in.) and a thickness of 0.7 mm (0.028 in.) with an aspect ratio of 70. Crimped fibers were used because of their good performance and workability [13]. Nominal maximum aggregate size was 10 mm (3/8 in.). All of the testing and construction was performed at the Bridge Engineering Laboratory of the University of Nevada, Reno. The footings were cast with the same materials used for the columns, but without silica fume slurry and steel fibers. Concrete for the footing was supplied by a local concrete ready mix company whereas column concrete was mixed in the laboratory. Target 56-day concrete strengths were 83 MPa (12 ksi) for the columns and 55 MPa (8 ksi) for the footings. Column concrete compressive strengths obtained from the testing of 150 mm x 300 mm (6 in. by 12 in.) cylinders are given in Table 1. Confining reinforcement more than specified by some codes was used in specimen A1 and A2 to attempt to obtaint the moment capacity predicted by the rectangular stress block and to improve ductility and drift (see Table 1). Reaching this level of load and high levels of ductility has been a problem with HSC columns without fibers. Specimen A3 was used to determine if a lower level of confinement with steel fibers would be adequate.

Table 1. Summary of Specimen Parameters

Specimen	f'c MPa (ksi)	Tie Spacing mm (in)	Steel Fiber %	% of Lateral Steel Compared to		
				ACI[2]	AASHTO[1]	CALTRANS[4]
A1	78.6 (11.4)	76 (3)	0	133	100	137
A2	82.1 (11.9)	76 (3)	0.75	127	95	133
A3	84.1 (12.2)	102 (4)	1.25	107	80	98

Longitudinal reinforcement type and arrangement were the same for the three specimens (see Figure 2). Grade 400 MPa - 19 mm (Grade 60 #6 bars) were used for longitudinal reinforcement. The average yield strength of the bars was found to be 480 MPa (69.6 ksi) by testing two bars in a uniaxial tensile testing machine. Grade 400 - 13 mm (Grade 60 #4) bars were used as lateral reinforcement. Square ties with 135 degree bends and cross ties with one end of 90 degree and the other with 135 degree bend were used. The hooks were alternated from side to side. Spacing of the lateral ties was varied from 76 mm (3 in.) to 102 mm (4 in.). The yield strength of the lateral bars was measured as 471 MPa (68.3 ksi), by testing two bars. Design clear cover was 19 mm (0.75 inches). Table 1 shows the summary of the specimen parameters.

Test Setup and Loading Configuration

Test setup and loading arrangements are shown schematically in Figure 3. The test setup for each specimen consisted of placing the test column in a vertical cantilever position. The footing was post-tensioned to the lab strong floor of the lab with four prestressing bars each with a force of 111 kN (25 kips). A steel I beam was placed on top of the column to transmit the axial and lateral loads. Axial load was applied with prestressing bars and two hydraulic jacks each with a capacity of 890 kN (200 kips). A hydraulic accumulator was connected to the hydraulic jacks to maintain a constant axial load during the test. Lateral loads were applied to the specimen with a 490 kN (110 kips) MTS hydraulic actuator with one end connected to the steel beam and the other end fixed to reaction blocks that were post-tensioned to the floor. The actuator had a stroke of $\pm$ 280 mm ($\pm$11 in.). The test setup was designed so that the specimen was placed in the middle of the actuator stroke. Therefore during the test, the maximum available displacement was $\pm$ 280 mm ($\pm$11 in.).

Instrumentation

Several types of instrumentation were used to obtain load-displacement, moment curvature, longitudinal bar strain profiles, hoop strains, cross tie strains and maximum concrete compressive strains. Actuator lateral load and horizontal displacement were measured with an actuator load cell and a linear variable differential transformer (LVDT) respectively. Axial load was measured with a load cell placed under one jack. Since the axial load jacks were identical and hooked to the same hydraulic system, one load cell was assumed sufficient. Internal measurements included strains on the longitudinal and lateral column reinforcement in the potential plastic hinge region. A total of 22 strain gages were applied to each specimen. LVDT's were used to measure axial deformation in and near the plastic hinge region. The LVDT's were attached to brackets secured on both sides of the specimen by a series of threaded rods embedded in the specimen. A strain distribution across the column section was obtained by considering a pair of LVDT's measuring longitudinal displacements at the same height on each side of the column. These measurements were used to calculate specimen rotation and curvature.

Test Procedure

Each column was subjected to a constant axial load and increasing levels of lateral displacements (see Figure 4). The axial load was applied first and kept constant at 1330 kN (300 kips), 20 percent of the column axial load capacity (P_o). The first loading cycle was used to establish the first yield displacement, Δ_y, as shown in Figure 5 . In the first lateral load cycle, the load was increased to the point where the applied moment at the column footing interface was 75% of the moment capacity of the column section calculated based on ACI318 [2] building code procedures, using a ϕ

factor of 1.0. The displacement at this load was then multiplied by 4/3 to obtain the yield displacement. Multiples of the first yield displacement (2 cycles at each displacement increment) were applied to the test columns to produce displacement ductility as shown in the loading schedule given in Figure 4. Displacement ductility at each cycle is defined as the ratio of peak lateral displacement divided by the first yield displacement. Testing was stopped when the specimen had a significant drop in lateral load capacity for a given displacement.

General Behavior

All three specimens exhibited good energy dissipation capabilities up to the last loading cycle. Specimens A2 and A3 (76.2 mm spacing) failed by fracturing of longitudinal bars and specimen A3 (102 mm spacing) by buckling of longitudinal bars. The plastic hinge formed just above the footing column interface. Length of the plastic hinge for specimen A1 (no steel fibers) was approximately 250 mm (10 in.), while for the two specimens with steel fibers (A2 and A3), it was approximately 200 mm (8 in.). With the addition of fibers, the spreading of the plastic hinge was reduced. This could help to reduce the level of maximum displacements. There was no damage to longitudinal or lateral reinforcement in the upper portion of the column beyond the plastic hinge length. In general, the first crushing of cover concrete was observed during the first cycle at a displacement ductility ratio of 1.5. At this ductility ratio, for the specimen without steel fibers, small pieces of cover concrete spalled in the plastic hinge region. In specimens with steel fibers, even though crushing was observed, the cover concrete stayed intact. Significant spalling of concrete cover was observed at ductility ratios of three and four. The amount of spalled concrete was much less in the two specimens with steel fibers compared to the specimen without steel fibers. Overall, there was a slight drop in horizontal load carrying capacity of the test columns as the cover concrete crushed. However, cover concrete crushing did not create a loud noise or brittle behavior as has been reported in the testing of HSC columns subjected to concentric axial loading alone [3,5]. In the specimens with steel fibers there was no significant opening of the peripheral hoops or cross ties after spalling. However, for the specimen without steel fibers, there was evidence of slight opening of peripheral hoops and cross ties at two layers by the conclusion of testing. A total of 14, 18 and 16 cycles of horizontal loading was applied to the specimens A1, A2 and A3 respectively.

Load versus Displacement

Figure 6 shows the lateral load versus horizontal displacement obtained for the specimen A2. In Figure 7, envelopes of the points corresponding to maximum lateral loads at increasing lateral displacement levels for all three specimens are shown. It can be seen from the envelopes, that the specimen with 0.75% steel fiber and minimum spacing sustains slightly more load and deflection. Specimen A3, with 1.25% of fiber and 102 mm spacing behavior is approximately the same as the specimen without steel fibers (102 mm spacing). It is also observed that the lateral load carrying capacity is almost the same for all the specimens.

Table 2 gives the first yield displacement (Δ_y) based on estimation as shown in Figure 5 , the maximum displacement (Δ_{max}) at which two complete cycles were applied when there was not a significant drop in capacity, and the maximum displacement ductility ratios (μ) for each column. The maximum ductility ranged from 5 to 6. It can be seen from the table that the estimated Δ_y using the procedure illustrated in Figure 5, is same for all the three specimens. Since the axial load is constant and the longitudinal reinforcement arrangement is typical for all three specimens, the estimated nominal moment is the same; this in turn gives approximately the same first yield displacement. Previous research has shown that the ACI 318 [2] provisions, assuming a rectangular stress block overestimates the horizontal load-carrying capacity (or, equivalently, the nominal moment capacity) for high strength concrete where the axial loads are at least $0.2P_o$[3]. No such reduction was observed since the concrete was highly confined. The highly confined high strength concrete has a stress-strain curve that is parabolic in shape. This parabolic shape and non brittle failure of the concrete allowed the rectangular stress block to conservatively predict the moment capacity. Therefore, Δ_y was based on Mn_{ACI} (255 kN-m, 139 kN) (188 ft-kips, 31 kips).

Table 2. Displacement and Displacement Ductility

Specimen	Δ_y mm (in)	$\overset{*}{\Delta}_y$ mm (in)	Δ_{max} mm (in)	$\mu = \Delta_{max}/\Delta_y$	$\overset{*}{\mu} = \Delta_{max}/\overset{*}{\Delta}_y$	Drift %
A1	31.2 (1.23)	37.1 (1.46)	177 (6.97)	5.6	4.8	9.6
A2	31.5 (1.24)	29.2 (1.15)	191 (7.52)	6.11	6.5	10.4
A3	31.5 (1.24)	31.5 (1.24)	192 (7.57)	6.00	5.7	10.5

* Strain Gage Measured Values

Table 2 compares the strain gage measured first yield displacements ($\overset{*}{\Delta}_y$) for these columns, which varied for each specimen, with those determined from the calculated moment capacity. It can be seen that the strain gage measured first yield displacement was higher for specimen A1 than specimen A2. Specimen A3 has a value between the other two specimens. The strain gage measured first yield displacement results in lowering the maximum displacement ductility ratio for specimen A1 and increasing the displacement ductility ratio for specimen A2. Also, the ductility ratio of specimen A3 is slightly reduced. Fibers seem to delay the first yield value. The is probably due to the bridging action of the strain gages over the cracks in the plastic hinge region.

Specimens A2 and A3 reached higher maximum displacement than specimen A1 (see Table 2). This behavior of specimen A2 was due to the combined effect of a high lateral steel percentage and steel fibers, which helped the material to behave in a ductile manner. It is also seen, that even after increasing the spacing for specimen A3, the column behaved similar to that of specimen A1 and A2, due to the presence of 1.25% fibers. All of the columns had excellent drift ratios.

Strain Distribution

The yielding of the lateral reinforcement was observed at a displacement ductility ratio of 3. The yield strain in the ties was reached at relatively high displacement (152 mm (6 in.)) almost at the end of the testing. The longitudinal bars yielded at a ductility ratio of 1. Table 3 gives the maximum compressive strain achieved for each test column prior to the first observed crushing of cover concrete. These strains are the calculated strains from the measured curvatures within the first 250 mm (10 in.) above the footing-column interface. Specimen A2 attained higher strains than specimen A1 due to the presence of steel fibers and a higher percentage of lateral reinforcement. Specimen A3 also attained higher strains than specimen A1 but less than that obtained by specimen A2.

Table 3.   Maximum Compressive Strain Just Prior to Cover Concrete Cracking

Specimen	Maximum Compressive Strain
A1	0.150
A2	0.200
A3	0.156

Shear Capacity

The columns did not show any shear distress. All of the columns had a maximum applied shear of approximately 142 kN (32 kips). According to a CALTRANS procedure that takes into account axial load and plastic hinging effects [5], the shear capacity of the column with 76 mm (3 in.)

spacing was 660 kN (148 kips), while the capacity was 500 kN (112 kips) for the column with 102 mm (4 in.) spacing. Only 27 kN (6 kips) of the shear capacity is provided by the concrete. The lateral steel was primarily for confinement, not shear.

Summary and Conclusions

Behavior of three confined high strength concrete columns with and without steel fibers, subjected to constant axial load and cyclic lateral loads was studied. The effects of confinement and influence of steel fibers on the strength and ductility of the material were examined. Based on the experimental studies, the following conclusions were drawn:

1.   By introducing steel fibers, partial replacement of confining reinforcement can be achieved; this will result in reducing congestion of reinforcement. Increasing the spacing of the lateral ties did not affect the overall capacity of the specimens but did change the failure mode. In the columns with the 76 mm (3 in.) tie spacing, the longitudinal bars fractured, while in the column with the tie spacing of 102 mm (4 in.), the longitudinal bars buckled. All of the test columns reached very high drift levels. More testing is needed to establish how much of an increase in spacing can be achieved for a given percentage of steel fibers and still have adequate performance.

2.   Introducing fibers improves the integrity, dimensional stability and performance of the material at higher stages of lateral loads. A slight increase in ductility was observed when steel fibers were added. It was observed that the spalling of cover concrete was delayed by introducing fibers.

3.   By using higher yield longitudinal bars in the higher lateral steel specimens, higher capacity could be achieved since the bars fractured. It is recommended that further research introducing different grade steels with different levels of axial loads and varying percentages of fibers be conducted.

Acknowledgments

Financial support from NSF is gratefully acknowledged under NSF Cooperative Agreement OSR-9353227. Student support from the Department of Energy is also acknowledged.

References

1.   American Association of State Highway and Transportation Officials

(AASHTO), *Standard Specifications for Highway Bridges*, Fifteenth Edition, Washington, D.C. 1992.

2. *American Concrete Institute Committee 318, Building Code Requirements for Reinforced Concrete and Commentary*, American Concrete Institute (ACI), Detroit, Michigan, 1992.

3. Aziznamini, A. et al, "Seismic Behavior of Square High-Strength Concrete Columns", *ACI Structural Journal*, Vol. 91, No. 3, May-June 1994, pp. 336-345.

4. California Department of Transportation, *Bridge Design Specification, 1983 AASHTO with Revisions by CALTRANS*, Sacramento, CA.

5. California Department of Transportation, *Memo to Designers*, March 1995, pp B11-B12.

6. Cusson, D. and Paultre, P., "Stress-Strain Model for Confined High-Strength Concrete", *ASCE Journal of Structural Engineering*, Vol. 121, No. 3, March 1995, pp. 468-477.

7. Fanella, D.A. and Naaman, A.E., "Stress-Strain Properties of Fiber Reinforced Mortar in Compression", *Journal of American Concrete Institute*, Vol. 182, No. 4, July-August 1985, pp. 475-483.

8. Fiorato, A.E, "PCA Research on High-Strength Concrete", *Concrete International*, Vol. 11, No. 4, April 1989, pp. 44-50.

9. Ganesan, N. and Ramana Murthy, J.V., "Strength and Behavior of Confined Steel Fiber Concrete Columns", *ACI Materials Journal*, Vol. 87, No. 3, May-June 1990, pp. 221-227.

10. Gefken, P and Ramey, M., "Increased Joint Hoop Spacing in Type 2 Seismic Joints Using Fiber Reinforced Concrete", *ACI Structural Journal*, Vol. 86, No. 2, March-April 1989, pp 168-172.

11. Sharma, A.K., "Shear Strength of Steel Fiber Reinforced Concrete Beams", *ACI Journal*, Vol 83, No. 4, July-August 1986, pp 624-628.

12. Smith, G.J. and Rad, F.N., "Economic Advantages of High Strength Concrete in Columns", *Concrete International*, Vol. 11, No. 4, April 1989, pp. 37-43.

13. Soroushian, P. and Bayasi, Z., "Fiber-Type Effects on the Performance of Steel Fiber Reinforced Concrete", *ACI Materials Journal*, Vol. 88, No. 2, March-April 1991, pp. 129-135.

14. Tan, K.H., Murugappan, K. and Paramasivam, P., "Shear Behavior of Steel Fiber Reinforced Concrete Beams", *ACI Structural Journal*, Vol. 90, No. 1, January-February 1990, pp 3-11.

15. Tang, J et al, "Seismic Behavior and Shear Strength of Framed Joint Using Steel-Fiber Reinforced Concrete", *Jounral of Structural Engineering*, ASCE, Vol. 118, No. 2, February 1992, pp 341-358.

16. Thomsen, J.H. and Wallace, J.W., "Lateral Load Behavior of Reinforced Concrete Columns Constructed Using High Strength Materials", *ACI Structural Journal*, September-October 1994, Vol. 91, No. 5, pp 605-615.

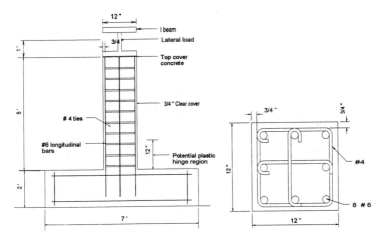

Figure 1. Test Specimen Elevation View

Figure 2. Cross Section

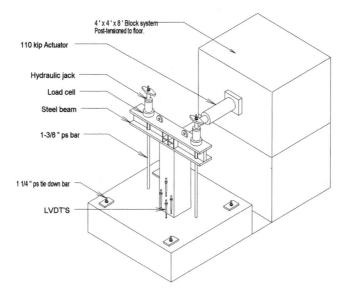

Figure 3. Test Setup

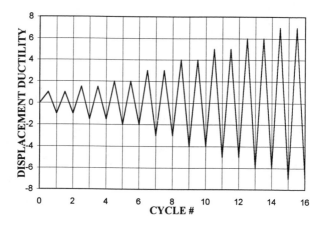

Figure 4. Lateral Loading Pattern for Specimen A2

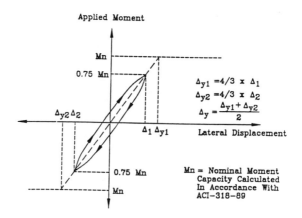

Figure 5. Definition of Yield Point [3]

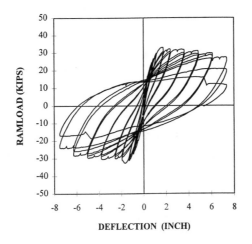

Figure 6. Hysterisis Curve for Specimen A2

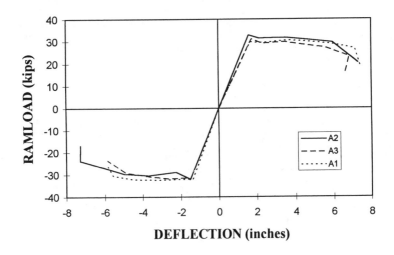

Figure 7. Displacement Envelops for All Specimens

Maximum Shear Strengths of Reinforced Concrete Structures

Li-Xin "Bob" Zhang[1] and Thomas T. C. Hsu[2]

Abstract

The maximum shear strength of reinforced concrete structures is related to the compressive strength of concrete, f_c'. How to express this relationship, however, has been controversial for the past 40 years. The ACI Code expresses the maximum shear strength as a function of $\sqrt{f_c'}$, but the equilibrium truss model (CEB-FIP Code, 1978) assumed the maximum shear strength to be directly proportional to f_c'. In order to establish the facts, full-size reinforced concrete membrane elements (panels) were tested at the University of Houston using concrete strengths of 42 MPa, 65 MPa and 100 MPa. These tests indicate that the softened coefficient of diagonal concrete struts is inversely proportional to $\sqrt{f_c'}$. This relationship can be used in the softened truss model to predict maximum shear strength as proportional to $\sqrt{f_c'}$. This conclusion supports the ACI Code.

Introduction

In 1962, ACI-ASCE Committee 326 proposed that the maximum shear strength, $v_{u, max}$, for a beam with web reinforcement is proportional to $\sqrt{f_c'}$, where f_c' is the compressive strength of standard concrete cylinders. This proposal was based on an extensive analysis of the results obtained from testing 166 beams with the concrete strengths from 10 to 50 MPa (1,500 to 7,000 psi). The committee then recommended conservative values for the maximum shear stresses $v_{u,max}$ as follows: $10\sqrt{f_c'}$ (f_c' in psi) for T-beams and beams with diagonal stirrups; and $8\sqrt{f_c'}$ for beams with rectangular cross sections and with vertical stirrups. Since both of these maximum shear stresses were conservative, a single value of $10\sqrt{f_c'}$ was adopted in the 1963 ACI Code and remained thereafter.

---

[1]Senior Structural Engineer, Barnett & Casbarian, Inc., 9225 Katy Freeway, Suite 307, Houston, TX 77024

[2]Professor, Department of Civil & Environmental Engineering, University of Houston, Houston, TX 77204

The 1960's also saw the development of the equilibrium truss model (Thurlimann, 1979) to analyze reinforced concrete structures. A reinforced concrete element may fail by the yielding of steel when the steel ratio is small or by the crushing of concrete when the steel ratio is large. The maximum shear strength, $\tau_{\ell tm}$, is then determined by assuming the concrete struts to fail before the yielding of steel. From the equilibrium condition, $\tau_{\ell tm}$ can be related to the peak compressive stress of concrete struts σ_p as follows:

$$\tau_{\ell tm} = -\sigma_p \sin\alpha\cos\alpha \tag{1}$$

Angle α in Eq. (1) is the angle between the longitudinal steel (ℓ-axis) and the post-cracking principal compressive stress of concrete (d-axis). Since the peak compressive stress of concrete struts σ_p was initially thought to be proportional to f'_c, Eq. (1) would then indicate that the maximum shear strength $\tau_{\ell tm}$ is also proportional to f'_c.

The equilibrium truss model was employed by many early investigators to analyze the web crushing of flanged beams. One of the earliest studies for beams with web crushing failure was reported by Robinson and Demorieux (1961). The peak compressive stress σ_p at web crushing failure was found to be $0.64f'_c$. The later investigations of Leonhardt and Walther (1964), Robinson and Demorieux (1968), Bennett and Balasooriya (1971), Lyngberg (1976) and Campbell et al. (1979) all expressed σ_p as f'_c times a constant. This approach, therefore, was adopted by the CEB-FIP Model Code (1978), and the constant was chosen to be 0.6.

The peak compressive stress σ_p in Eq. (1) can be expressed as f'_c times a coefficient ζ:

$$\sigma_p = \zeta f'_c \tag{2}$$

ζ is called the softened coefficient in the softened-truss model (Hsu, 1993). This softened coefficient ζ is a constant in the 1978 CEB-FIP Model Code, but is inversely proportional to $\sqrt{f'_c}$ in the ACI Code. This difference created a controversy that continued up to the present time.

The ACI Code provision received additional support from Placas and Regan (1971), who carried out extensive studies on the shear resistance of T-beams, I-beams, and rectangular beams, all made of normal strength concrete. From the specimens that failed in web crushing, they derived the softened coefficient ζ as:

$$\zeta = \frac{25 + 500r}{\sqrt{f'_c}} \quad (f'_c \text{ in psi}) \tag{3}$$

It is notable that the coefficient ζ in Eq. (3) is inversely proportional to $\sqrt{f'_c}$. The presence of the variable r is not surprising, because it is an indirect reflection of the principal tensile strain ε_r. At failure of the beam, the severity of cracking can be expected to decrease (lower ε_r) with the increase of web reinforcement (higher r).

Nielsen et al. (1978) studied the correlation between the softened coefficient and the web width, reinforcement details, and the concrete strength. Their test results showed that there was a minor influence on the softened coefficient from the width of web and reinforcement details, but a large influence from the concrete strength. For practical purposes, therefore, the softened coefficient was expressed only as a function of the concrete strength. An expression for evaluating the average value of ζ was given by

$$\zeta = 0.7 - \frac{f_c'}{200} \quad (f_c' \text{ in MPa}) \tag{4}$$

In Equation (4) the softened coefficient ζ is a function of the concrete strength f_c'. This equation was adopted in 1988 by the Architectural Institute of Japan for the earthquake design of reinforced concrete buildings (A.I.J., 1988) and in 1991 by the Eurocode (EC2, 1991).

All the softened coefficients ζ mentioned above were determined from web crushing tests of beams subjected not only to shear but also to bending, axial loads and boundary restraints. This study pursues a clearer understanding of shear behavior alone by subjecting reinforced concrete membrane elements to uniform shear stresses. Here, the softened coefficient ζ will first be evaluated from shear elements (panels). Then the new softened coefficient ζ will be used in the softened truss model to predict the maximum shear strength of elements and beams.

Compressive Stress-Strain Curve of Normal Strength Concrete

At the University of Houston, a parabolic equation was proposed (Pang and Hsu, 1995) for the compressive stress-strain curve of normal strength concrete:

$$\sigma_d = \zeta_\sigma f_c' \left[2\left(\frac{\varepsilon_d}{\zeta_\varepsilon \varepsilon_0}\right) - \left(\frac{\varepsilon_d}{\zeta_\varepsilon \varepsilon_0}\right)^2 \right] \quad \frac{\varepsilon_d}{\zeta_\varepsilon \varepsilon_0} \le 1 \tag{5a}$$

$$\sigma_d = \zeta_\sigma f_c' \left[1 - \left(\frac{\varepsilon_d/\zeta_\varepsilon \varepsilon_0 - 1}{2/\zeta_\varepsilon - 1}\right)^2 \right] \quad \frac{\varepsilon_d}{\zeta_\varepsilon \varepsilon_0} > 1 \tag{5b}$$

where ε_0 is the concrete cylinder strain corresponding to the cylinder strength f_c'. The symbols ζ_σ and ζ_ε are the stress-softened coefficient and the strain-softened coefficient, respectively. Equation (5) with different values of stress-softened coefficient ζ_σ and strain-softened coefficient ζ_ε is sketched in Fig. 1.

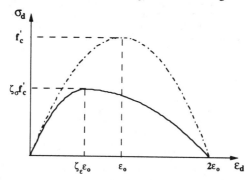

Figure 1 Proposed Stress-Strain Curve of Concrete

The stress-softened coefficient ζ_σ is defined as $\zeta_\sigma = \sigma_p/f_c'$, i.e. the ratio of the peak compressive stress σ_p of the panel to the companion cylinder strength f_c'. The strain-softened coefficient ζ_ε is defined as $\zeta_\varepsilon = \varepsilon_p/\varepsilon_0$, i.e. the ratio of the peak compressive strain ε_p (corresponding to σ_p) of the panel to the peak compressive strain ε_0 of the companion cylinder. These two softened coefficients, ζ_σ and ζ_ε, have been determined from tests of reinforced membrane elements (or panels) with longitudinal reinforcement perpendicular to the direction of principal compressive stress (90° panels) as shown in Fig. 2.

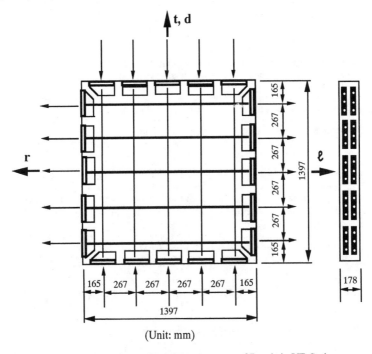

(Unit: mm)

Figure 2 Dimension and Steel Arrangement of Panels in VE-Series

The two softened coefficients, ζ_σ and ζ_ε, had been obtained for normal strength concrete under sequential loading and proportional loading (Belarbi and Hsu, 1995). The effect of load paths on the concrete compressive softening was observed by comparing two series of 90° panels. In one series of panels loaded sequentially, the principal tensile stress was first applied up to a target level, then followed in sequence by the principal compressive stress up to failure. In another series of panels loaded proportionally, the principal tensile stress and the principal compressive stress were increased in a proportional manner up to failure. It was found that the loading path does have a significant effect on the softening behavior:

Sequential Loading

$$\zeta_\sigma = \frac{0.9}{\sqrt{1 + 250\varepsilon_r}} \tag{6}$$

$$\zeta_\varepsilon = 1 \tag{7}$$

Proportional Loading

$$\zeta_\sigma = \zeta_\varepsilon = \zeta = \frac{0.9}{\sqrt{1 + 400\varepsilon_r}} \tag{8}$$

where ε_r is valid up to 0.04. Equations (6) and (7) show that the stress was softened in sequential loading, but the strain-softening did not occur. In proportional loading, however, Equation (8) shows that both the stress and the strain were softened in a proportional manner. Therefore, the stress-softened coefficient ζ_σ and the strain-softened coefficient ζ_ε can be replaced by one coefficient ζ. The use of one coefficient ζ significantly simplifies the stress-strain curve of Eq. (5).

High-Strength Concrete Panels Under Sequential loading

To determine the softening behavior of high-strength concrete of 100 MPa, a test program of five full-size reinforced panels (VE-series) was carried out. The dimension and steel arrangement for these five 90° panels are shown in Fig. 2. The test panels had a size of 1397 x 1397 x 178 mm (55 x 55 x 7 in.), except that the panel VE0 had a thickness of 152 mm (6 in.) to accommodate the capacity limit of the test facility.

Table 1   Mechanical Properties of Test Panels in VE-Series

PANEL	CONCRETE		STEEL					
	f'_c (MPa)	ε_o	Rebars	ρ_ℓ	$f_{\ell y}$ (MPa)	Rebars	ρ_t	f_{ty} (MPa)
VE0	98.2	0.00240	15M@ 267mm	0.0098	409	10M@ 267mm	0.0049	445
VE1	103.1	0.00245	15M@ 267mm	0.0084	409	10M@ 267mm	0.0042	445
VE2	110.4	0.00250	15M@ 267mm	0.0084	463	10M@ 267mm	0.0042	445
VE3	104.3	0.00245	20M@ 267mm	0.0126	455	10M@ 267mm	0.0042	445
VE4	97.6	0.00240	25M@ 267mm	0.0211	470	10M@ 267mm	0.0042	445

ρ_ℓ = steel ratio in the longitudinal direction.
ρ_t = steel ratio in the transversal direction.

This series of panels was tested in the Universal Panel Tester (Hsu, Belarbi and Pang; 1995) with a newly installed servo-control system to provide strain-control procedure (Hsu, Zhang and Gomez, 1995). Tensile loading in the horizontal direction was first applied up to a desired average tensile strain. Then compressive loading was gradually applied in the vertical direction until the specimen reached failure, while at the same time maintaining the desired constant tensile strain. The main variable in this series was the principal tensile strains, which were selected to be 0.0065, 0.010, 0.020, and 0.030, for panels VE0, VE1 (and VE2), VE3, and VE4, respectively. The mechanical properties of test panels are given in Table 1.

The concrete compressive stress-strain curves of panels in the VE-series are plotted and compared with the concrete stress-strain curves of their companion cylinders in Fig. 3. It can be seen from these figures that the peak stresses have been softened as compared to the corresponding standard cylinder strengths.

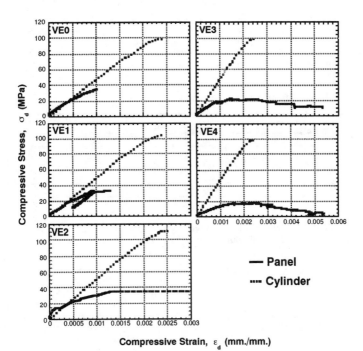

Figure 3 Compressive Stress-Strain Curves of Panels and Corresponding Cylinders

Panel VE1 was identical to panel VE2, except that the compression loading of VE1 in the second phase was controlled by strain and VE2 by load. A comparison of these two panels shows that the control mode has no effect on the stress softening. Panels VE0, VE1 and VE2 with the tensile strain ε_r less than 0.015 failed in a brittle manner with the peak strain considerably less than that for the cylinders (the descending portion of panel VE2 indicated by the dashed curve in Fig. 3 was obtained in a very short period of time and may not be reliable). In contrast, panels VE3 and VE4 with the tensile strain ε_r greater than 0.015 failed in a ductile manner with a gentle descending branch. The strain at peak point is close to that for the cylinders.

This series of tests shows that ζ_σ given by Eq. (6) for normal strength concrete can be generalized to include high strength concrete. This can be done by replacing the constant 0.9 by a variable $R(f_c')$ which is a function of concrete cylinder strength f_c'. The generalized form of the stress-softened coefficient ζ_σ can be written as

$$\zeta_\sigma = \frac{R(f_c')}{\sqrt{1 + 250\varepsilon_r}} \qquad (9)$$

The $R(f_c')$ values in Eq. (9) are calculated from the measured ζ_σ and ε_r for the five panels in VE-Series in Table 2. Similar calculations for $R(f_c')$ were made for the fifteen panels in E-Series with 42 MPa concrete (Belarbi and Hsu, 1995) and the three panels in HE-Series with 65 MPa concrete (Zhang, 1992). All the $R(f_c')$ values are plotted in Figure 4.

Table 2 Stress Softening for High Strength Concrete

PANEL	f_c' (MPa)	ε_r (x 10^{-3})	ζ	$R(f_c')$	C
VE0	98.2	6.81	0.357	0.587	5.82
VE1	103.1	10.18	0.304	0.572	5.81
VE2	110.4	11.05	0.290	0.562	5.91
VE3	104.3	20.60	0.236	0.585	5.98
VE4	97.6	30.50	0.196	0.576	5.69

Figure 4 shows that $R(f_c')$ is inversely proportional to $\sqrt{f_c'}$ and the proportional constant, represented by the slope of the straight line C, is equal to 5.8:

$$R(f_c') = \frac{5.8}{\sqrt{f_c'}} \qquad (f_c' \text{ in MPa}) \qquad (10)$$

In Fig. 4 the points for panels with normal strength concrete ($1/\sqrt{f_c'} \approx 0.16$) are quite scattered. This is because these panels are tested to explore several variables with wide ranges: the steel ratio varied from 0.6% to 3.0%, the spacing of steel bars from 2.625 in. (66.7 mm) to 10.5 in. (266.7mm), and the principal tensile strains from 0 to 0.03.

Substituting Eq. (10) into Eq. (9) gives the general equation of stress-softened coefficient ζ_σ for the sequential loading:

$$\zeta_\sigma = \frac{5.8}{\sqrt{f_c'}} \; \frac{1}{\sqrt{1 + 250\varepsilon_r}} \qquad (f_c' \text{ in MPa}) \qquad (11)$$

It should be noted that the variable $R(f_c')$ becomes 0.9 for concrete with a normal strength of 42 MPa (6,000 psi). In other words, Eq. (11) reduces to Eq. (6) for the normal strength concrete. When the concrete strength is less than 42 MPa, the stress-softened coefficient ζ_σ in Eq. (9) should be limited to 0.9. Also, in the case of 100 MPa concrete strength, the calculated softened coefficient is 0.353 when ε_r has the lowest value of 0.00681 in panels VE0. It is quite possible that Eq. (11) is too conservative when ε_r is less than 0.00681. Panels in this region would have failure loads exceeding the capacity of the test facility.

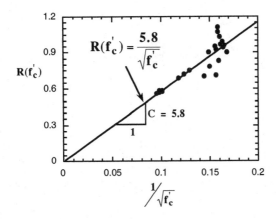

Figure 4 Relationship Between $R(f_c')$ and $1/\sqrt{f_c'}$

Effect of Concrete Strength on Softened Coefficient for Proportional loading

Preliminary tests of panels with 100 MPa concrete under proportional loading indicates that $R(f_c')$ defined by Eq. (10) is also valid for proportional loading. Pang and Hsu (1995) also found that the orientation of steel had little effect on the compressive softening of concrete in the principal direction. Therefore, Eq. (10) proposed for the 90° panels under sequential loading can be used for the non-90° panels under proportional loading.

To generalize the softened compressive stress-strain curves of concrete for proportional loading, the constant 0.9 in Eq. (8) is replaced by $R(f_c')$ given in Eq. (10), resulting in

$$\zeta = \frac{5.8}{\sqrt{f_c'}} \; \frac{1}{\sqrt{1 + 400\varepsilon_r}} \qquad (f_c' \text{ in MPa}) \qquad (12)$$

<u>Maximum Shear Strength of Beams</u>

Substituting Eqs. (12) and (2) into Eq. (1), the maximum shear strength of beams with vertical web reinforcement can be written as

$$\tau\ell_{tm} = \frac{5.8\sqrt{f_c'}}{\sqrt{1 + 400\varepsilon_r}} \sin\alpha\cos\alpha \quad (f_c' \text{ in MPa}) \tag{13}$$

At the maximum shear stress the principal tensile strain ε_r is close to 1% (Zhang, 1995). Substituting $\varepsilon_r = 1\%$ into Eq. (13), we have the following simplified equations:

$$\tau\ell_{tm} = 2.6\sqrt{f_c'} \sin\alpha\cos\alpha \quad (f_c' \text{ in MPa}) \tag{14a}$$

$$\tau\ell_{tm} = 31\sqrt{f_c'} \sin\alpha\cos\alpha \quad (f_c' \text{ in psi}) \tag{14b}$$

It is remarkable that the constant 31 in Eq. (14b) is very close to the constant 30 proposed by Ramirez and Breen (1991). The angle α in Eq. (14) depends on the ratio $\eta = \rho_t f_{ty}/\rho_\ell f_{\ell y}$, where ρ_t and ρ_ℓ are the steel ratios in the t- and ℓ- directions, respectively; and f_{ty} and $f_{\ell y}$ are the steel yield stresses in the two directions (Zhang, 1995). When the ratio η is in the range of 1/3 -1/2, commonly used in beams, the angle α reaches its lower ACI limit of 30°. Substituting $\alpha = 30°$ into Eq. (14), $\tau\ell_{tm}$ becomes $1.1\sqrt{f_c'}$, where f_c' is in MPa, and $13\sqrt{f_c'}$, where f_c' is in psi.

When a shear element is applied to the web of a beam, the maximum shear strength $\tau\ell_{tm}$ in Eq. (14) is based on V_u/bd_v, where d_v is the distance between the compressive stringer and the tensile stringer. If shear stress in a beam is expressed in terms of $v_u = V_u/bd$, where the effective depth of a beam is $d = 1.1d_v$, then the maximum shear stress $v_{u, max} = \tau\ell_{tm}/1.1$ can be expressed as

$$v_{u, max} = \sqrt{f_c'} \quad (f_c' \text{ in MPa}) \tag{15a}$$

or

$$v_{u, max} = 12\sqrt{f_c'} \quad (f_c' \text{ in psi}) \tag{15b}$$

The current ACI shear provisions for beams subjected to shear and flexure limit the maximum shear stress $v_{u, max}$ to

$$v_{u, max} = v_c + v_s = \frac{1}{6}\sqrt{f_c'} + \frac{4}{6}\sqrt{f_c'} = \frac{5}{6}\sqrt{f_c'} \quad (f_c' \text{ in MPa}) \tag{16a}$$

or

$$v_{u, max} = v_c + v_s = 2\sqrt{f_c'} + 8\sqrt{f_c'} = 10\sqrt{f_c'} \quad (f_c' \text{ in psi}) \tag{16b}$$

The maximum shear stress calculated by the ACI provision, Eq. (16), is somewhat less than that calculated by the rotating-angle softened-truss model, Eq. (15). Equations (15) and (16) are compared in Fig. 5 to the experimental values of partially under-reinforced panels B4, B5 and VB3 with low η values (α angle much less than 45°). It can be seen that the maximum shear strengths, $v_{u, max}$, calculated by Eq. (15) agree better with the test results than those calculated by Eq. (16). However, ACI equation, Eq. (16), provides an approximate lower bound for the maximum shear strengths.

<u>Balanced Condition in Shear Elements</u>

The balanced condition in shear defines a mode of failure where both the longitudinal and the transverse steel yield simultaneously as the concrete crushes. The balanced steel ratio ρ_b, therefore, divides the under-reinforced element from the over-reinforced element under the equal strain condition. It is used as the maximum limit of steel ratio to prevent a brittle failure due to crushing of concrete. In short, ρ_b in shear serves the same purpose as the maximum shear strength.

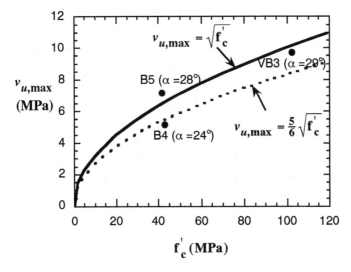

Figure 5 Comparison of Maximum Shear Strengths

Based on the softened-truss model, a solution procedure for the balanced condition of a pure shear element was given (Zhang, 1995). Using this procedure, the balanced steel ratios ρ_b are calculated for concrete strengths f_c' of 42 MPa, 65 MPa and 100 MPa. These three ρ_b values are plotted against the parameter $\sqrt{f_c'}/f_y$ as shown in Fig. 6. It can be seen that ρ_b can be expressed as a function of the square root of the concrete strength $\sqrt{f_c'}$:

$$\rho_b = 1.89\frac{\sqrt{f_c'}}{f_y} \qquad (f_c' \text{ in MPa}) \qquad (17)$$

Figure 6 show that Eq. (17) closely represents the values calculated by the softened-truss model.

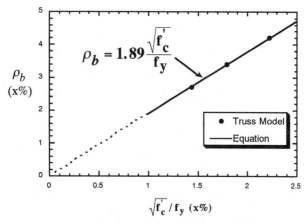

Figure 6 Balanced Reinforcement Ratios

In the design of a reinforced concrete structure, we can check the minimum cross section of members either by the balanced steel ratio in shear or by the maximum shear strength. Both ways are intended to prevent the brittle crushing of concrete before the yielding of steel. It is, therefore, not surprising to see that both the balanced steel ratio in shear and the maximum shear strength are proportional to the square root of the concrete strength, $\sqrt{f_c'}$.

Conclusions

Full-size reinforced membrane elements (panels) with concrete strengths of 100 MPa were tested. When compared to the previous tests with concrete strengths of 42 MPa and 65 MPa, the softened coefficient ζ of diagonal concrete struts was found to be inversely proportional to $\sqrt{f_c'}$. In other words, the peak strength of concrete struts $\zeta f_c'$ is proportional to $\sqrt{f_c'}$ in a shear element. Using this new softened coefficient, the softened truss model will predict that the balanced steel ratio is proportional to $\sqrt{f_c'}$ in a shear element.

Applying such a shear element to the web of a beam, the softened truss model will predict that the maximum shear strength of a beam is proportional to $\sqrt{f_c'}$ for f_c' up to 100 MPa. This conclusion supports the ACI Code, in which the maximum shear strength of a reinforced concrete beam is taken to be $10\sqrt{f_c'}$ (f_c' in psi).

References

ACI Committee 318 (1963; 1989), "Building Code Requirement for Reinforced Concrete," ACI Standard 318-89, American Concrete Institute, Detroit.

ACI-ASCE Committee 326 (1962), "Shear and Diagonal Tension," Journal of the American Concrete Institute, Proceedings Vol. 59, pp. 1-30, pp. 277-334, pp. 352-396.

A.I.J. (1988), "Design Guidelines for Earthquake Resistant Reinforced Concrete Buildings Based on Ultimate Strength Concept," with Commentary, Architectural Institute of Japan, 337 pp.

Belarbi, A. and T. T. C. Hsu. (1995), "Constitutive Laws of Softened Concrete in Biaxial Tension-Compression," Structural Journal of the American Concrete Institute, Vol. 92, No. 5, Sept.-Oct.

Bennett, E. W. and Balasooriya, B. M. A. (1971), "Shear Strength of Prestressed Beams with Thin Webs Failing in Inclined Compression," Journal of the American Concrete Institute, Proceedings Vol. 68, No. 3, March, pp. 204-212.

CEB-FIP Model Code (1978), "International System of Unified Standard Codes of Practice of Structures," CEB-FIP International Recommendations, 4th Edition.

Campbell, T. Ivan., Batchelor, B. deV., and Chitnuyanondh, L. (1979), "Web Crushing in Concrete Girders with Prestressing Ducks in the Web," Journal of Prestressed Concrete Institute, Vol. 24, No. 5, Sept. Oct. pp. 70-88.

Robinson, J. R. and Demorieux, J. M. (1968), "Poutres en Double Te en Beton Arme," Annales de l'Institut Technique du Batiment et des Travaux Publics, Vol. 21, No. 246, Paris, pp. 931-934.

EC2 (1991): Eurocode No. 2 "Design of Concrete Structures, Part I, General Rules and Rules for Buildings."

Hsu, T. T. C., Unified Theory of Reinforced Concrete, CRC Press Inc., Boca Raton, FL, 1993.

Hsu, T. T. C., Belarbi, A., and Pang, X. (1994), "A Universal Panel Tester," Journal of Testing and Evaluation, ASTM, Vol. 22, No.1, pp. 41-49.

Hsu, T. T. C., Zhang, L. X., and Gomez, T. (1995), "A Servo-Control system for Universal Panel Tester," Journal of Testing and Evaluations, ASTM, Vol. 23, No. 6, Nov.-Dec.

Leonhardt, F and Walther, R. (1964), "The Stuttgart Shear Test, 1961," Library Translation No. 111, Cement and Concrete Association, London, England, 134 pp.

Lyngberg, B. S. (1976), "Ultimate Shear Resistance of Partially Prestresed Reinforced Concrete I-Beams," Journal of the American Concrete Institute, Proceedings Vol. 73, No. 4, April, pp. 214-222.

Nielsen, M. P., Braestrup, M. W., Jensen, B. C., and Bach, F. (1978), "Concrete Plasticity - Beam Shear - Shear in Joints - Punching Shear," Danish Society for Structural Science and Engineering, Technical University of Denmark, Lyngby, Copenhagen, 129 pp.

Pang, X. B. and Hsu, T. T. C. (1995), "Behavior of Reinforced Concrete Membrane Elements in Shear," Structural Journal of the American Concrete Institute, Vol. 92, No. 6, Nov.-Dec.

Placas, A. and Regan, P. E. (1971), "Shear Failure of Reinforced Concrete Beams," Journal of the American Concrete Institute, Proceedings Vol. 68, No. 10. pp. 763-773.

Ramirez, J. A. and Breen, J. E. (1991), "Evaluation of a Modified Truss-Model Approach for Beams in Shear," Structural Journal of the American Concrete Institute, Vol. 88, No. 5, pp. 562-571.

Robinson, J. R. and Demorieux, J. M. (1961), "Essais a l'effort tranchant de Poutres a Ame Mince en Beton Arme," Annales des Ponts et Chaussees, Vol. 131, pp. 225-255.

Thurlimann, B. (1979). "Shear Strength of Reinforced and Prestressed Concrete - CEB Approach," ACI-CEB-PCI-FIP Symposium, Concrete Design: U.S. and European Practices, ACI Publication SP-59, American Concrete Institute, Detroit, MI, pp. 93-115.

Zhang, L. X. (1992), "Constitutive Laws of Reinforced Elements with Medium-High Strength Concrete," M. S. Thesis Presented to the Faculty of the Department of Civil Engineering, University of Houston, 214 pp.

Zhang, L. X. (1995), "Constitutive Laws of Reinforced Elements with High Strength Concrete," Ph. D. Dissertation Presented to the Faculty of the Department of Civil Engineering, University of Houston, 303 pp.

Behavior of High-Strength Concrete Beam-Column Joints

Michael E. Kreger[1] and Elias I. Saqan[2]

Abstract

Current seismic-resistant design provisions for beam-column connections were developed using data from cyclic load tests performed on beam-column joints constructed with concrete strengths of 41.4 MPa (6000 psi) or less. Results of twenty-six beam-column connection tests conducted in Japan and the U.S. are used to evaluate current provisions for use in design of exterior and interior beam-column connections constructed with concrete strengths exceeding 41.4 MPa.

Introduction

During a strong-motion earthquake, beam-column connections in reinforced concrete moment-resisting frames can experience severe cyclic loads. If the connections do not possess adequate strength, the overall strength and stiffness of the frame system can be adversely affected. Current provisions in the ACI Building Code (1995) and recommendations developed by ACI Committee 352 (1985) for design of beam-column connections were developed primarily from results of tests conducted on specimens constructed with concrete strengths of 41.4 MPa (6000 psi) or less. Although it is now technically and economically feasible to commercially produce concrete with a compressive strength of 80 MPa (approximately 12,000 psi) or higher, it is not economical to use this concrete in beams and slabs. However, concrete compressive strengths exceeding 80 MPa have already been used in columns to reduce creep deformations and cross-section dimensions. Because the

---

[1]   Associate Professor, Department of Civil Engineering, The University of Texas at Austin, Austin, TX, 78759.

[2]   Graduate Research Assistant, Department of Civil Engineering, The University of Texas at Austin, Austin, TX, 78759.

column concrete may be cast through the joint region, it is possible that concrete strengths exceeding 40 MPa will be used in future beam-column connections.

Results of cyclic load tests conducted in Japan and the United States on 26 beam-column connection specimens constructed with concrete strengths ranging from 41.4 to 107 MPa (6000 to 15,500 psi) were used to evaluate current ACI Building Code and Committee 352 joint-shear strength provisions for use in design of interior and exterior connections constructed with high-strength concrete.

Overview of Current Design Provisions

Current design provisions for joint-shear strength compare horizontal shear in the joint, V_u, with a shear strength, ϕV_n, that is a function of the geometry of the beam-column connection. The horizontal shear in the joint is computed as shown in Fig. 1. The nominal joint-shear strength is computed as

$$V_n = 0.083 \gamma \sqrt{f_c'} \ (MPa) \ b_j h \qquad (1)$$

where b_j is the effective joint width, h is the column depth in the direction of loading, and γ is a constant equal to 12, 15, or 20, depending whether the beam-column connection is classified as a corner, exterior, or interior connection, respectively.

The effective width, b_j, is the average of the column width and the beam width, b_b, in the direction of loading, but not greater than the beam width, b_b, plus one half the column depth, h, on each side of the beam.

Figure 2 illustrates the various connection geometries associated with interior, exterior, and corner connections. Connections classified as "interior" have beams framing into all sides of the connection. Exterior connections have at least two beams framing into opposite sides of the column. In order to satisfy the requirements for number of beams framing into the column, beams must be at least three-quarters the width of the column, and the depth of the shallowest beam must be no less than three-quarters the depth of the deepest beam framing into the connection. Connections not satisfying these requirements are classified as corner connections. The value of γ used in the joint-shear strength relation (Eq. 1) is an indicator of the confinement provided by horizontal members framing into a connection. It is believed that more pairs of horizontal members framing into opposite sides of a joint result in enhanced joint-shear strength.

In addition to horizontal members framing into joints, transverse reinforcement is also required by the 352 provisions for confinement of beam-column joints. Single hoops, overlapping hoops, or hoops with crossties are provided so that the area of steel in each direction is at least equal to

$$A_{sh} = 0.3 \frac{s_h h'' f_c'}{f_{yh}} \left(\frac{A_g}{A_c} - 1 \right) \qquad (2)$$

but not less than

$$A_{sh} = 0.09 \frac{s_h h'' f_c'}{f_{yh}} \qquad (3)$$

where s_h is the vertical spacing of the hoops, h'' is the core dimension, f_{yh} is the yield strength of the hoops, A_g is the gross area of the column section, and A_c is the area of the column core. The constant (0.09) in Eq. 3 is equal to 0.12 in the comparable equation in the ACI Code. Vertical spacing of hoops may not exceed one-quarter the minimum column dimension, six times the diameter of column longitudinal bars, or 6 inches.

Evaluation of Joint-Shear Strength Data

Joint shear strength provisions are evaluated using results of tests performed by Sugano, Nagashima, Kimura, and Ichikawa (1991), Noguchi and Kashiwazaki (1992 & 1993), Oka and Shiohara (1992), Kitayama, Lee, Otani, and Aoyama (1992), and Guimaraes, Kreger and Jirsa (1992). Connections considered in this evaluation had concrete compressive strengths of at least 41.4 MPa (6000 psi), generally incorporated details similar to those used in U.S. practice, and in all but two cases, experienced joint-shear failure.

Story shear versus drift histories were available for all tests except those performed by Oka and Shiohara (1992). Maximum story shears experienced by specimens for drift ratios of 2% or less were used to estimate maximum joint shears. The maximum computed joint shears were then compared with joint-shear strengths obtained using provisions from ACI 352 (1985) and 318 (1995) (Eq. 1). Oka and Shiohara reported maximum story shears, but not the drift ratios at which these story shears occurred.

Maximum joint shear was computed for each specimen by first determining the applied beam moments from the maximum story shear

$$M_{beam} = \frac{V_{story}H}{L}\left(\frac{L-h}{2}\right) \qquad (4)$$

where H is the height of the column, L is the beam span, and h is the depth of the column in the loading direction. The joint shear, V_j, was computed using the free body in Fig. 1 that was discussed earlier. Note that the joint shear in Fig. 1 is denoted by V_u. Resultant compression and tension forces in the beams at the vertical faces of the joint were approximated by M_{beam}/j_b, where j_b was taken as $(7/8)d$. The internal moment arm, j_b, was approximated because insufficient information about stresses in the longitudinal reinforcement was provided in the references to accurately assess its value. The value of j_b used here was also used by many of the researchers listed in the references. The resulting equation used to estimate joint shear was

$$V_j = V_{story}\left[\frac{H}{j_b}\left(1-\frac{h}{L}\right)-1\right] \qquad (5)$$

Table 1 lists the specimen identifier, concrete compressive strength, joint shear, ratio of estimated to computed joint-shear strength, ratio of joint reinforcement provided to that required by Eqns. 2 and 3, and failure mode for 22 exterior beam-column connections. The notation "J" or "BJ" in the failure mode column indicates whether joint-shear failure occurred without beam yielding (J) or following beam yielding (BJ).

Only two of the 22 exterior beam-column connections tested yielded joint-shear strengths less than those predicted by the joint-shear strength relationship presented in Eq. 1. These two tests (MKJ-1 and MKJ-3) were part of the second series of tests performed by Noguchi and Kashiwazaki (1993), and the four tests in this series produced four of the five lowest joint-shear strength ratios for the 22 tests considered. Even though the exterior dimensions of specimens in the MKJ series were identical to those for the OKJ series (also conducted by Noguchi and Kashiwazaki), changes in the joint core dimensions and longitudinal reinforcing bars likely contributed to the lower joint-shear strength ratios. The joint core width for the MKJ specimens was 40 mm less than for the OKJ specimens, resulting in 20 mm of additional concrete cover on side faces of the joint. Bond stresses for longitudinal bars that passed through the joint were higher for the MKJ series than for most other specimens because large-diameter bars (as large as 22 mm) were used. Because high bond stresses in the joint region are known to result in deterioration of joint-shear strength, the ACI 352 provisions (1985) place a limit of 20 on the ratio of column depth to beam longitudinal bar diameter, and the ratio of beam depth to column longitudinal bar diameter. The column-depth to beam-bar-diameter ratio for each specimen in the MKJ series was 13.6 or 15.7 compared with 23.6 for specimens tested in the OKJ series.

The mean joint-shear strength ratio was 1.31 for all 22 exterior-joint specimens, and was 1.39 when the MKJ series of tests was excluded. The latter mean value suggests a joint-shear coefficient, γ, of 20.8, compared with the value of 15 recommended for exterior connections in the ACI 352 provisions and ACI Building Code. The mean joint-shear strength ratio for connections that experienced joint-shear failure without beam yielding was slightly higher than for connections (excluding the MKJ series) that experienced beam yielding prior to joint-shear failure (1.48 versus 1.34). Classification of most of the connections as "exterior" is inappropriate if current design provisions are interpreted rigorously. Current provisions require that a minimum of two beams on opposite sides of the column "mask" at least 75% of the column width in order to be referred to as exterior connections. All "exterior" connections considered here, with the exception of those tested by Oka and Shiohara, had beam to column-width ratios of 0.67 and 0.68. So, even though the specimens had the appearance of exterior connections, according to current design provisions, they should actually be treated as "corner" connections with a joint-shear coefficient of 12. Improved confinement provided by beams that masked a higher percentage of the column width (80%) in specimens tested by Oka and Shiohara resulted in apparent increases in joint-shear strength; the average joint-shear strength ratio for specimens tested by Oka and Shiohara was 1.45, compared with 1.36 for the remaining 11 exterior-joint specimens (excluding the MKJ series of tests).

Details about joint reinforcement varied for the tests considered in this study. Transverse reinforcement strengths were not reported by Oka and Shiohara, and placement of transverse reinforcement within each joint was generally ambiguous. Despite this, joint reinforcement quantities were estimated as best as possible with the information given.

Transverse reinforcement quantities in the joints varied from 0.07 to 2.02 times the amount required by Eqns. 2 and 3. There was no discernible correlation between the amount of transverse reinforcement in connections and failure modes or joint-shear strengths inferred from the test data. Only five specimens contained more joint reinforcement than required by ACI 352 (from 1.50 to 2.02 times more). The average ratio of transverse reinforcement provided in the remaining exterior-joint specimens (once again excluding the MKJ series) was 0.47. Considering these specimens attained, on average, 1.42 times the computed joint-shear strength, the data strongly suggest that joint reinforcement could be reduced for beam-column connections constructed with high-strength concrete.

The specimen identifier, concrete compressive strength, joint shear, ratio of estimated to computed joint-shear strength, ratio of joint reinforcement provided to that required by Eqns. 2 and 3, and failure mode are presented in Table 2 for four interior beam-column connections. The notation "B" in the failure mode column indicates beam yielding without joint-shear failure. The two interior connections tested by Guimaraes et al. (1992) experienced joint-shear failures at loads exceeding strengths specified by ACI 352 and 318. The remaining two connections tested by Kitayama et al. (1992) and Oka and Shiohara (1992) also developed joint shears that exceeded specified capacities, even though the actual strength of the joint in each of these connections was never mobilized. Furthermore, these connections attained specified joint-shear strengths despite three of the four specimens having from 22 to 64% of the joint reinforcement required by ACI 352, and despite the "interior" connection tested by Kitayama et al. having beams that masked only 67% the width of the column (instead of the minimum 75% required to be considered an interior connection).

Conclusions

The primary objective of this study was to evaluate current joint-shear strength provisions for design of joint regions in beam-column connections constructed with high-strength concrete. As a secondary consideration related to joint shear strength, transverse reinforcement requirements for the joint region were also examined. The observations and conclusions that follow were based on the results of tests conducted on 22 exterior beam-column connections and four interior connections.

1. The joint-shear strength provisions for exterior beam-column connections were evaluated using 18 of the 22 tests considered. Four specimens in the MKJ series tested by Noguchi and Kashiwazaki (1993) had joint core dimensions and longitudinal reinforcement sizes that were outside the norms of standard practice. It is quite likely that these contributed significantly to the lower joint-shear strengths observed, and as a result, these tests were not used in the evaluation.

2. Current ACI 352 (1985) and 318-95 joint-shear strength provisions for exterior beam-column connections are valid for concrete strengths up to 107 MPa. The mean joint-shear coefficient, γ, inferred from 18 exterior connection specimens tested by Sugano et al. (1991), Oka and Shiohara (1992), Kitayama et al. (1992), and Noguchi and Kashiwazaki (1992) was 20.8, compared with 15 recommended by ACI 352 and 318.

3. Current ACI 352 and 318-95 joint-shear strength provisions also appear to be valid for interior beam-column connections constructed with concrete strengths as high as 98.8 MPa. The mean joint-shear coefficient, γ,

inferred from four interior connection specimens tested by Guimaraes et al. (1992), Oka and Shiohara, and Kitayama et al. was 23.8, compared with 20 recommended by ACI 352 and 318. It is quite likely that use of this value will result in underestimated joint-shear strengths because two of the four test results used in computing this value were from specimens that did not experience joint-shear failure.

4. Quantities of transverse reinforcement required by ACI 352 (and ACI 318-95) in the joint region of beam-column connections can likely be reduced for connections constructed with high-strength concrete (f_c' > 41.4 MPa). Thirteen of the 18 exterior connections considered in the joint-shear strength evaluation contained substantially less joint reinforcement (on average, 47%) than required by ACI 352. The mean joint-shear strength for these 13 specimens was 1.42 times the strength predicted by current ACI 352 and 318 provisions. Similarly, three of the four interior connections contained, on average, 39% of the joint reinforcement required by ACI 352, while developing 1.15 times the joint-shear strength predicted by ACI 352 and 318.

Acknowledgements

The work summarized in this paper was originally presented at the second U.S.-Japan-New Zealand-Canada multilateral meeting on "Structural Performance of High-Strength Concrete in Seismic Regions," which was held in Honolulu, Hawaii from 29 November to 1 December 1994. Travel to this meeting was funded through a National Science Foundation grant to the University of Minnesota.

Table 1 Evaluation of joint shear strength data for exterior beam-column connections					
Specimen ID	f_c' (MPa)	V_j (kN)	V_j(est.)/ V_j(ACI 352)	A_{sh}(prov.)/ A_{sh}(352)	Failure Mode
Sugano, Nagashima, Kimura, and Ichikawa (Note: b_b/b_c = 0.68)					
J6-0	60.5	1970	1.25	2.02	BJ
J6-1	62.3	1970	1.24	1.97	BJ
J8-0	77.6	1990	1.12	1.58	BJ
J8H-0	80.1	2740	1.51	1.53	BJ
Oka and Shiohara					
J-1	81.2	1180	1.29	0.21	BJ
J-2	81.2	1270	1.39	0.79	J
J-3	81.2	1320	1.45	1.50	J
J-4	72.8	1220	1.41	0.24	BJ
J-5	72.8	1370	1.59	0.24	J
J-6	79.2	1260	1.40	0.07	BJ
J-8	79.2	1430	1.59	0.27	J
Kitayama, Lee, Otani, and Aoyama (Note: b_b/b_c = 0.67)					
I1	98.8	1320	1.44	0.22	BJ
I3	41.4	710	1.19	0.36	BJ
Noguchi and Kashiwazaki (1992) (Note: b_b/b_c = 0.67)					
OKJ-1	70.0	1120	1.44	0.75	BJ
OKJ-3	107	1220	1.26	0.49	J
OKJ-4	70.0	1107	1.42	0.75	BJ
OKJ-5	70.0	1190	1.53	0.75	J
OKJ-6	53.5	1070	1.56	0.98	J
Noguchi and Kashiwazaki (1993) (Note: b_b/b_c = 0.67)					
MKJ-1	84.4	667	0.78	0.41	BJ
MKJ-2	84.4	925	1.08	0.46	BJ
MKJ-3	98.5	785	0.85	0.38	BJ
MKJ-4	98.5	1050	1.13	0.42	BJ

Specimen ID	f_c' (MPa)	V_j (kN)	V_j(est.)/ V_j(ACI 352)	A_{sh}(prov.)/ A_{sh}(352)	Failure Mode
Table 2 Evaluation of joint shear strength data for interior beam-column connections					
Guimaraes, Kreger, and Jirsa					
J5	78.0	4530	1.33	1.10	BJ
J6	92.2	4260	1.15	0.64	BJ
Oka and Shiohara					
J-9	79.2	1350	1.13	0.22	B
Kitayama, Lee, Otani, and Aoyama (Note: $b_b/b_c = 0.67$)					
I2	98.8	1440	1.16	0.30	B

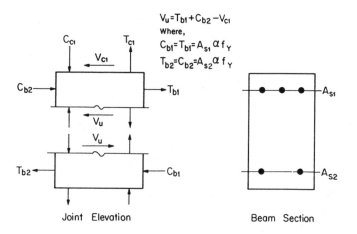

Joint Elevation　　　　Beam Section

$V_u = T_{b1} + C_{b2} - V_{c1}$

Where,

$C_{b1} = T_{b1} = A_{s1}\,\alpha\,f_Y$

$T_{b2} = C_{b2} = A_{s2}\,\alpha\,f_Y$

Fig. 1 Computation of horizontal joint shear
[from ACI-ASCE 352 (1985)]

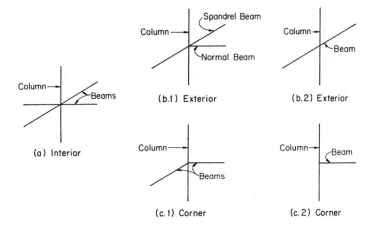

Fig. 2 Geometric description of joints
[from ACI-ASCE 352 (1985)]

References

"Building code requirements for reinforced concrete," (1995). *ACI 318-95*, Am. Concrete Inst. (ACI), Detroit.

Guimaraes, G.N., Kreger, M.E., and Jirsa, J.O. (1992). "Evaluation of joint-shear provisions for interior beam-column-slab connections using high-strength materials," *ACI Struct. J.*, Am. Concrete Inst. (ACI), 89(1), 89-98.

Kitayama, K., Lee, S., Otani, S., and Aoyama, H. (1992). "Behavior of high-strength R/C beam-column joints," *Proc. of 10WCEE*, Madrid, Spain, 3151-3156.

Noguchi, H., and Kashiwazaki, T. (1992). "Experimental studies on shear performances of RC interior column-beam joints with high-strength materials," *Proc. of 10WCEE*, Madrid, Spain, 3163-3168.

Noguchi, H., and Kashiwazaki, T. (1993). "Capacity of interior column-beam joints with high-strength concrete," *Proc., First Meeting of Multilateral Program on High-Strength Concrete in Earthquake-Resistant Design*, 26.

Oka, K., and Shiohara, H. (1992). "Tests of high-strength concrete interior beam-column-joint subassemblages," *Proc. of 10WCEE*, Madrid, Spain, 3211-3216.

"Recommendations for design of beam-column joints in monolithic reinforced concrete structures," (1985). *ACI-ASCE 352-R85, ACI Struct. J.*, Am. Concrete Inst. (ACI), 82(3), 266-283.

Sugano, S., Nagashima, T., Kimura, H., and Ichikawa, A. (1991). "Behavior of beam-column joints using high-strength materials," *ACI Spec. Publ., SP 123-13*, Am. Concrete Inst. (ACI), Detroit, Mich., 359-377.

TENSILE RESPONSE OF REINFORCED
HIGH STRENGTH CONCRETE MEMBERS

S. P. Shah[1] and C. Ouyang[2]

Abstract

Effects of various parameters such as reinforcement ratio and distribution of steel bars were experimentally examined for high strength concrete tensile members. A fracture energy approach is proposed to predict the response of the members. It was found that high strength concrete members require greater minimum reinforcement ratio than normal strength concrete members.

Introduction

For the design of reinforced concrete structures, one frequently needs to know its tensile response. Since high strength concrete is normally more brittle than normal strength concrete, the former may have somewhat different response compared to the latter. A study on tensile behavior of reinforced high strength concrete members is presented. Tests were performed to examine the effects of several variables on the tensile response. A theoretical model is proposed to predict this response. It was found that high strength concrete members require greater minimum reinforcement ratio than normal strength concrete members.

Test Program

The reinforced high strength concrete panels tested had a gross cross-sectional area of 127 mm x 50.8 mm and had notched (a_0 = 10 mm) on the center of the 50.8 mm thick sides. The length of the specimen was 686 mm. They were reinforced with deformed rebars of 6-9 mm in diameters. Reinforcement ratios were in the range of 0.78% to 3.3%. The mix proportions for the high strength concrete by weight were 1 : 0.5 : 2.67 : 2.83 and 1 : 0.26 : 1.31 : 1.68 (cement

---

[1] NSF Center for Advanced Cement Based Materials, Northwestern University, 2145 Sheridan Road, Evanston, IL 60208.

[2] Iowa Department of Transportation, Office of Materials, 800 Lincoln Way, Ames, IA 50010.

: water : sand : coarse aggregate), respectively. Silica fume in the form of slurry and naphthalene based superplasticizer were added. All specimens were cured in a waterbath until the time of testing to avoid cracking due to shrinkage stresses.

All uniaxial tension tests on the panels were carried out in a servo-hydraulic MTS testing machine as shown in Fig. 1. Two identical loading fixtures were used to transfer the load from the machine to the specimen. One fixture was mounted on to the machine's actuator and the other one was connected to the load cell at the machine cross-head. The loading fixtures were connected through a sliding steel block and four hardened steel pins to four steel plates which were glued to the specimen ends. The tests were conducted with actuator stroke as the feedback control signal with a rate of 0.2 μm/sec.

Experimental Results

Typical stress-strain curve of the tensile member is shown in Fig. 2. The member was initially uncracked, and stress and strain increased linearly. The use of the notch assured that the first crack nucleated within the measurement length of the LVDT's. The axial stiffness of the member decreased gradually with each additional crack until a stable cracked pattern developed. From this point onwards, the average stress-strain curve of the member and that of a bare rebar had almost identical slope. The difference between the two curves approximately represented the contribution of the concrete matrix.

The effect of number of rebars (i. e. spacing) on first cracking strength is shown for the panels with a reinforcement ratio of approximately 3% in Fig. 3a. Here f_{tm} is the first cracking strength of the concrete matrix which was obtained by subtracting the rebar stress from the total stress on the panel based on the rule of mixture, and f_{tc} is the tensile strength of the corresponding plain concrete. The calculated matrix strength is about 9% higher in the panels with seven rebars compared to the panels with three rebars. The first cracking strength is plotted against reinforcement ratio in Fig. 3b. When the reinforcement ratio changes from 0.78% to 3.07%, the average first cracking strength increases about 6%.

Theoretical Model

A reinforced concrete tensile member containing multiple cracks is shown in Fig. 4. The crack width is denoted as w_0. The average concrete and steel strains at the loading ends are ϵ_c and ϵ_s, respectively, where ϵ_c does not include the crack width while ϵ_s does. The relationship between ϵ_c and ϵ_s is given as

$$\epsilon_s = \epsilon_c + \frac{Nw_0}{L} \qquad (1)$$

where L is the length of the member and N is the number of cracks.

Fig.1 : Experimental set-up

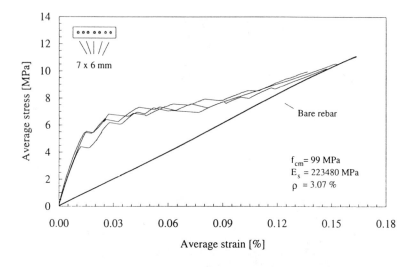

Fig. 2 Total stress-strain curve for reinforced high strength concrete panel

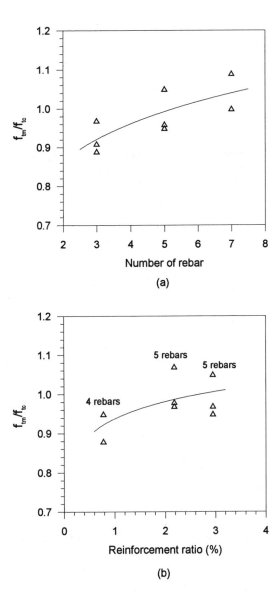

(a)

(b)

Fig. 3 Effects of number of rebar and reinforcement ratio
on the first cracking strength

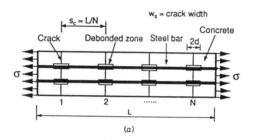

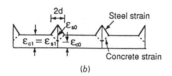

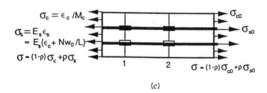

Fig. 4   Fracture states and distributions of stresses and strains:
(a) fracture of a reinforced concrete member, (b) strain
distributions in steel bar and concrete, and (c) stress in
steel and concrete at a cracked section.

The unknown average concrete strain ϵ_c in Eq. (1) can be determined from the total energy dissipations during the cracking process. Based on fracture mechanics, energy equilibrium during cracking requires that the strain energy release rate φ_c, should be equal to the sum of all dissipated energies and the fracture resistance R_{Icf} of concrete. Dissipated energies during cracking are mainly attributed to debonding and sliding effects at the interface of reinforcing bars and concrete. Since sliding only happens at the debonded interface, the interfacial slip between the reinforcing bar and the concrete is small. Therefore, the sliding energy may be neglected for simplicity. As a result, energy equilibrium during cracking can be written as below,

$$-\frac{1}{bt}\frac{\partial \varphi_c}{\partial N} = R_{Icf} + \frac{1}{bt}\frac{\partial \varphi_d}{\partial N} \tag{2}$$

where b and t are the specimen width and thickness, respectively, φ_c is the strain energy of concrete containing N cracks, φ_d is the total debonding energy on all debonded interface associated with N cracks, R_{Icf} is the fracture resistance of a plain concrete member which has the same dimension as the corresponding reinforced concrete member.

In order to use Eq. (2), values of R_{Icf} , φ_c and φ_d should be known. The strain energy φ_c and the debonded energy φ_d are given as (Ouyang and Shah, 1994):

$$\varphi_c = \frac{L\ b\ t\ (1-\rho)\ \sigma_c\ \varepsilon_c}{2} = \frac{L\ b\ t\ (1-\rho)\ \varepsilon_c^2}{2\ M_N} \tag{3}$$

and

$$\varphi_d = \frac{4\ N\ \rho\ d\ \gamma_d\ b\ t}{r} \tag{4}$$

where M_N is the effective compliance of plain concrete with N cracks, ρ is the reinforcement ratio, γ_d is the debonded energy on unit area of interface between steel and concrete, and r is the radius of the steel bar. The effective compliance, M_N, was expressed as (Ouyang et al., 1995)

$$M_N = \frac{1 + \omega\ \xi\ N}{1 - \xi\ N\ (1-\omega)}\ M_o \tag{5}$$

where

$$\xi = \frac{2}{Lb} \left[\int_0^{a_{cf}} a\ F^2(a)\ da + (b - a_{cf})\ a_{cf}\ F^2(a_{cf}) \right] \tag{6}$$

and the geometry function F(a/b) for a double-edge notch tensile specimen which was used in this investigation, is given by Tada et al. (1985),

$$F_I(a/b) = \frac{1.122 - 0.561a/b - 0.205(a/b)^2 + 0.471(a/b)^3 - 0.190(a/b)^4}{\sqrt{1-a/b}} \tag{7}$$

The values of ω is between 0 and 1 and will be determined later.

Substituting Eqs. (3), (4) and (5) into Eq. (2) results in

$$\frac{L\,(1-\rho)\,\epsilon_c^2\,\xi}{2\,M_o\,(1+\omega\xi N)^2} - \frac{4\,\rho\,d\,\gamma_d}{r} - R_{Icf} = 0 \tag{8}$$

After knowing the matrix strain ϵ_c, the average concrete stress σ_c and the average steel stress σ_s can be computed as :

$$\sigma_c = \frac{\epsilon_c}{M_N} = \epsilon_c\,\frac{1-\xi N(1-\omega)}{(1+\omega\xi N)\,M_o} \tag{9}$$

and

$$\sigma_s = E_s\,\epsilon_s = E_s\left[\epsilon_c + \frac{N\,w_o}{L}\right] \tag{10}$$

The value of the crack width w_0 of concrete can be determined

$$\frac{w_o}{d} = \frac{\left[\dfrac{1-\rho}{\rho E_{cM_c}} + 1\right]\epsilon_c - \eta}{1 - \dfrac{Nd}{L} - \dfrac{\eta d}{w_c}} \tag{11}$$

where η is given by

$$\eta = \frac{k\,(1-\rho)\,f_t}{\rho\,E_s} + \frac{k\,f_t}{E_c} \tag{12}$$

Constant k will be determined later.

The theoretical and experimental comparison is shown in Fig. 5. In the theoretical calculation, the initial notch length was $a_o = 10$ mm. The fracture resistance of concrete, $R_{Icf} = 0.03$ N/mm, was determined from R-curves (Ouyang

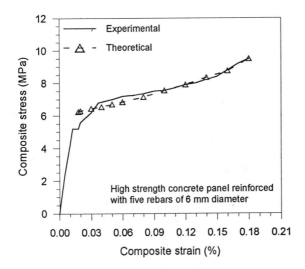

Fig. 5 Comparison of experimental and theoretical composite
     stress-strain curve

and Shah, 1994) based on the measured values of $K^s_{Ic} = 35.9$ MPa mm$^{1/2}$, CTOD$_c$ = 0.0066 mm and E_c = 36.6 GPa. Values of w_c = 0.12 mm, k = 0.15, γ_d = 0.01 N/mm and ω = 0.86 were assumed for the high strength concrete. The value of d = r/(20ρ) was used for the deformed rebar (Gilbert, 1992). It is demonstrated that the proposed model predicts reasonable response of the reinforced high strength concrete tensile member compared to the experimental measurement.

Minimum Reinforcement Ratio

The minimum reinforcement ratio is determined from stress equilibrium immediately after a single crack has developed. At this point the sum of the average tensile stresses in concrete and reinforcing steel bars should be equal to the yield strength of the steel bars at the crack such that:

$$(1-\rho_{min})\sigma_{cr} + \rho_{min}\sigma_{sr} = \rho_{min} f_{sy} \tag{13}$$

where σ_{cr} and σ_{sr} are the tensile stresses in concrete and in steel bars away from the crack and f_{sy} is the yield strength of the steel bar. The concrete strain is related to the steel strain through Eq. (1). Immediately after the crack forms, it may be assumed for simplicity that the debonded length and the crack width are both equal to zero. Under these specific conditions the strains in concrete and steel are identical and can be obtained as:

$$\varepsilon_{cr} = \varepsilon_{sr} = \left[\frac{2R_{Icf}}{L\xi(1-\rho_{min})E_c}\right]^{1/2} (1+\xi\omega) \tag{14}$$

Substituting Eq. (14) into Eqs. (9) and (10), respectively, results in σ_{cr} and σ_{sr}. These values of σ_{cr} and σ_{sr} are used in Eq. (13) for determining the minimum reinforcement ratio. This leads to

$$\rho_{min} = \frac{E_c(1-\xi+\xi\omega)}{f_{sy}\sqrt{\dfrac{L\xi E_c}{2R_{clf}}} - E_s(1+\xi\omega)} \tag{15}$$

Notice that the approximation of $(1-\rho_{min}) \approx 1$ was used in Eq. (15).

The predicted minimum reinforcement ratios for normal and high strength concrete members with different width b are shown in Figs. 6a and b. The specimen length (L = 1000 mm) and thickness (t= 50.8 mm), the ratio a_0= b/6 and ω = 0.86 were kept constant. The selected properties of steel were E_s = 210 GPa and f_{sy} = 500 MPa. Fig. 6a shows that ρ_{min} increases with increasing concrete strength and decreases with increasing specimen width. The qualitative comparison

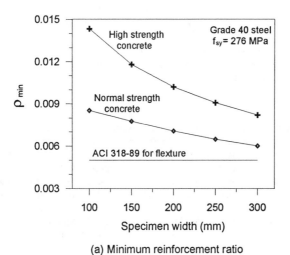

(a) Minimum reinforcement ratio

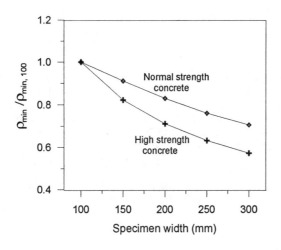

(b) Relative minimum reinforcement ratio

Fig.6 Minimum reinforcement ratios for high strength
and normal strength concretes

with the ACI requirement for flexural members shows the significance of the specimen geometry and the concrete strength regarding ρ_{min}. Fig. 6b shows that the size dependency of ρ_{min} is more pronounced for high strength concrete members. It can be shown that the increase of ρ_{min} with decreasing specimen width is higher in high strength concrete members than in normal strength concrete members. The stronger size effect may be attributed to the more brittle material behavior of high strength concrete.

Summary

Experimental results revealed that the first cracking strength of the members was affected by rebar distribution and reinforcement ratio. A previously proposed fracture energy approach was modified for better predicting experimentally measured composite stress-strain curves. The proposed fracture energy approach shows a good agreement with experimentally measured stress-strain curves of reinforced concrete tensile members. The model is also able to predict the minimum reinforcement ratios in tensile members. The predicted minimum reinforcement ratio increases with increasing concrete strength, as well as indicates a stronger size dependency for the high strength concrete members.

Acknowledgment

Support from the NSF Center for Science and Technology of Advanced Cement-Based Materials is appreciated.

References

Gilbert, R.I., (1992), "Shrinkage Cracking in Fully Restrained Concrete Members", ACI Structural Journal, Vol. 89, No.2, pp. 141-149.

Ouyang, C., Wollrab, E., and Shah, S. P., (1995), "Prediction of Cracking Response of Reinforced Concrete Tensile Members," Submitted to publication, Journal of Structural Engineering, ASCE.

Ouyang, C., and Shah, S.P., (1994), "Fracture Energy Approach for Predicting Cracking of Reinforced Concrete Tensile Members," ACI Structural Journal, Vol.91, No. 1, pp. 71-82.

Tada, H., Paris, P.C., and Irwin, G.R., (1985), Stress Analysis of Cracks Handbook, 2nd Edition, Paris Productions, St. Louis.

COMPRESSION FAILURE IN REINFORCED CONCRETE COLUMNS AND SIZE EFFECT

ZDENĔK P. BAŽANT[1], Fellow ASCE, and YUYIN XIANG[2]

Abstract

The size effect in the failure of columns or other reinforced concrete compression members is explained by energy release due to transverse propagation of a band of axial splitting cracks. According to their stress and strain states, three regions are distinguished in the column: cracking, unloading and invariable zones. The microslabs of the material between the axial splitting cracks are considered to buckle and undergo post-critical deflections. Based on the equality of the energy released from these regions and the energy consumed by formation of the axial splitting cracks in the band, the failure condition is formulated. It leads to a closed-form expression relating the characteristic size and the nominal strength of the structure. The results of laboratory tests of reduced-scale columns reported previously by Bažant and Kwon (1994) are analyzed according to the proposed formulation and are shown to be described by the proposed theory quite well. Although the present theory is formulated for a reinforced concrete column, it can be adapted to compression failures of other quasibrittle materials such as rocks, ice, ceramics or composites.

Introduction

Compression failure of quasibrittle materials has been studied extensively and important results have been achieved (Biot, 1965; Ashby and Hallam, 1986; Batto and Schulson, 1993; Horii and Nemat-Nasser, 1985, 1986; Kendall, 1978; Bažant, 1967; Bažant et al., 1991; Sammis and Ashby, 1986; Shetty et al., 1986). However, attention has been focused primarily on the microscopic

---

[1]Walter P. Murphy Professor of Civil Engineering and Materials Science, Northwestern University, Evanston, IL 60208.

[2]Graduate Research Assistant, Northwestern University.

mechanisms that initiate the compression failure rather than on the final global mode of failure and the size effect. Structures made of quasibrittle materials, such as concrete, rock, ice, ceramics and composites, are known to exhibit a significant size effect (Bažant, 1993a). The nominal strength of structure is not constant, as predicted for materials following yield or strength failure criteria, but decreases with an increasing size of structure. Previous studies have demonstrated that such a size effect exists in various tensile and shear failures, including the diagonal shear failure of reinforced concrete beams, torsional failure, punching shear failure, pullout of bars and anchors, etc.

The size effect on nominal strength of structure is due to the release of energy due to propagation of fracture or damage bands. Such a size effect must also be expected for compression failures in which the energy release depends on the structure size, for example, reinforced concrete columns. A general analysis of the size effect for compression failures has already been outlined in general terms (Bažant, 1993b). However, detailed formulas for the size effect have not been derived and comparisons with test results have not been made. This paper presents a more detailed theoretical analysis of the size effect in concrete columns and comparison with the test results. In full detail, the analysis of size effect in compression members will be presented in a forthcoming journal article.

Analysis of Energy Release

Consider a prismatic compression member (a column) loaded by axial compressive force P with eccentricity e (shown in Fig. 1). The column has length L, width D (taken as the characteristic dimension), and unit thickness $b = 1$. If one end cross section is subjected to axial displacement u and rotation θ and the other is fixed, linear elastic solution indicates that the initial normal stress in the cross sections before fracturing is

$$\sigma_0(x) = -\frac{E}{L}\left[u + \theta\left(\frac{D}{2} - x\right)\right], \tag{1}$$

in which E = Young's elastic modulus, and x = transverse coordinate measured from the compressed face (Fig. 1a).

Assume now that, at a certain moment of loading, axial cracks of spacing s and length h, forming a band as shown in Fig. 1a,b,c, suddenly appear, and that the microslabs of the material between the axial cracks, behaving as beams of depth s and length h, lose stability and buckle. This kind of buckling can happen in any of the three configurations shown in Fig. 1a,b,c. For all of them, the present type of approximate solution turns out to be identical if the length of the cracks in the pair of inclined bands in Fig. 1c is denoted as $h/2$. During the buckling of the microslabs, certain zones in the column undergo unloading. For the sake of simplified analysis, we assume that the column can be divided

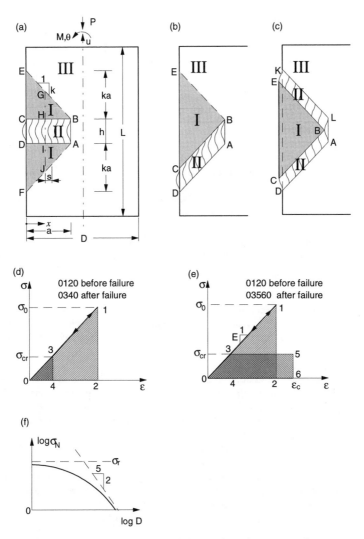

Figure 1: (a-c) Compression splitting cracks, (d,e) energy release, (f) size effect

into three regions (Fig.1): Region I—the elastic unloading region, shown as the shaded triangular area; region II—the slab buckling band, in which the microslabs are undergo post-critical deflections; region III—the region with no cracking nor unloading.

The key idea for failure analysis of the column is to calculate the change of stored strain energy caused by buckling (Bažant, 1993b). For region III, this calculation is not needed because no energy loss takes place. We need to calculate the energy losses only in the first two regions. The critical stress for buckling of the microslabs is

$$\sigma_{cr} = -\frac{\pi^2 E s^2}{3h^2}. \tag{2}$$

Region I is bounded by the so-called "stress diffusion lines" of slope k, whose magnitude is close to 1. The precise value of k could be obtained from the full elastic solution, but is not needed for the present purposes. For the analysis of size effect the important fact is that k is a constant if geometrically similar columns are considered. Under the assumption that the stress in the shaded triangular stress-relief zones is reduced all the way from the initial stress $\sigma_0(x)$ to σ_{cr}, the strain energy density before and after fracture is given by the areas of the triangles 0120 and 0340 in Fig. 1d. So the loss of strain energy density along a vertical line of horizontal coordinate x (Fig. 1a) is

$$\Delta \bar{\Pi}_r = \frac{\sigma_0^2(x)}{2E} - \frac{\sigma_{cr}^2}{2E}. \tag{3}$$

The situation is more complicated for the crack bands. The microslabs buckle, and the energy associated with the postbuckling behavior must be taken into account. Before the buckling of microslabs, the strain energy density is given by the area 0120 in Fig. 1a,e. After the buckling, the stored elastic strain energy is the area 03560. The analysis of postbuckling behavior of columns (Bažant and Cedolin, 1991, Sec. 1.9 and 5.9) indicates that the stress in the axis of the microslabs follows, after the attainment of the critical load, the straight line 35 which has a very small positive slope (precisely equal to $\sigma_{cr}/2$), compared to the slope E before buckling, and can therefore be neglected. Thus, line 35 can be considered to be horizontal. The triangular area 0340 in Fig. 1e represents the axial strain energy density of the microslabs, and the rectangular area 35643 represents the bending energy density. Because the microslabs remain elastic during buckling, the stress-strain diagram 035 is fully reversible and the energy under this diagram is the stored elastic strain energy. From the foregoing analysis, the change in strain energy density in Region II is the difference of areas 0120 and 03560 in Fig. 1e, that is,

$$\Delta \bar{\Pi}_c = \frac{\sigma_0^2(x)}{2E} - \left[\sigma_{cr} \epsilon_c(x) - \frac{\sigma_{cr}^2}{2E} \right] \tag{4}$$

where $\epsilon_c(x)$ is the axial strain of the microslabs in the crack band after buckling. It is important to note that $\epsilon_c(x)$ is generally not equal to 04 nor 02 in Fig. 1e, but can be determined from the compatibility condition. The stress in region III of the column, as an approximation, may be assumed to be unaffected by introduction of the crack and thus constant during failure, and so region III behaves as a rigid body. Consequently, the line segment GJ in Fig. 1a at any x does not change length. Expressing the change of length of this segment on the basis of σ_{cr}, ϵ_c and $\sigma_0(x)$, and setting this change equal to zero, one obtains

$$\epsilon_c(x) = \frac{\sigma_0(x)}{Eh}\left[h + 2k(a-x)\right] - \frac{2k}{h}(a-x)\frac{\sigma_{cr}(x)}{E} \tag{5}$$

Integrating (3) and (4) over their specific area yields the total loss of potential energy at constant u and θ:

$$\Delta\Pi = \int_0^a \left(\frac{\sigma_0^2(x)}{2E} - \frac{\sigma_{cr}^2}{2E}\right) 2k(a-x)dx + \int_0^a \left\{\frac{\sigma_0^2(x)}{2E}\left[\sigma_{cr}\epsilon_c(x) - \frac{\sigma_{cr}^2}{2E}\right]\right\} h dx \tag{6}$$

where a = horizontal length of the crack band (Fig. 1a,b,c). When the column is failing, $\Delta\Pi$ must be equal to the energy consumed by the formation of the surfaces of all the axial splitting cracks. Assuming that there is no other energy dissipation but fracturing, we may write the energy balance criterion of fracture mechanics as:

$$-\left[\frac{\partial\Delta\Pi}{\partial a}\right]_{u,\theta} = \frac{\partial}{\partial a}\left(G_f h\frac{a}{s}\right) = \frac{h}{s}G_f \tag{7}$$

where G_f is the fracture energy of the axial splitting cracks, assumed to be a material property.

It can be shown (Bažant, 1993b) that the foregoing equations yield a failure criterion of the form

$$F(k, a, s, h, G_f, \sigma_N) = 0, \qquad \sigma_N = \frac{P}{bD}\left(1 + \frac{6e}{D}\right) \qquad (b=1) \tag{8}$$

in which P = maximum load, and σ_N = nominal strength of the compression member. The value of the diffusion slope k can be approximately estimated as the k-value which gives the exact energy release rate for an edge-cracked tensile fracture specimen according to linear elastic fracture mechanics. The unknown spacing of the axial cracks, s, can be determined from the condition that load P be minimized, which requires that $\partial(\delta F)/\partial(\delta s) = 0$. It can further be shown that this condition indicates s to increase with size D as $D^{-1/5}$.

In the case of slender columns which undergo global buckling, the effect of slenderness on σ_N needs to be taken into account. This can be done by using

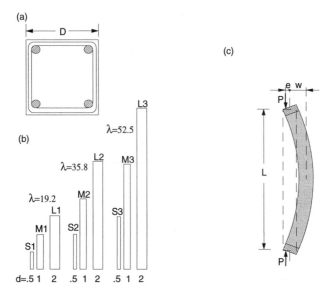

Figure 2: Reduced scale columns tested by Bažant and Kwon (1994)

one of the two concepts proposed in Bažant (1993, Eqs. 45–50). The simpler of these two concepts is based on modifying eccentricity e in Eq. 8 on the basis of the magnification factor for global buckling of columns (e.g., Bažant and Cedolin, 1991, Chapter 1).

Assuming the ratio a/D for failures of compression members of various sizes to be the same, the following relationship between the characteristic dimension D and the nominal strength σ_N can be deduced from the foregoing formulation.

$$D = k\frac{(\sigma_0 - \sigma_{cr} + \sigma_N)(\sigma_0 + \sigma_{cr} - \sigma_N)}{(\sigma_N - \sigma_{cr})}, \qquad \sigma_N = \frac{P}{bD} \quad (b = 1) \qquad (9)$$

in which σ_0, σ_{cr} = constants = critical normal compressive stress for the microslab buckling, which is the same as the intrinsic compression strength of the material.

Comparison to Test Results

The present formulation has been compared and calibrated with the test data for failure of reduced-scale reinforced concrete columns (Fig. 2) reported in

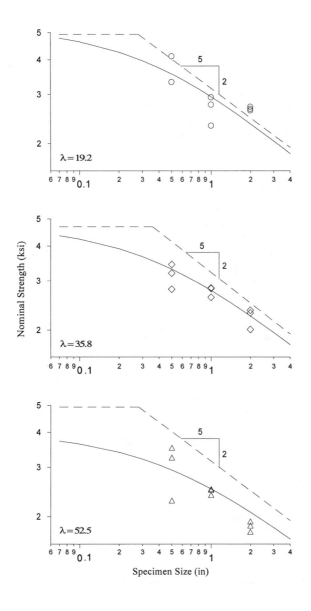

Figure 3: Size effect from experiment and from theory

Bažant and Kwon (1994). Fig. 3 shows the comparison between the experimental data and the theoretical results from Eq. 8. The data points are the experimental results for various column slendernesses $\lambda = \ell/r$; $r =$ radius of gyration of the cross section and $\ell =$ effective length of the column. The predictions according to the present theory are indicated by the curves. The horizontal asymptote of the size effect curve according to the strength theory and the inclined asymptote according to linear elastic fracture mechanics (which has the slope $-2/5$) are also marked in the figures. As one can see, reasonable agreement with the test results has been achieved.

It should be noted that the existing code procedures for concrete structures give no size effect, which is contradicted by the present test results.

Conclusion

The mechanism of compression failure of quasibrittle materials can be described as transverse propagation of a band of axial microcracks. Assuming the axial stress transmitted by the band to be limited by buckling of the microslabs of the material between the axial splitting racks, the failure loads can be calculated on the basis of the energy release. This calculation predicts a size effect which is in reasonable agreement with available reduced-scale laboratory tests.

ACKNOWLEDGEMENT: Partial financial support under NSF Grant MSS-9114426 to Northwestern University is gratefully acknowledged.

References

Ashby, M.F., and Hallam, S.D. (1986). The failure of brittle solids containing small cracks under compressive stress states. *Acta Metall.*, Vol. 34, No. 3, 497–510.

Barenblatt, G.I. (1979). *Similarity, self-similarity, and intermediate asymptotics.* Consultants Bureau, New York and London.

Barenblatt, G.I. (1987). *Dimensional analysis*, Gordon and Breach, New York.

Batto, R.A., and Schulson, E.M. (1993). On the ductile-to-brittle transition in ice under compression, *Acta metall. mater.*, 41(7), 2219–2225.

Bažant, Z.P. (1967). Stability of continuum and compression strength, (in French), *Bulletin RILEM*, Paris, No. 39, 99-112.

Bažant, Z.P. (1993a). Scaling Law in Mechanics of Failure, *J. of Engineering Mechanics ASCE* 119, in press.

Bažant, Z.P. (1993b). "Size effect in tensile and compressive quasibrittle failures." Proc., *JCI International Workshop on Size Effect in Concrete Structures*, held at Tohoku University, Sendai, Japan, October, 141–160.

Bažant, Z.P., and Cedolin, L. (1991). *Stability of Structures: Elastic, Inelastic,*

Fracture and Damage Theories (textbook and reference volume), Oxford University Press, New York, 1991 (984 + xxvi pp.).

Bažant, Z.P., Lin, F.-B., and Lippmann, H. (1991). Fracture energy release and size effect in borehole breakout, Structural Engineering Report 91-11/457f, Northwestern University; also *Int. J. of Num. and Anal. Methods in Geomechanics*, 17, 1-14.

Biot, M.A. (1965). *Mechanics of Incremental Deformations*, John Wiley & Sons, New York.

Horii, H., and Nemat-Nasser, S. (1985). Compression-induced microcrack growth in brittle solids: Axial splitting and shear failure, *J. of Geophys. Res.*, Vol. 90, 3105-3125.

Horii, H., and Nemat-Nasser, S. (1986). Brittle failure in compression, splitting, faulting and brittle-ductile transition, Phil. Tran. Royal Soc. London, 319(1549), 337-374.

Kendall, K. (1978). Complexities of compression failure, Proc. Royal Soc. London., A., 361, 245-263.

Sammis, C.G., and Ashby, M.F. (1986). The failure of brittle porous solids under compressive stress state, *Acta Metall*, 34(3), 511-526.

Shetty, D.K., Rosenfield, A.R., and Duckworth, W.H. (1986). Mixed mode fracture of ceramics in diametral compression, *J. Am. Ceram. Soc.*, 69(6), 437-443.

Constitutive Relationships of Mortar–Aggregate Interfaces in High Performance Concrete

Oral Büyüköztürk[1] and Brian Hearing[2]

Abstract

The fracture behavior of high performance concrete is often governed by the characteristics of mortar–aggregate interfaces. This study presents a fracture mechanics methodology to develop strain–softening constitutive relationships for mortar–aggregate interfaces and investigates the effect of mortar properties and aggregate surface conditions on the relationship. Physical interface specimens consisting of aggregate inclusions in mortar matrices are tested under mode I and mixed mode loading conditions. Interfaces with different properties are introduced by varying the strength of the mortar and the surface roughness of the aggregate inclusions. High strength mortars are prepared with silica fume and water–reducing admixtures. Fracture mechanics parameters for the different material property combinations are calculated from the results. Interfacial fracture energies and bond strengths are found to increase with mortar strength and surface roughness. These parameters are used to simulate the interfacial deformation behavior of the tested specimens in a fracture model. The fracture model is a cohesive type where the cohesive forces are related to the crack opening displacement through an interfacial constitutive relationship. Two constitutive relationships are investigated and the load/deflection curves obtained with the fracture models are compared to those obtained during physical specimen testing. A linear strain–softening relationship is found to best model cohesive behavior during fracture of mortar–aggregate interfaces in mode I and mixed mode crack propagation.

[1]Professor, Department of Civil and Environmental Engineering, Massachusetts Institute of Technology, Cambridge, MA 02139-4307

[2]Research Asst., Dept. of Civil and Environmental Eng., Massachusetts Institute of Technology, Cambridge, MA 02139-4307

Introduction

Considerable research has been conducted to study the damage process, deformation behavior, and failure mechanisms of concrete. The development of bond cracks at the mortar–aggregate interface often influence microcracking and failure of concrete; it has been shown that interfacial zones are the "weak link" in crack formation (Büyüköztürk et al., 1971; Büyüköztürk et al., 1972). It is generally aggreed that propagation and joining of cracks lead to failure of the material (Struble et al., 1980). Interest in the study of mortar-aggregate interfaces has increased with the development of high–performance concrete.

Several investigators and models have described the complex failure mechanisms of concrete. Perhaps one of the simplest yet widely accepted model of crack propagation is Hillerborg's single discrete crack model (Hillerborg et al., 1976). Using this model, many fracture mechanics approaches have described the nonlinear strain–softening behavior of concrete and its constitutive materials during fracture (Foote et al., 1986). It has been used to characterize the fracture behavior of concrete as a homogeneous material (Llorca and Elices, 1990) and as a composite where crack propagation occurs through its constitutive materials (Kitsutaka et al., 1993). However, few quantitative investigations of the interface between these materials have been performed. Such information is essential in the understanding of concrete composite fracture behavior and in developing high performance concrete materials.

This study investigates crack propagation within the interface of mortar and aggregate in concrete. At issue are the fracture properties of the interface and how they are affected by concrete characteristics such as mortar strength and aggregate surface roughness. A variety of interfacial combinations are tested through mortar beams with embedded granite inclusions. Two different strengths of mortar and two aggregate surface roughnesses are tested to compare differences in fracture behavior.

A modified Hillerborg model is employed to investigate the cohesive behavior of the tested interfaces during failure. The model uses experimental interfacial constitutive relationships to simulate the failure behavior and the load/load–line displacement curves of the tested composite beams. Two strain–softening material relationships are examined in this simulation to model the behavior of the interfacial region during failure.

Experimental Program

Sandwiched beam specimens were tested under the three point bending geometries shown in Figure 1 to investigate interfacial fracture under mode I and mixed mode loading conditions. Two different mortar strengths were combined with smooth and sandblasted granite surfaces to create a variety of concrete interfaces.

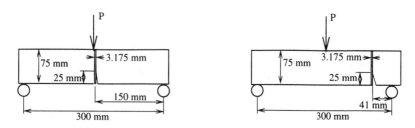

Figure 1: Three point bending for mode I and mixed mode loading.

Table 1: Material properties

Material	σ_c [MPa]	E [GPa]	ν
Normal strength mortar	40.0	27.8	0.2
High strength mortar	83.8	33.3	0.2
Granite	123.0	42.2	0.16

Materials

Two different strength mortars were combined with granite slices to create the interfacial specimens. The mortars were tested for compressive strength and modulus of elasticity. Poisson's ratio was assumed from previous research (Büyüköztürk and Lee, 1993). The properties of the tested mortars are reported in Table 1. All specimens were made using Type III cement to produce high early strength mortars. High strength mortars were manufactured using a naphtalene sulfonate type superplasticizer and condensed silica fume in slurry form.

The two roughnesses of the granite were characterized optically. Smooth surfaces were achieved by a diamond saw cut; the surface was smooth to the touch with no visible scratches. The sandblasted surface achieved a pit depth of about 0.5 mm with an average grain diameter of about 3.0 mm.

Scope of Tests

For each series of specimens six beams were cast. The relative initial crack length was held constant at 1/3 the beam depth; this crack was created by epoxy resin hardened onto the aggregate surface before the mortar was cast. The specimens were tested after 7 days on a 1–kip Instron testing machine and the load/load–line deflection curves from this machine were recorded. Peak failure loads were used to calculate fracture parameters of the various interfaces.

Table 2: Fracture loads and energy release rates

	Mode I testing			Mixed mode testing			
Series	P_u [kN]	G_f [J/m^2]	f'_t [MPa]	P_u [kN]	G_f [J/m^2]	f'_t [MPa]	f'_s [MPa]
Normal strength	0.215	1.423	1.501	0.904	2.146	1.682	0.350
High strength	0.310	2.949	2.160	1.207	3.824	2.245	0.468
High strength/ sandblasted	0.331	3.362	2.306	1.250	4.101	2.325	0.484

Fracture Parameters of Tested Specimens

Average failure loads P_u for the beams tested in each category are reported in Table 2. Fracture toughnesses for the interfaces are calculated from these loads and the stress intensity factors of the test specimens. For mode I loading, the three point bending stress intensity factor is calculated (Srawley, 1976) as

$$K_I = \frac{3SP}{2tW^2}\sqrt{a}f(\alpha), \qquad \alpha = a/W$$

where $f(1/3) = 1.08228$, a/W is the relative crack length, t is the specimen thickness, W is the specimen height, and S is the distance between supports.

For mixed mode loading, both mode I and mode II fracture toughnesses are computed. The stress intensity factors are analyzed (Hua et al., 1982) as

$$K_I = \frac{M}{tW^{3/2}}f_b$$

and

$$K_{II} = \frac{Q}{tW^{1/2}}f_s$$

where the factors f_b and f_s are computed for a relative crack length of $a/d = 1/3$ as $f_b = 7.03$ and $f_s = 1.00567$. Here M is the bending moment and Q is the shear force at the crack tip.

The critical interface fracture energy release rate is computed by

$$G_f = G_I + G_{II} = (K_I^2 + K_{II}^2)/E$$

The interface energy release rate G_f is computed for each category from the measured failure loads to compare the effects of different material combination, aggregate surface roughness, and loading ratio.

The ultimate tensile bond strengths f'_t and ultimate shear bond strengths f'_s of the interfaces are also reported in Table 2. They are calculated by

$$f'_t = \frac{M_{cr}(\frac{W}{2})}{I_{cr}}$$

and

$$f'_s = \frac{V_{cr}}{A_{cr}}$$

where M_{cr} and V_{cr} are the critical moment and shear at failure and I_{cr} and A_{cr} are the moment of inertia and area of the section reduced by the initial crack.

Results of Experimental Program

Fracture energies and bond strengths of the interfaces shown in Table 2 are found to increase with mortar strengths. Similarly, the fracture energies and bond strengths also increase with rougher aggregate surfaces. Load/load–line displacement curves for the tested beams were also recorded. Sample curves are shown in Figure 2. The post–peak failure shape of these curves provide a qualitative comparison of the interfaces. When compared to normal strength mortar, interfaces with high strength mortars appear more brittle. Interfaces with normal strength mortars in mixed mode appear more brittle than in mode I loading conditions. However, the interfaces of high strength mortar and sandblasted aggregate surfaces appear more ductile when compared to high strength mortar with smooth surfaces. This is more prominent in mixed mode loading samples and may be attributable to the sandblasted surface's greater capacity for aggregate interlock in shear.

Analytical Model

Cohesive models have been used to model fracture in mortars, aggregates, and concretes as homogeneous materials (Kitsutaka et al., 1993; Llorca and Elices, 1990). The fracture process is modeled with cohesive forces trailing the crack tip to simulate material plasticity, as illustrated in Figure 3. The strain–softening constitutive relationship of the material determines the cohesive forces in the fracturing material behind the crack tip. It has been concluded that a nonlinear strain–softening relationship best models fracture in most concretes.

Because fracture of the interfacial specimens appear brittle, this investigation examines both a bilinear and a linear strain–softening relationship to model the interfacial behavior. Figure 4 shows the relationships that are tested to relate cohesive forces with crack opening displacements during fracture. These constitutive relationships are used in a Hillerborg–type single discrete crack model to simulate the cohesive behavior in the interface of the tested specimens. This model is modified to investigate interfacial crack propagation and applied to

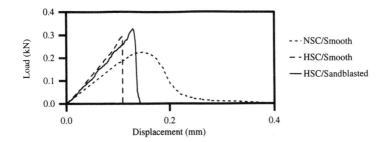

(a) Interface specimens in mode I loading

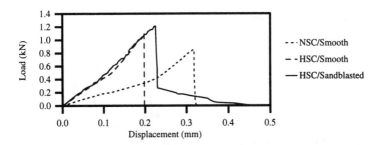

(b) Interface specimens in mixed mode loading

Figure 2: Sample load/load-line displacement curves of interfacial specimens

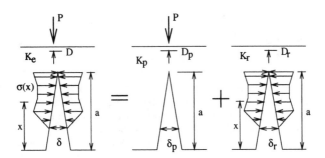

Figure 3: Mode I K-superposition

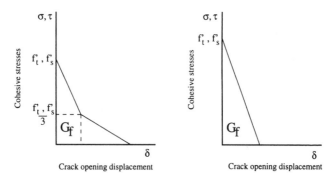

Figure 4: Bilinear and linear constitutive models

both the mode I and mixed mode testing configurations.

<u>Mode I</u>

Using an iterative computer model, simulated load/load–line displacement diagrams of the tested interfacial specimens are generated. The simulations allow a comparison of modeling accuracy to determine which constitutive relationship best models cohesive behavior of the interface.

The relation between load and load–line displacement is obtained by solving the equations of the net stress intensity factor (K_e), the equilibrium of crack opening displacement (COD) at the crack surface, and the constitutive function of the COD and cohesive stresses. The crack opening displacements are calculated by a K–superposition method. Figure 3 shows that there are two contributions to K_e. One is the stress intensity factor due to the applied load (P), K_p; the other, K_r, is due to the cohesive forces $\sigma(x)$ acting across the crack faces in the process zone.

$$K_e = K_p + K_r = 0$$

The equilibrium of crack opening displacement at the crack surface is given by

$$\delta = \delta_p + \delta_r$$

where δ_p is the COD due to the three point bending load and δ_r is the COD due to the cohesive stress. By solving the stress intensity equilibrium for the three point bending load and substituting it into the COD equilibrium, a relationship between the COD and the cohesive stresses is obtained.

A constitutive relationship for the material is needed to solve the problem.

It is given by an equation relating the COD to the cohesive stresses

$$\sigma = f(\delta)$$

By substituting this relationship into the COD/cohesive stress equation the crack opening displacements, $\delta(a, x)$, can be solved.

Corresponding cohesive forces found from the constitutive model are compared to assumed initial forces; when the COD and cohesive forces agree, the crack zone is in equilibrium. The external force P and load–line displacement D can be computed.

Mixed Mode

Hillerborg observed that the model is also valid for mixed mode loading (Hillerborg et al., 1976). A similar iterative computer model simulates the load/load–line deflection diagram of the sandwich specimen in mixed mode loading. The K–superposition includes both tensile and shear cohesive forces and may be expressed as

$$K_e = (K_{p_I} + K_{r_I}) + i(K_{p_{II}} + K_{r_{II}})$$

and the COD is given by

$$\delta = (\delta_{P_I} + \delta_{r_I}) + i(\delta_{P_{II}} + \delta_{r_{II}})$$

Utilizing Castigliano's theorem equations are derived relating mixed mode three point bending load to the cohesive stresses. By substituting the stress intensity equilibrium into the COD equilibrium a relationship between the COD and cohesive stresses is obtained.

Because both mode I and mode II stresses are present two constitutive relationships for the material are needed. The tensile relationship (similar to the relationship used in the mode I analysis) and the shear relationship are given by

$$\sigma = f(\delta_I)$$

and

$$\tau = f(\delta_{II})$$

The bilinear and linear relationships of figure 4 were used with calculated shear fracture properties to investigate shear behavior. Again by substituting these relationships into the COD/cohesive stress equations the crack opening displacements can be solved.

The solution is iterative in the same manner as the mode I simulation. An agreement is found between the crack opening displacements and the cohesive stresses. In equilibrium the external three point bending load is found as the sum of the load creating mode I stresses and the load creating mode II stresses.

$$P = P_I + P_{II}$$

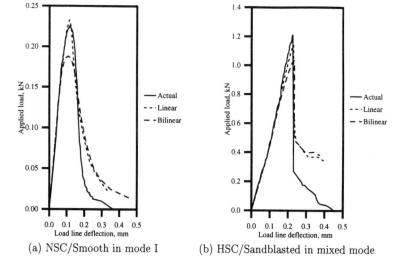

(a) NSC/Smooth in mode I   (b) HSC/Sandblasted in mixed mode

Figure 5: Experimental and analytical load/load-line deflection diagrams.

Similarly the three point bending displacement is found as the sum of the displacement due to mode I tractions and the displacement due to mode II tractions.

$$D = D_I + D_{II}$$

Results of Analytical Model and Discussion

Figure 5 shows sample results for the mode I and mixed mode simulations. The results of the simulations are plotted with the experimental results previously shown in Figure 2. A qualitative comparison shows that the behaviorial trends are similar to the experiments; i.e. relative post peak ductility is decreased with higher strength mortar and increased with surface roughness.

A comparison of the constitutive models show the linear relationship achieved a more accurate simulation of interfacial fracture behavior. The mode I simulation demonstrates that a linear strain-softening relationship best models tensile failure of the interface. Similarly, the mixed mode simulation demonstrates that combined tensile and shear failure in the interface is also best modeled with a strain-softening relationships.

Conclusion

The testing program of the interfacial specimens demonstrates that fracture energies and bond strengths of mortar–aggregate interfaces increase with mortar strength. Fracture energies and bond strengths also increase with rougher aggregate surfaces. Brittleness is observed to increase with higher mortar strengths and to decrease with rougher aggregate surface.

The investigation of strain–softening behavior resulted in successful modeling of interfacial crack propagation for both tensile (mode I) stresses and shear (mixed mode) stresses. Use of a linear strain–softening relationship is proven more accurate than bilinear in modeling both tensile and shear interfacial cohesive forces during fracture.

References

Büyüköztürk, O. and Lee, K. M. (1993). Assessment of interfacial fracture toughness in concrete composites. *Cement and Concrete Research*, 15.

Büyüköztürk, O., Nilson, A. H., and Slate, F. O. (1971). Stress-strain response and fracture of a concrete model in biaxial loading. *ACI Journal*, 68(8).

Büyüköztürk, O., Nilson, A. H., and Slate, F. O. (1972). Deformation and fracture of particulate composite. *Journal of Engineering Mechanics, ASCE*, 98(6).

Foote, R. M. L., Mai, Y.-W., and Cotterell, B. (1986). Crack growth resistance curves in strain–softening materials. *Journal of Mechanical Physics of Solids*, 34(6).

Hillerborg, A., Modéer, M., and Petersson, P.-E. (1976). Analysis of crack formation and crack growth in concrete by means of fracture mechanics and finite elements. *Cement and Concrete Research*, 6(6).

Hua, G., Brown, M. W., and Miller, K. J. (1982). Mixed–mode fatigue thresholds. *Fatigue of Engineering Materials and Structures*, 5(1).

Kitsutaka, Y., Büyüköztürk, O., and Lee, K. M. (1993). Fracture behavior of high strength concrete composite models. *Submitted for Publication in ASCE Journal of Engineering Mechanics*.

Llorca, J. and Elices, M. (1990). A simplified model to study fracture behavior in cohesive materials. *Cement and Concrete Research*, 20(1).

Srawley, J. E. (1976). Wide range stress intensity factor expressions for ASTM E 399 standard fracture toughness specimens. *International Journal of Fracture*, 12.

Struble, L., Skalny, J., and Mindess, S. (1980). A review of the cement–aggregate bond. *Cement and Concrete Research*, 10(2).

Size effects in the fracture of fiber reinforced materials

Roberta Massabó[1] and Alberto Carpinteri[2]

Abstract

A nonlinear fracture mechanics model is applied to analyze the flexural behavior of a beam made of a fibrous brittle matrix composite. The two assumptions, of a vanishing and a nonvanishing crack tip stress intensity factor, are examined, and the influence of matrix fracture toughness on mechanical behavior is investigated. Different types of response are predicted by varying dimensionless parameters, which depend on the mechanical and geometrical properties of the composite beam. Two transition in the flexural behavior, i.e. brittle–to–ductile and then the reversal ductile–to–brittle, are found when the beam depth is increased.

Introduction

The mechanical behavior of fiber reinforced materials undergoes substantial alterations when the mechanical and geometrical properties of the structural components are varied. The shape of the constitutive flexural relationship of a fiber reinforced cementitious material, for instance, can vary from strain-softening to strain-hardening when the fiber volume ratio is increased (Jenq and Shah, 1986). Moreover, variations in the characteristic dimensions of the element can induce modifications in the structural responses. The results of tests on high-strength fiber-reinforced concrete beams under three-point bending show that the flexural response changes from brittle to ductile when the beam depth increases (Gopalaratnam et al., 1995).

In order to explain these experimental results and to predict the behavior of the real–size structural components, a nonlinear fracture mechanics model is applied. The model has been formulated by Carpinteri and Massabó (1995.a,b) for the evaluation of the constitutive flexural relationship of brittle matrix materials with uniformly distributed reinforcements. Crack propagation in a cross section under bending is analyzed. In accordance with the Barenblatt-Dugdale model the bridged zone, where the fibers cross and bridge the crack

[1] Assistant Prof., Ist. Scienza Costruzioni, Universitá di Genova, Via Montallegro 1, 16145, Genova. [2] Professor, Dip. Ingegneria Strutturale, Politecnico di Torino, Corso Duca degli Abruzzi 24, 10129, Torino, ITALY.

faces, is replaced by a fictitious crack and a closing traction distribution. The two assumptions, of a vanishing and a non vanishing crack tip stress intensity factor, are examined, and the influence of matrix fracture toughness on macrostructural response is investigated.

The model

Carpinteri and Massabó (1995.a,b) formulated a nonlinear fracture mechanics model for the analysis of brittle–matrix composites. The model examines the evolutive process of crack propagation in a composite beam in bending, and defines the constitutive flexural relationship which characterizes the mechanical response of the component. The basic assumptions are briefly reviewed here.

Consider the schematic description of the cracked cross section of a composite beam, shown in Fig.1.a. The beam depth and thickness are h and b, respectively, and M is the applied bending moment. In accordance with the models of Barenblatt (1959) and Dugdale (1960), the crack of length a consists of a traction–free part of length a_r and a fictitious part of length a_f, acted upon by closing tractions σ_0. The fictitious crack can represent either a microcracked process zone ahead of a macrocrack or a macrocrack wake bridged by reinforcing elements. The reinforcement is assumed to be uniformly distributed and taken into account in the post–cracking loading phase through the closing tractions σ_0. The closing tractions are linked to the crack opening displacement w, according to an assigned relationship $\sigma_0(w)$, which has to be derived either from micromechanics models or experimental tests. If the reinforcement is present in a low volume ratio, the pre–cracking response of the composite coincides with that of the matrix, which is assumed to be linear–elastic. Reference is made to the two–dimensional single–edge notched–strip solutions (Tada et al. 1985) to define the fracture mechanics parameters.

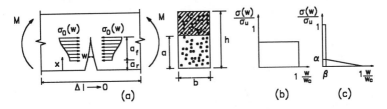

Fig.1.: (a) Schematic description of the cracked cross section in bending; (b) rectilinear bridging law; (c) two part cohesive law.

At the tip of the crack a global stress intensity factor (SIF), $K_I = K_{IM} - K_{I\sigma}$, can be defined by means of the superposition principle, K_{IM} and $K_{I\sigma}$ being the stress intensity factors due to the applied bending moment M and to a distribution of opening tractions σ_0, respectively.

Two different crack growth criteria are applied. If the matrix is assumed to be brittle and K_{IC} is the fracture toughness, a singular stress field exists at the tip of the crack, which starts propagating when K_I reaches the critical value K_{IC}. In this case the closing tractions σ_0 (bridging tractions), which control crack opening, are governed by the properties of the reinforcing phase and by its interaction with the matrix. On the other hand, a finite stress field can be assumed in the crack tip vicinity, provided the closing tractions σ_0 (cohesive tractions) represent the combined restraining action of the matrix and the secondary phases on crack propagation. In this case the crack propagates when the global crack tip stress intensity factor K_I vanishes. The matrix toughness is then combined with the toughening mechanism of the secondary phases. The damage process producing the propagation of the crack is the same as that governing the opening process along the process zone. The two crack growth criteria are given by

$$K_I = K_{IM} - K_{I\sigma} = \begin{cases} K_{IC} & nonvanishing\ SIF \\ 0 & vanishing\ SIF \end{cases} \tag{1}$$

By means of eq.1, the crack–propagation moment M_F, which defines the load–carrying capacity of the system, can be derived for each crack length. The corresponding localized rotation ϕ is calculated using Clapeyron's Theorem. The problem is a nonlinear statically indeterminate problem, the indeterminate closing tractions σ_0 depend on the unknown crack opening displacement w. A numerical iterative procedure has been formulated to evaluate the beam configuration satisfying both equilibrium and kinematics compatibility.

Dimensionless parameters

In accordance with Buckingham's II Theorem of Dimensional Analysis (Buckingham, 1915), the theoretical results can be synthesized by means of a functional relationship between the physical variables involved in the flexural response of the composite material. This relationship brings out the dimensionless parameters controlling the structural behavior. Physical similitude in the response is predicted when the mechanical and geometrical properties vary, provided the dimensionless parameters are kept constant.

Nonvanishing crack tip stress intensity factor

A nonvanishing stress intensity factor is assumed at the tip of the crack in the cross section of Fig.1 and the closing tractions σ_0 are defined as a power law of the crack opening displacement w, i.e. $\sigma_0 = \rho\sigma_u(w/w_c)^n$. Based on the first assumption, the toughening mechanisms proper of the brittle matrix and the toughening mechanisms of the secondary phases are separately modeled. The first set of mechanisms is controlled by the single parameter K_{IC} or matrix fracture toughness, whose physical dimensions are $[F][L]^{-1.5}$. The second set of mechanisms is controlled by the ultimate stress $\rho\sigma_u$, $[F][L]^{-2}$, the critical crack opening displacement w_c, $[L]$, and the exponent n, which describe the bridging relationship. The ultimate stress $\rho\sigma_u$ is given by the product of the fiber volume ratio ρ and the ultimate strength of the fibers σ_u (i.e., yielding or pulling–out strength). The crack propagation phenomenon as well as the constitutive flexural response of the composite beam, depend on the two toughening mechanisms and on the geometry of the component.

Let us assume the matrix toughness K_{IC} and the beam depth h as the fundamental set of dimensionally independent variables. The constitutive flexural relationship, linking the crack–propagation moment M_F to the localized rotation ϕ, can be given the following general form:

$$f\left(\frac{M_F}{K_{IC}h^{1.5}b}, \phi, \frac{\rho\sigma_u h^{0.5}}{K_{IC}}, \frac{Eh^{0.5}}{K_{IC}}, \frac{w_c}{h}, n; r_i\right) = f\left(\tilde{M}, \phi, N_P, \tilde{E}, \tilde{w}_c, n; r_i\right) = 0$$
(2)

whose terms are the dimensionless products obtained by multiplying the different variables involved by a suitable combination of the fundamental set: $\tilde{M}$ is the dimensionless crack–propagation moment, N_P is the dimensionless ultimate stress, $\tilde{E}$ is the dimensionless Young's modulus, $\tilde{w}_c$ is the normalized critical crack opening displacement and the r_i are the geometrical ratios describing the shape of the cross section. The Poisson ratio has been assumed to be negligible. The dimensionless parameter N_P is the brittleness number, formerly defined by Carpinteri for the description of the failure mechanisms in reinforced concrete beams (Carpinteri, 1984).

The dimensionless variables $\tilde{E}$ and $\tilde{w}_c$ control the propagation of the traction – free crack a_r, and they appear in the analytical formulation only as a product (see Carpinteri and Massabó, 1995.b). Therefore, if we fix the beam geometrical ratios and the exponent n of the bridging law, the structural behavior is controlled by two dimensionless parameters, namely N_P and $\tilde{E}\tilde{w}_c$:

$$f\left(\tilde{M}, \phi, N_P, \tilde{E}\tilde{w}_c\right) = 0,$$
$$N_P = \frac{\rho\,\sigma_u\,h^{0.5}}{K_{IC}},$$
$$\tilde{E}\tilde{w}_c = \frac{Eh^{0.5}}{K_{IC}}\frac{w_c}{h} = \frac{Ew_c}{K_{IC}h^{0.5}}.$$
(3)

Both the dimensionless parameters depend on the beam depth h, and consequently they do not characterize the composite material but the structural component.

These arguments hold for a generic bridging power law $\sigma_0(w)$. Different bridging mechanisms give rise to a different number of dimensionless parameters. Let us consider, for instance, a composite whose bridging mechanism can be modeled through a rigid–perfectly plastic bridging law $\sigma_0 = \rho\sigma_u$, i.e., a cementitious material continuously reinforced with long ductile wires (ferrocement). In this case the problem can be directly solved through the verification of the equilibrium condition (1) and the dimensionless parameter N_P proves to be the sole parameter controlling the behavior. The dimensionless diagram in Fig.2.a shows the theoretical moment versus rotation relationships of three beams having different brittleness numbers, that is N_P =0.3, 1.1 and 2.5. The responses have been evaluated by following the propagation process of a crack extending from a_0=0.1h to $a = 0.95h$. The initial crack of length a_0 has been assumed in this application, as well as in the others of the section, as a matrix crack crossed by the reinforcing elements. The horizontal dashed lines represent the ultimate moments for the totally disconnected sections, $\tilde{M}_u$=0.5N_P, and are asymptotes for the theoretical curves.

The model predicts a strain–softening response for the lower brittleness number N_P=0.3, a nearly perfectly plastic response for N_P=1.1 and a strain-hardening response for the greater brittleness number, $N_P = 2.5$. According to eq.(3), this brittle–to–ductile transition can be due to an increase in the fiber volume ratio ρ or in the ultimate stress σ_u, the beam depth h being kept unchanged, or to an increase in the beam depth h, the material properties being kept unchanged. Note that all the curves in the diagram are in the large–scale bridging regime, as the crack in the cross section is fully crossed by the reinforcing elements up to total disconnection.

The beam with $N_P = 0.3$ (curve A) shows a hyper–strength phenomenon, which is an indication of a peak loading capacity $\tilde{M}_a$ (circled in Fig.2.a) that is greater than the ultimate loading capacity $\tilde{M}_u$. The response of this beam in the first post–cracking phase is strongly affected by the matrix fracture toughness, which prevails over the secondary phase toughening action controlling the ultimate loading capacity (see next section). When the brittleness number increases, this effect turns into a snap–through instability, as shown in curve B. This discontinuity would be represented by a jump at constant load in a load–controlled process. Subsequently, as the brittleness number further increases, the local discontinuity disappears and the structural response is essentially controlled by the secondary–phase toughening mechanism, which gives rise to strain–hardening responses. Note that the hyper–strength phenomenon and the snap–through instability would disappear if the initial matrix crack length were increased and, for a_0 greater than a limit value depending on N_P, the structural response would result totally in strain–hardening. Likewise, an initial notch, without any restraining between the crack faces, would smooth the local discontinuities.

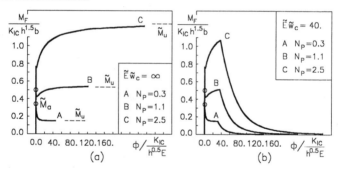

Fig.2: Dimensionless crack–propagation moment versus localized rotation diagrams for a composite cross section with $\tilde{E}\tilde{w}_c = \infty$ (a) and $\tilde{E}\tilde{w}_c = 40$ (b), as the brittleness number N_P varies.

If the bridging mechanism is characterized by a critical crack opening displacement value w_c, beyond which the bridging tractions vanish, the constitutive flexural response follows the functional relationship (3). Let us consider, for instance, a composite material whose bridging mechanism can be represented by a rectilinear law with a limit value of the crack opening displacement, $\sigma_0(w) = \rho\sigma_u$ if $w \leq w_c$, and $\sigma_0(w) = 0$ if $w > w_c$ (Fig.1.b). This law might represent the pullout of low–resistance ductile short fibers embedded in a cementitious matrix. The structural response of a beam with an initial crack of depth $a_0 = 0.1h$ has been examined.

The dimensionless moment versus rotation diagram in Fig.2.b has been obtained for $\bar{E}\bar{w}_c = 40$ and three different values of the brittleness number N_P, namely $N_P = 0.3$, 1.1 and 2.5. Note that the diagram of Fig.2.a, which has been previously discussed, describes a limit behavior of this material, for very high values of $\bar{E}\bar{w}_c$. Comparison between the two diagrams highlights the effects of $\bar{E}\bar{w}_c$ on the structural behavior. All of the curves in Fig.2.b define unstable structural responses, and are characterized by three branches. Consider, for instance, the beam with $N_P = 1.1$ (curve B). The initial linear branch describes the response until the crack starts to propagate. The second branch describes the large–scale bridging regime, when the bridging zone is invading the cross section and the crack is fully crossed by the fibers. After the cuspidal point, the beam reaches the small–scale bridging regime characterized by the third unstable branch. It points to a catastrophic crack propagation due to the advancement of the traction–free crack. This branch can be theoretically or experimentally predicted by reducing the applied moment, in a deformation–controlled process. The three different curves, with increasing N_P, might represent the constitutive flexural responses of a beam of constant depth when the fiber volume ratio increases, the fiber–bridging law and the other material properties being kept unchanged.

The dimensionless moment versus rotation diagrams shown in Fig.3 have been obtained by assuming three different values of the brittleness number, $N_P = 0.5$, 1.3 and 2.5 (Figs 3.a, 3.b and 3.c), and by varying, for each value of N_P, the dimensionless parameter $\bar{E}\bar{w}_c$.

For each fixed value N_P, the four diagrams I, II, III and IV show different type of behaviors which represent a ductile–to–brittle transition when the parameter $\bar{E}\bar{w}_c$ decreases. In Figure 3.b the global response of the beam with $N_P = 1.3$ is strain–hardening if $\bar{E}\bar{w}_c$ is greater than 250 (curve IV), while it is strain–softening if $\bar{E}\bar{w}_c$ is lower than 1.0 (curve I). The intermediate curves show more complex responses characterized by the three previously discussed branches. A somewhat analogous behavior is shown in Fig.3.c by the beam with $N_P = 2.5$, but in this case the local discontinuities, which follow the linear–elastic branches in the diagrams of Fig.3.b, are not found. The lower brittleness number $N_P = 0.5$, in Fig.3.a, gives rise to strain–softening responses. On the other hand, whereas the beam with the greater value of $\bar{E}\bar{w}_c$ (curve IV) is characterized by an ultimate value of the resistance moment greater than zero, albeit lower than the peak loading capacity, the other curves present a zero ultimate capacity.

Diagrams IV in the three sequences 3.a,b,c reproduce the constitutive flexural response of the composite beam when the dimensionless parameter $\bar{E}\bar{w}_c$ is very high. These curves coincide with the ones of a composite whose reinforcement presents a rigid–perfectly plastic bridging law, $\bar{E}\bar{w}_c \to \infty$ (Fig.2.a). Similar responses have been noticed in the tests carried out by Gopalaratnam et al. (1995) on fiber–reinforced concrete beams in bending.

For this particular composite material we cannot generally state, as we did for a rigid–plastic bridging law, that the mechanical behavior ranges from brittle to ductile when the structural size is increased. Let us consider the three curves in Figs. 3.a.IV, 3.b.IV and 3.c.III for $N_P = 0.5$, 1.3 and 2.5, respectively. These curves might characterize the constitutive flexural behavior of three beams made with the same composite material and presenting different depths. If we assume $w_c = 4mm$, $E = 40000 Nmm^{-2}$, $\sigma_u = 200 Nmm^{-2}$, $\rho = 0.02$, and $K_{IC} = 50 Nmm^{-1.5}$, the curves in Figs. 3.a.IV, 3.b.IV and 3.c.III are

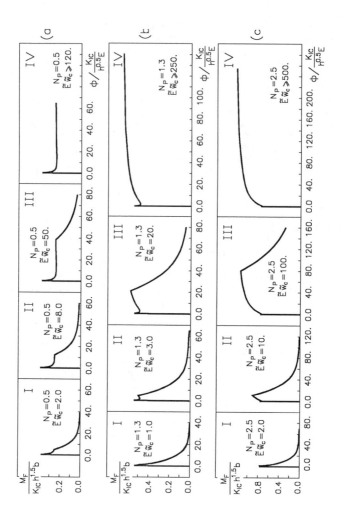

Fig.3: Dimensionless crack–propagation moment versus localized rotation diagrams for a composite cross section with N_P=0.5, 1.3 and 2.5, as the dimensionless parameter $\tilde{E}\tilde{w}_c$ varies.

respectively related to three beams of depths $h \simeq 40mm$, $h \simeq 260mm$, and $h \simeq 980mm$. Therefore the theoretical model predicts a brittle–to–ductile transition in the structural response when the beam depth is increased from $40mm$ to $260mm$, but an inversion in this trend, i.e., a ductile–to–brittle transition, is predicted when the beam depth is further increased in the range from $260mm$ to $980mm$.

Note that if a different power law were assumed in order to describe different kinds of bridging mechanism (e.g., a second order power law), the curves in Figs.2 and 3 would be generally smoothed, but the same transition regimes would be predicted.

Vanishing crack tip stress intensity factor

A vanishing stress intensity factor is now assumed at the tip of the crack. In this case the crack growth in the cross section is controlled by the combined matrix–secondary phase toughening mechanism. This mechanism is represented by the shielding effect that the cohesive tractions develop on the crack tip. Let us assume the cohesive traction $\sigma_0(w)$, characterizing the homogenized composite, as a function of the composite ultimate stress σ_u, the critical crack opening displacement w_c, and the exponent n, according for instance to the power law $\sigma_0(w) = \sigma_u(1 - (w/w_c)^n)$. Based on these assumptions the homogenized fracture toughness of the composite, K_{IC}, can be defined as a function of the area beneath the cohesive curve using the relationships $\mathcal{G}_{IC} = K_{IC}^2/E$ and $\mathcal{G}_{IC} = \int_0^{w_c} \sigma(w)dw$, $\mathcal{G}_{IC}$ being the composite fracture energy.

On assuming geometrical similarity, and K_{IC} and h as a basic set of dimensionally independent variables, and if the exponent n of the power law is kept constant, the functional relationship linking the cross–sectional resistance moment to the localized rotation can be given the following form:

$$f\left(\frac{M_F}{K_{IC}h^{1.5}b}, \phi, \frac{\sigma_u h^{0.5}}{K_{IC}}, \frac{Eh^{0.5}}{K_{IC}}\frac{w_c}{h}\right) = f\left(\tilde{M}, \phi, s, \tilde{E}\tilde{w}_c\right) = 0 \qquad (4)$$

in which s is the brittleness number, early defined by Carpinteri (1981) to describe the failure mechanism in a brittle homogeneous material. By equating the two above mentioned expressions of the fracture energy, the dimensionless parameters $\tilde{E}\tilde{w}_c$ and s and the exponent n of the cohesive law can be related. For the assumed power law, for instance, the relationship is $\tilde{E}\tilde{w}_c=(n + 1)s/n$. The dimensionless functional relationship (4) becomes:

$$f\left(\tilde{M}, \phi, s\right) = 0, \qquad s = \frac{K_{IC}}{\sigma_u h^{0.5}} \qquad (5)$$

in which s is the single governing parameter. Analogous arguments can be produced for any shape of the cohesive law.

The flexural behavior of cementitious composite beams has often been analyzed based on the assumption of a vanishing stress intensity factor (cohesive–crack model). The flexural response of a beam in three–point bending has been previously studied by Carpinteri (1989), on assuming a linearly decreasing cohesive law, and a ductile–to–brittle transition has been predicted as the beam depth increases. Such modeling of the cohesive law could be used for the description of a self–reinforced material, e.g., concrete.

In order to represent the flexural behavior of a fibrous composite, i.e., a fiber reinforced concrete, the proposed model is applied and the cohesive tractions are defined by the two part cohesive law, shown in Fig.1.c, with $\alpha = 0.1$ and $\beta = 0.001$. The first part of the relationship describes the matrix toughness and the second part the toughening mechanism of the secondary phases.

The moment versus localized rotation dimensionless diagrams in Fig.4 are obtained by analyzing a cross section with an initial crack depth of $a_0 = 0.05h$, for six different brittleness numbers, i.e., $s=5.0$, 1.0, 0.7, 0.5, 0.3 and 0.2. The moment versus rotation response changes substantially according to the variations in the brittleness number s. For $s = 5.0$ the cross section is characterized by a strain–softening response, for $s = 0.7$ a nearly strain–hardening response is predicted and for brittleness numbers lower than $s = 0.5$ the behavior again exhibits strain–softening. The model, therefore, reproduces the size–scale effect previously observed in Fig.3, according to which a double transition is predicted when the beam depth is increased.

The double transition, the local discontinuities and the hyper–strength phenomena shown by the curves in Fig.4 are due to the particular choice of the cohesive law, and these phenomena would not be predicted if the cohesive relationship were characterized by a generic power law.

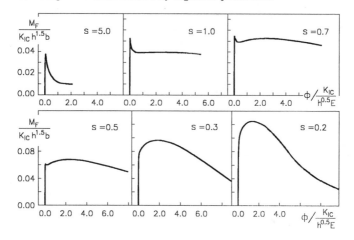

Fig.4: Dimensionless crack–propagation moment versus localized rotation diagrams for a composite cross section as the brittleness number s varies.

Influence of matrix toughness on structural response

Under particular conditions, the influence of matrix toughness on the overall behavior of the composite beam may be neglected. In this situation the application of the model based on the two different assumptions of a vanishing and a nonvanishing stress intensity factor leads to the same global results. Let us consider a composite material whose secondary–phase bridging mechanism can be simulated through a rigid–plastic bridging relationship $\sigma_0(w)=\rho\sigma_u$. The mechanical behavior of a cross section in bending, with an initial matrix crack of depth $a_0 = 0.1h$, is evaluated on assuming a singular stress field at the tip of the crack. Three different brittleness numbers are considered, namely $N_P = 0.5, 1.1$ and 2.5. Then, the theoretical model is applied based on the assumption of a finite stress field and with a cohesive law coincident with the previously assumed bridging law. By means of this second application, the flexural response of the composite material can be studied disregarding the effects of the matrix toughness.

In order to make a direct comparison of the results, a new set of dimensionally independent variables, $\rho\sigma_u$ $[F][L]^{-2}$ and h $[L]$, is chosen, and the dimensionless constitutive relationship links $M/(\rho\sigma_u h^2 b)$ to $\phi/(\rho\sigma_u E^{-1})$.

The predicted dimensionless relationships are shown in the diagram of Fig.5. The curves obtained based on the assumption of a vanishing stress intensity factor define strain–hardening response, whereas the curves obtained based on the assumption of a nonvanishing stress intensity factor show the brittle–to–ductile transition previously discussed in Fig.2.a, when the dimensionless parameter N_P is increased.

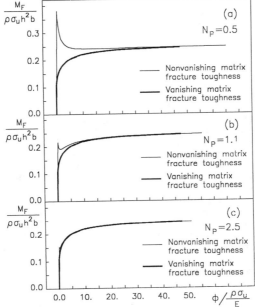

Fig.5: Dimensionless crack–propagation moment versus localized rotation diagrams for a composite cross section with N_P=0.5, 1.1 and 2.5, and a rigid–plastic bridging relationship. Comparison between the constitutive curves predicted by considering or not the matrix fracture toughness.

The ultimate moments, for the totally disconnected sections, are the same for both the curves in each diagram, as they are not affected by the matrix toughness which, on the other hand, strongly controls the remaining part of the curves. For a high value of N_P, the two curves define almost the same global response. This result indicates that the structural response is merely controlled by the toughening action of the reinforcing elements. This situation can be obtained when the material properties are characterized by a low ratio $K_{IC}/(\rho\sigma_u)$, and also for very deep beams. On the other hand, the two curves in each diagram profoundly differ for the lower values of N_P. In this case the matrix toughness strongly affects the structural performance during the entire loading phase. This situation characterizes composites with a high ratio $K_{IC}/(\rho\sigma_u)$, but also shallow beams. A different choice of the bridging law would modify the shape of the curves shown in Fig.5 but the general considerations continue to hold.

In conclusion, disregarding the matrix toughness does not affect the quality of the results in the theoretical simulation of the mechanical behavior of a composite element with an high value of the dimensionless parameter N_P. On the other hand, the same assumption in a composite with a low N_P, induces an underestimation of the cross section loading capacity and an erroneous prediction of the shape of the flexural constitutive curve.

It is worth noticing that a perfect agreement between the flexural relationships predicted based on the two assumptions would be obtained, for each value of N_P, provided the cohesive law were properly defined to represent the toughening mechanism of the homogenized composite (Carpinteri and Massabó, 1995.b).

Conclusions

The flexural behavior of a fibrous brittle–matrix composite has been analyzed by means of a nonlinear fracture mechanics model previously formulated by Carpinteri and Massabó (1995.a,b). The two assumptions, of a vanishing and a nonvanishing crack tip stress intensity factor, have been examined. In the first case the composite is modeled as homogeneous. A global toughening mechanism, which describes the combined restraining action of the matrix and the secondary phases on the crack propagation, is considered, and a single dimensionless parameter is found to control the macrostructural behavior. In the second case the composite is modeled as a biphase material, the two toughening mechanisms of the matrix and the fibers are separately taken into account, and two dimensionless parameters are found to control the behavior.

Two different regimes are predicted in the flexural response of a beam when the depth is increased. In the first regime, a brittle to ductile transition, typically encountered on fiber reinforced concrete beams of limited depth, is observed. In this regime the beams are characterized by a bridged zone which covers the total crack length up to the complete disconnection of the beam.

On increasing the beam depth further, a second transition, from ductile to brittle responses, is observed. This is characterized by the propagation of a traction–free crack. Quasi–brittle materials, such as concrete, typically show this transition in the range of depths normally covered by the laboratory specimens, due to the weak bridging mechanism developed by the aggregates. Such behavior has not been observed in fiber reinforced concrete specimens, due to the fact that the ductile–brittle transition occur at a beam depth much larger than those normally tested in the laboratory.

The beam depth at which the two transitions occur can be theoretically defined as a function of the mechanical properties of matrix and fibers, and of the properties of the fiber–matrix interface. The proposed model can be used to "engineer" the composite material, so that unstable responses will be avoided in the flexural behavior of real–size structural components.

References

Barenblatt, G.I. (1959), The formation of equilibrium cracks during brittle fracture. General ideas and hypotheses. Axially–symmetric cracks, *J. of Applied Mathematics and Mechanics*, **23**, 622–636.

Buckingham, E. (1915), Model experiments and the form of empirical equations, *Trans. ASME*, **37**, 263–296.

Carpinteri, A. (1981), Static and energetic fracture parameters for rocks and concretes, *Materials and Structures*, **14**, 151–162.

Carpinteri, A. (1984), Stability of fracturing process in r.c. beams, *J. Struct. Engng.*, **110** (3), 544–558.

Carpinteri, A. (1989), Cusp catastrophe interpretation of fracture instability, *J. Mech. Phys. Solids*, **37**, 567–582.

Carpinteri, A. and Massabó, R. (1995.*a*), Nonlinear fracture mechanics models for fibre reinforced materials, Proc. Int. Symp. on *Advanced Technology on Design and Fabrication of Composite Materials and Structures*, (eds. Sih, G.C., Carpinteri, A. and Surace, G.), Torino.

Carpinteri, A. and Massabó, R. (1995.*b*), Bridged versus cohesive crack and double brittle–ductile–brittle size–scale transition in flexure, submitted for publication.

Dugdale, D.S. (1960), Yielding of steel sheets containing slits, *J. Mech. Phys. Solids*, **8**, 100–104.

Gopalaratnam, V.S., Gettu, R., Carmona, S. and Jamet D. (1995), Characterization of the toughness of fiber reinforced concretes using the load–cmod response, in *Fracture Mechanics of Concrete Structures*, proc. FRAMCOS–2, ed. F.H. Wittman, Aedificatio Publ., Freiburg, 769–782.

Jenq, Y.S. and Shah, S.P. (1986), Crack propagation in fiber–reinforced concrete, *J. Engng. Mech.*, **112**, 19–34.

Tada, H., Paris, P.C. and Irwin, G. (1985), *The Stress Analysis of Cracks Handbook*, Paris Productions Incorporated (and Del Research Corporation), St. Louis, Missouri.

Matrix First Cracking Strength in Continuous Fiber Cement Composites

Jamil M. Alwan[1] and Antoine E. Naaman[2]

Abstract

The fundamental concept behind the fracture of a brittle matrix subjected to a tensile load is the initiation and propagation of small cracks and flaws which extend and inter connect until irreversible crack propagation leading to rupture occurs. Cracking behavior of composites consists of a combination of sequence of fracture of individual constituents or interface separation between reinforcement and matrix. If these fractures are locally restrained from extending into adjacent materials, then first cracking strength as well as the resistance of the material in the post cracking stage can be significantly improved. The increase in the first cracking strength of continuous fiber cement composites have been reported in many studies as a function of the spacing between wires which causes a reduction in the length of critical flaws. In this study a fracture mechanics concept is deployed to model the physical interconnection of microcracks in the process of forming larger macrocracks, and thus determining the matrix first cracking strength in the presence of fibers. Therefore, the occurrence of the first composite crack can be referred to microcracks joining and connecting in a plane normal to the direction of reinforcement. The fracture mechanics model is based on the elastic solution of *Westergaard* (1939), *Irwin* (1958) and *Koiter* (1959) for an infinite row of evenly spaced collinear cracks in an isotropic material. A parametric analysis is carried out in which the first cracking strength of a cement matrix is computed for various steel fiber diameters, different combination of reinforcement layers, as well as fiber volume fraction. A very good correlation is observed between model predictions and experimental data.

Introduction

A composite material consists of a matrix and reinforcement which act together to form a new material designed to exhibit better characteristics than either of its constituents alone. Examples include cement based composites. The reinforcement of such composites can be classified into two types. Continuous long fibers (wires or meshes) and discontinuous short fibers. The mechanical response of

---

[1]Research Engineer/Analyst, CAEtech Inc., Computer Aided Engineering Technology, 38701 West Seven Mile Rd., Suite 130, Livonia, MI 48152.

[2]Professor of Civil Engineering, University of Michigan, Dept. of Civil and Environmental Engineeirng, 2305 G.G. Brown Bldg., Ann Arbor, MI 48109.

cement composites is greatly affected by the type and amount of fibers used. Continuous metallic fibers embedded in a brittle cementitious matrix cause the latter to assume ductile characteristics similar to those of the reinforcement. This type of behavior is refered to as the brittle matrix-ductile fiber behavior. Reinforcing cement matrices with discontinuous fibers is also another alternative to enhance the mechanical response of such composites. However, it was only recently with the introduction of High Performance composites like SIFCON (*Lankard* and *Nowell* 1984), that short fiber cement composites were shown to exhibit tension stiffening and multiple cracking. Regular short fiber composites are still labeled with the tension softening behavior. Moreover, the amount of energy absorbed and crack width are more easily controlled with continuous reinforcement.

Mechanical Response Model

A number of analytical models have been developed in an attempt to describe the mechanical response in tension of continuous fiber reinforced cement composites. These models are based on composite mechanics (*Naaman* 1970, and *Somayaji* 1979), energy approach (*Aveston, Cooper*, and *Kelly* 1971; *Aveston*, and *Kelly* 1973), and the inclusion method (*Yang, Mura* and *Shah* 1991; *Li, Mura*, and *Shah* 1992).

In *Alwan* 1994, a new model is introduced to predict the tensile stress-strain response of continuous fiber reinforced cement composites. The model utilizes principles of fracture mechanics to determine first cracking strength of the cement composite that is assumed to depend on fiber reinforcing parameters. This technique yields the experimentally observed increase in the first cracking strength of the matrix due to an increase in the fiber volume fraction V_f and specific surface of the reinforcement S_R (lateral surface of fibers per unit volume of composite). A composite mechanics approach is then used to compute internal stresses due to external loading of the composite. During multiple cracking, and as the composite is divided into smaller and shorter segments, the matrix cracking strength is updated using Weibull statistics to account for one dimensional size effect phenomena. Comparisons of model output with experimental results from the technical literature is carried out. Good agreement between model output and experimental results is observed (Fig.1). The objective of this paper is to present the adopted technique in modeling the enhancement of the first cracking strength of the composite in presence of fibers.

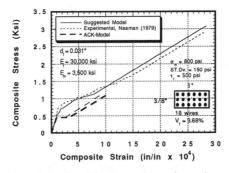

Fig. 1 - Comparison of model output and experimental results.

<u>Theoretical Consideration</u>

A typical tensile stress-strain response of a brittle matrix reinforced with ductile continuous fibers is shown in Fig. 2. The response consists of three regions. The first region is characterized by an elastic response, where the load is directly proportional to the composite deformation. The second region is characterized by a multiple cracking stage with a slight increase in the composite load-carrying capacity.

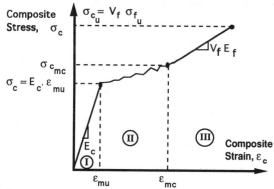

Fig. 2 - Typical stress-strain response of a continuous fiber reinforced cementitious composite under tension.

Finally, the third region known as the post-multiple cracking region is characterized by a linear response where the load-carrying capacity of the composite is determined only by the reinforcing fibers. Eventually, the composite fails as the reinforcing fibers are stretched out to their ultimate strength. Given such a response, one can clearly identify key points needed to model the composite behavior in tension. The key parameters are

E_c = the composite elastic modulus

σ_{mu} = the composite stress at first crack

ε_{mu} = the composite strain at first crack

ε_{mc} = the composite strain at the end of multiple cracking

σ_{mc} = the composite stress at the end of multiple cracking.

σ_{cu} = the composite ultimate stress.

and,

E_m, E_f = the matrix and fiber elastic moduli respectively.

Among other assumptions, under tensile stresess, structural cracking in the matrix is assumed to occur when all the minor cracks are joined to form a larger macrocrack crossing the entire width of the specimen (Fig. 3). This leads to the following discussion.

<u>Matrix First Cracking Strength</u>

The fundamental concept behind the fracture of a brittle matrix subjected to a tensile load is the initiation and propagation of small cracks and flaws which extend and inter connect until the whole internal structure is completely disrupted. Cracking behavior of composites consists of a combination of sequence of fracture of individual constituents or interface separation between reinforcement and matrix. If

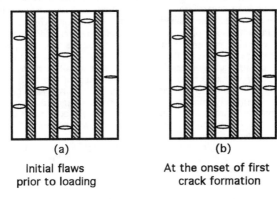

(a)   (b)

Initial flaws   At the onset of first
prior to loading   crack formation

Fig. 3 - Inter connectivity of microcracks. [Ref. *Shah (1991)*]

these fractures are locally restrained from extending into adjacent materials, then first cracking strength as well as the resistance of the material in the post cracking stage can be significantly improved provided sufficient length and volume fraction of reinforcing fibers are embedded in the matrix.

The increase in the first cracking strength of continuous fiber cement composites have been reported in many studies. *Romualdi* and *Batson* (1963, and 1964) tried to link the increase of matrix first cracking strength to the spacing between wires which causes a reduction in the length of critical flaws. Since that work, considerable research has been undertaken, where the strain at first cracking has been carefully measured. Based on these tests, it is now generally accepted that the type of fibers currently used do not significantly affect the first cracking stress of fiber reinforced composites. However, the first cracking stress is affected by the amount and geometry of reinforcing fibers. *Naaman* (1970) reported a linear relationship between composite first cracking strength and the specific surface of reinforcement S_R (lateral surface of steel per unit volume of composite) for continuous fiber composites. Similar results were reported by *Somayaji* and *Shah* (1981), and *Shah* (1991). Because the crack size was constant (i.e. these were structural cracks), this contradicts the first cracking stress definition of Romualdi and Batson which is based on Grffith's Formula for a single crack,

$$\sigma = \sqrt{\frac{G_c E}{(1-v^2)\pi a}} \tag{1}$$

where, σ = average tensile stress at failure
$\quad G_c$ = critical elastic energy release rate
$\quad E$ = modulus of elasticity
$\quad v$ = poisson's ratio
$\quad a$ = half the length of the critical flow.

One shall note here that G_c is the critical rate of releasing elastic energy per unit area of crack propagation, and that it is different than G_f which represents the total fracture energy of the material (the mattrix in this case). G_f is mobilized at displacements several times larger than that at the peak load, while G_c represents the energy release rate at the oncet of crack propagation. However, the idea of relating the matrix first cracking strength to the matrix critical fracture energy is very appealing and physically correct. This justifies the use of linear elastic fracture mechanics concepts to model the physical interconnection of microcracks in the process of forming larger macrocracks, and thus determining the matrix first cracking strength. In Fig. 3, *Shah* (1991) shows the cracking process where small microcracks have eventually joined together and formed a large macrocrack which represents the composite first effective crack. Therefore, the occurrence of the first composite crack can be referred to microcracks joining and connecting in a plane normal to the direction of reinforcement.

Westergaard (1939), *Irwin* (1958) and *Koiter* (1959) gave solutions to an infinite row of evenly spaced collinear cracks in an isotropic material as depicted in Fig.4. The applied tensile stress is defined as

$$\sigma = \frac{\sqrt{G_c E}}{\sqrt{\pi l} \sqrt{\frac{2}{\pi} tan(\pi a/2l)}} \tag{2}$$

where, G_c = matrix critical elastic energy release rate (as in Eq. 1)
$2l$ = center-to-center distance between adjacent cracks
$2a$ = crack length.
Using the same assumption as *Romualdi* and *Batson* (1963), and assuming matrix isotropy, we can say that the cracks in Eq. (2) are separated by the fibers as in Fig. 3(b). Thus, the first cracking strength of the matrix is given by Eq. (2) where,
$2l$ = center to center distance between fibers.
$2a$ = distance between fiber edges = $(2l - d_f)$, where d_f = fiber diameter.
and, E = composite elastic modulus, approximated by the rule of mixtures.

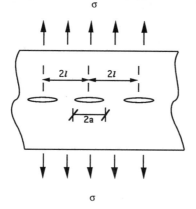

Fig. 4 - Infinite plate with collinear cracks.

Fig. 5 explains the concept of superposing the reinforcement geometry on the infinite plate with collinear cracks. However, this crack configuration can occur in one plane of reinforcement at a time. To obtain the total first cracking strength due to multiple layers of reinforcement, one has simply to add the contribution of each layer by multiplying the one-layer cracking strength by the number of reinforcement layers.

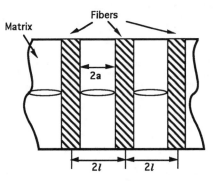

Fig. 5 - Assumed cracking spacing at first macrocrack.

Fig. 6 shows the extension of the one-layer strength to multiple layers of reinforcement. Therefore, the matrix first cracking strength of a continuous fiber composite reinforced with multiple layers is given by,

$$\sigma = \frac{\sqrt{G_c E}}{\sqrt{\pi l} \sqrt{\frac{2}{\pi} \tan(\pi a / 2l)}} \times \textit{(number of layers)} \qquad (3)$$

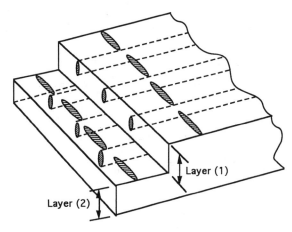

Fig. 6 - Macrocracks in each reinforcement layer.

A parametric analysis was carried out in which the first cracking strength of a cement matrix was computed for various steel fiber diameters, different combination of reinforcement layers, as well as fiber volume fraction. The matrix fracture energy was considered the same for all cases to minimize variation. All the numerical data is summarized in Table 1. Fig. 7 shows a plot for all simulated data in open symbols. Solid symbols represent experimental measurement obtained by *Somayaji* and *Shah* (1981), and *Yang, Mura*, and *Shah* (1991). A very good correlation is observed between Eq. (3) predictions and experimental data. Fig. 8 shows another plot of the same data presented in Table 1. The plots, however, are grouped according to the number of reinforcement layers. It is found that the matrix first cracking strength increases almost linearly with the increase of the ratio of specific surface of reinforcement to the fiber spacing. This means smaller flaw size density in the virgin composite. This also seems to confirm observations on the first cracking strength of ferrocement composites.

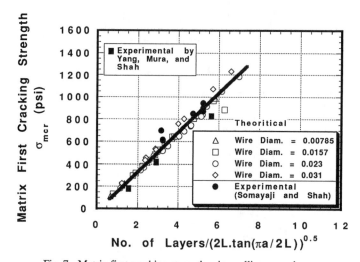

Fig. 7 - Matrix first cracking strength using collinear cracks concept.

Conclusion

This study shows how fracture mechanics is used to evaluate a key parameter, here the matrix first cracking strength in the behavior of cement composites. This can be later integrated in a composite-mechanics-based model to predict the mechanical response of cement based composites subjected to uniaxial tension. The evaluation of the first cracking strength of the cementitious matrix by the fracture-mechanics-based model came in good agreement with experimental data reported in the literature.

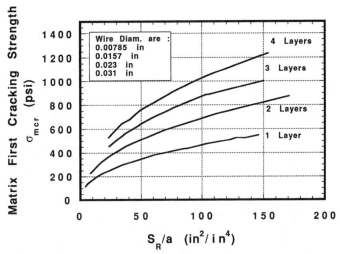

Fig. 8 - Matrix first cracking strength vs. specific surface of reinforcement to fiber spacing ratio.

Acknowledgement

This research was supported in part by a grant from the National Science Foundation to the NSF Center for Advanced Cement Based Materials. The center is a consortium of five institutions: Northwestern University, University of Illinois at Urbana Champaign, University of Michigan, Purdue University, and the National Institute of Standards and Technology. Any opinions, findings and conclusions expressed in this study are those of the authors and not necesserily reflect the views of NSF or the ACBM center.

Table. 1 - Matrix average cracking strength for different fiber volume fractions, fiber diameters, and number of reinforcement layers.

Fiber Diameter (in)	No. of fibers per layer	No. of layers	Total No. of fibers	V_f (%)	S_R (in^2/in^3)	S_R/a (in^2/in^4)	σ_{matrix} crack (psi)
	5	1	5	0.06	0.33	4.14	113
	10	1	10	0.13	0.65	15.8	208
	5	2	10	0.13	0.65	8.28	226
	10	2	20	0.26	1.31	31.67	416
0.00785	6	4	24	0.31	1.57	23.30	529
	7	3	21	0.27	1.38	23.57	453
	10	3	30	0.39	1.97	47.5	625
	8	4	32	0.40	2.1	40.75	681
	10	4	40	0.50	2.6	63.34	833
	12	4	48	0.62	3.15	91.39	985
	10	3	30	1.54	3.19	105.0	886
	8	2	16	0.82	2.1	44.0	481
	6	3	18	0.92	2.36	37.20	560
	5	2	10	0.51	1.31	17.40	319
	5	4	20	1.03	2.63	34.0	639
	7	3	21	1.08	2.76	50.50	641
0.0157	8	4	32	1.65	4.2	88.20	963
	9	3	27	1.39	3.55	84.20	804
	7	1	7	0.36	.93	16.80	213
	4	1	4	0.20	.52	5.70	133
	10	1	10	0.51	1.30	35.0	297
	13	1	13	0.67	1.70	61.0	381
	18	1	18	0.93	2.37	128	527
	5	1	5	0.55	0.96	13.4	184
	5	2	10	1.10	1.92	26.8	368
	5	4	20	2.20	3.85	53.6	737
	10	2	20	2.21	3.85	113.2	689
	7	3	21	2.32	4.04	79.3	742
	8	3	24	2.65	4.62	104	838
0.023	6	3	18	1.99	3.46	57.8	647
	15	1	15	1.67	2.89	146.3	518
	12	2	24	2.65	4.62	171.0	824
	8	2	16	1.77	3.08	70.0	559
	10	1	10	1.10	1.92	56.70	344
	12	1	12	1.33	2.30	85.75	436
	13	1	13	1.44	2.50	103.0	473
	14	1	14	1.55	2.70	123.0	510
	6	4	24	2.65	4.60	77.0	914
	7	4	28	3.10	5.40	105.0	1048
	8	4	32	3.54	6.16	140.0	1184

Cont'd

Cont'd

	5	1	5	1.00	1.29	19.2	228
	5	2	10	2.00	2.60	38	456
	4	4	16	3.20	4.15	49.2	757
	6	2	12	2.40	3.11	55.7	535
0.031	6	3	18	3.60	4.67	83.5	803
	12	1	12	2.40	3.10	135	525
	10	1	10	2.00	2.60	86.7	434
	9	4	24	4.80	6.23	111	1071
	8	4	32	5.60	7.27	154	1232

References

Alwan, J. M., "Modeling of The Mechanical Behavior of Fiber Reinforced Cement Based Composites Under Tensile Loads", *Ph. D. Thesis*, Dept. of Civil and Envir. Engineering, University of Michigan, Ann Arbor, 1994.

Aveston, J., and Kelly, A.,"Theorey of Multiple Fracture of Fibrous Composites", *Journal of Materials Science*, vol. 8, 1973, pp. 352-362.

Aveston, J., Cooper, G. A., and Kelly, A.,"Single and Multiple Fracture. The Properties of Fiber Composites", *Conference Proceedings of National Physical Laboratory*, IPC Science and Technology Press Ltd., 1971, pp. 15-25.

Irwin, G. R., Fracture, Handbuch der Physik, vol. vi, Springer 1958, pp. 551-590.

Koiter, W. T.,"An Infinite Row of Collinear Cracks in an Infinite Elastic Sheet", Ingenieur-Archiv, vol. 28, 1959, pp. 168-172.

Lankard, D. R., and Nowell, J. K.,"Preparation of highly reinforced steel fiber reinforced concrete composites", *Fiber Reinforced Concrete (ACI-SP 81)*, American Concrete Institute, Detroit, 1984, pp. 286-306.

Li, S. H., Mura, T., and Shah, S. P.,"Multiple Fracture of Fiber-Reinforced Brittle Matrix Composites Based On Micromechanics", *Engineering Fracture Mechanics*, vol. 43, no. 4, 1992, pp. 561-579.

Naaman, A. E.,"Reinforcing Mechanisms in Ferro-Cement", *Master of Science Thesis*, Department of Civil Engineering, Massachusetts Institute of Technology, 1970.

Romualdi, J. P., and Batson, G. B.,"Mechanics of Crack Arrest in Concrete", *Journal of Engineering Mechanics, ASCE*, February 1964, pp. 167-173.

Romualdi, J. P., and Mandel, J. A.,"Tensile Strength of Concrete Affected by Uniformly Distributed and Closely Spaced Short Lengths of Wire Reinforcement", ACI Journal, Proceedings, vol. 61, no. 6. June 1964, pp. 657 -670.

Shah, S. P.,"Do Fibers Increase The Tensile Strength of Cement-Based Matrixes?", *ACI Materials Journal*, vol. 88, no. 6, November-December 1991, pp. 595-602.

Somayaji, S., and Shah, S. P.,"Bond Stress-Slip Relationship and Cracking Response of Tension Members",*ACI Journal*, vol. 78, no. 20, 1981, pp. 217-225.

Westergaard, H. M.,"Bearing Pressures and Cracks", *Journal of Applied Mechanics*, vol. 61, 1939, pp.A49-53.

Yang, C. C., Mura, T., and Shah, S. P.,"Micromechanical Theory and Uniaxial Tensile Tests of Fiber Reinforced Cement Composites", *J. of Materials Research Society*, vol. 6, no. 11, November 1991, pp. 2463-2473.

Damage Caused by Projectile Impact to High Strength Concrete Elements

A.N. Dancygier[1]

Abstract

According to common penetration equations the resistance of concrete elements to projectile impact increases with increasing concrete strength. Therefore it is expected that using high strength concrete (HSC) will enhance the performance of elements that are designed to resist projectile impact. Normal strength concrete (NSC) and HSC plates were impacted, in an experimental study, by 160 g. (including the sabot) cylindrical projectiles, with a conical nose, at velocities that ranged between 100 - 250 m/sec. The experiments were done at NBRI laboratory with a gas gun that accelerated hard steel projectiles, whose impact velocities were measured by an electro-optical system. This paper describes a part of the study, in which craters at the rear and front faces of plate specimens were examined. The differences between the response of the NSC and of the HSC specimens under similar dynamic loading conditions (i.e., same projectile and similar impact velocities) were indicated by the fracture path in the rear face craters (around or through the aggregates). Other parameters that were examined were the crater volume and dimensions.

Introduction

Design of concrete elements in protective structures, or in nuclear power plants structures, requires the consideration of impact loading, caused by accelerated projectiles (for each category of structures there is a different range of projectile weights and velocities). This part of the design is commonly based on formulae that predict the element thickness, required to prevent rear face scabbing, or perforation, for a given projectile weight, velocity, shape, and other pertinent parameters of the target element, such as the concrete strength. These formulae are based on physical considerations (e.g., modified NDRC - Kennedy, 1976; or Hughes, 1984), but their parameters are adjusted to fit empirical results. Among these formulae, the classical

---

[1]Lecturer, Faculty of Civil Engineering, National Building Research Institute, Technion - Israel Institute of Technology, Technion City, Haifa 32000, Israel

modified NDRC set of equations (Kennedy, 1976) is repeatedly cited in the literature as a reliable way to represent a "normally reinforced" concrete element, with a relatively wide range of relevant data (Haldar and Hamieh, 1984; Sliter, 1980; Williams, 1994). Another set of equations which is mentioned recently in the literature are Barr's equations (Barr, 1990; Williams, 1994), which consider also the effect of the reinforcement ratio and spacing. Most of the formulae account for the contribution of the concrete strength to its resistance against projectile impact by including a term that relates the perforation resistance to f_c^{α}, where f_c is the concrete compressive strength and α is an empirical parameter (e.g. $\alpha=1/2$ in the NDRC and in Barr's equations). With the introduction of High Strength Concrete (HSC) the required resistance of the designed elements may be achieved with reduced thickness. However, because the existing formulae are mostly based on empirical database from normal strength concrete (NSC) results, it is not clear whether they can be applied to the design of HSC elements. For example, Barr's equations are limited to NSC by their definition. Furthermore, the increased brittleness (Holand, 1989) of HSC may cause a level or type of damage, that is different from NSC. Observations from an experimental study indicate the effect of concrete strength on the response of a plate subjected to hard projectile impact, in view of the above considerations. Note that in the following text (and commonly in the literature) *perforation* is defined as the case when the projectile fully penetrates the plate, while in some cases the projectile may cause *punching* through the plate without perforation.

Description of the experiments

Experimental Setup. The tests were conducted in the laboratory setup described in Fig. 1. The impact load was generated by a gas gun system that accelerated steel projectiles. The 120 g. 25 mm diameter projectiles, described in Fig. 2, had a conical nose, and were made of high yield steel to prevent their deformation in the tests. They were accelerated through the 50 mm diameter of the gun barrel by two leading rings (sabot) made of hard plastic, that weighed ~40 g. (the total weight of the projectile and the sabot was ~160 g.). Measurements of the velocity were performed by an electro-optical device, which consisted of two infra-red light emitting diodes (LED) and two matching receivers, located at a fixed distance of 0.5 m from each other. When the LED beams were cut by the moving projectile they produced a signal that showed the time interval between the two crossings. That information was translated into velocity data.

Test Plan. NSC and HSC panels were tested at impact velocities that ranged between 100 and 250 m/s. The panels were 40x40 cm plates, that were 4 and 5 cm in thickness. Some of the plates were unreinforced (except for a pair of 5 mm ties along their perimeter), some had steel fibers (60 Kg/m^3) and others were reinforced with steel meshes near their rear face, with a reinforcement ratio that ranged between 0.16% and 1.82%. Several types of reinforcing steel meshes were used: smooth bars (f_y=470~530 MPa); woven, small diameter wires (f_y=380 MPa); and woven fence meshes (f_y=180~300 MPa). The NSC plates had a compressive strength of 34-39

MPa, and the HSC plates had a 104-112 MPa compressive strength. In a separate set of tests, 5 cm thick, NSC and HSC plates with a reinforcement ratio of 0.20% (smooth bars, $\phi 3.25 @ 104$ mm e.w.) were tested also under static punching loads. In these tests the projectile was loaded by a standard compressive testing machine. In order to maintain the local nature of shear penetration the bending resistance was eliminated by a round steel support that had a 20 cm diameter (Fig. 3).

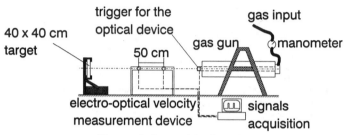

Figure 1: Impact tests set-up

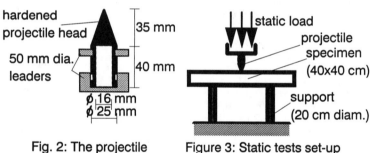

Fig. 2: The projectile
(W = 160 ~ 165 gr
including plastic leaders)

Figure 3: Static tests set-up

The NSC concrete was made of ordinary Portland cement with a nominal fly ash content of 10%, and water/cement ratio of 0.64. The aggregates were dolomitic with a particle size of 4.75 - 9.5 mm, crushed dolomitic aggregate with a maximum particle size of 4.75 mm, and natural siliceous sand (s.g. 2.63). The HSC concrete was made of ordinary Portland cement with pure silica fume that was 14.5% of the cement (by weight), and with a superplasticizer (trade mark "Rheobuild 2000") that was 5% of the cement (by weight). The water/(cement + silica fume) ratio was 0.28, and the aggregates were dolomitic with a particle size of 4.75 - 9.5 mm, 1.18 - 4.75 mm, and natural siliceous sand (s.g. 2.63).

Results

The following discussion refers to the results of the specimens that were punched through their thickness. Their impact velocities were measured together with their rear crater volumes, and their front and rear craters gross dimensions (length and width according to the longer axis of the crater and to a perpendicular axis, respectively). The volumes were measured by pouring dry sand into the craters. In order to account for the influence of the different amounts and spacing of the reinforcement on the plates resistance, a 'reinforcement parameter' was used, r_{BARR}, which was calculated according to Barr's equation for the resistance of reinforced concrete elements against perforation of hard (non-deformed) projectiles (Barr, 1990). Two additional parameters were used to describe the rear craters: D_{eq} is an equivalent measure of the crater diameter, evaluated by $D_{eq} = \sqrt{D_1 \cdot D_2}$, where D_1 and D_2 are the measured gross dimensions of the crater. A_{eq} is an equivalent measure of the rear crater surface area, assuming an equivalent crater shape of a cone, hence $A_{eq} = \pi \left(D_{eq} + d/2\right)\sqrt{\left(D_{eq} - d/2\right)^2 + h_{eq}^2}$, where D_{eq} is the diameter of the cone base (i.e., at the plate rear face), h_{eq} is its height, and d is the top diameter of the cone. Assuming a similar slope for the front and for the rear craters (see Fig. 4), where V is the measured crater volume, h_{eq} is given by $h_{eq} = (3/\pi) \cdot 4V \big/ \left[D_{eq}^2 + d\left(D_{eq} + d\right)\right]$. A_{eq} of the punched specimens are plotted against D_{eq} in Fig. 5, which together with the above equation of A_{eq} shows that h_{eq} of these plates hardly influenced the area of the cracked surface. Furthermore, the measured craters volumes of the NSC and of the HSC specimens were similar, for the different amounts of reinforcement, and for the impact velocities that were tested. Hence, the rear face damage level is evaluated by the equivalent diameters of the craters, D_{eq}.

Effects of Concrete Strength. The impact velocities in the tests of the 50 mm thick plates that were perforated, are shown in Fig. 6 for the various amounts of reinforcement, r_{BARR}. In the range of the projectile velocities that was studied, Fig. 6 shows that the HSC plates required higher projectile velocities to cause perforation as compared to similar NSC plates. The damage levels of these plates are evaluated by their rear craters equivalent diameter in Figs. 7 and 8, for the 40 mm and 50 mm thick plates, respectively. As can be seen in these figures, D_{eq} of the HSC specimens was higher than that of the NSC specimens, indicating also a higher area of their failure surface, A_{eq}. Note that the scatter in Figs. 6 to 8 represents not only the usual statistics of experimental results, but it is also a result of the different *types* of reinforcement.

Effect of the Fibers. The perforation resistance of the 40 mm plates with steel fibers (60 Kg/m^3, "Dramix ZP 30/.50") was similar to that of the 40 mm specimens that were reinforced with a mesh of steel rebars (ϕ5 @ 100 mm). However, the rear craters dimensions of the NSC and of the HSC specimens with fibers were smaller than those of the other specimens (see D_{eq} of NSC and HSC in Fig. 7), indicating that the fibers reduced the damage at the rear face.

Strength and Loading Rate Effect. Figs. 9 and 10 show typical cracked surface areas of NSC and HSC rear face craters, respectively. As can be seen in Fig. 9, most of the aggregates at the NSC craters remained uncracked, and the cracked crater surface passed in the concrete matrix around the aggregates. This result was observed in both the dynamic and the static tests (Fig. 9), and it is therefore not in agreement with the theory that relates the increase in concrete tensile strength at high loading rates to a different crack path in the two loading conditions (Zielinski and Reinhardt 1982). On the other hand, in HSC the aggregates ruptured, both under static and impact conditions (Fig. 10).

Fig. 11 shows the crater dimensions (represented by D_{eq}) and volumes of the NSC and of the HSC plates, at static and dynamic punching. They show that at static loading the craters were approximately the same, while under projectile impact the NSC craters were smaller than the craters that were developed at the HSC plates rear face. When the crater dimensions and volumes are considered, the impact loading affected the results of the NSC specimens, whose crater dimensions were reduced. On the other hand, the impact loading hardly affected the HSC crater volumes (Fig. 11a), and increased their equivalent diameter (Fig. 11b).

It was also observed, in both NSC and HSC plates that there were no front craters in the static loading, while they developed in the impact tests. This result is due to higher friction force that develops between the projectile and the concrete in the static loading. The inclination of the projectile during impact penetration, compared with a straight path (perpendicular to the plate's surface) in static loading, may also be related to this result.

Concluding Remarks

Results of several experimental series show that the response of NSC plates to hard projectile impact is different than that of HSC plates. The experimental results showed that within the velocity range that was tested the *perforation resistance* of the HSC plates was higher than that of NSC plates. This is an expected result, because whether it is the concrete compressive or tensile strength that governs the perforation resistance, the strength parameters of HSC are higher than those of NSC.

The performance of HSC versus NSC was characterized by larger dimensions of the HSC rear face craters, indicating a higher level of damage. This observation and the similar crater volumes of the NSC and HSC specimens, also indicate a different surface areas of the conical failure surface (i.e., of the craters), implying a different energy dissipation in the NSC and HSC elements. Yet another observation was that there was a smaller number of larger fragments of the HSC craters, than in the NSC craters that broke into a larger number of smaller fragments. Therefore, the effect of the concrete type on its ductility, under impact loading should be further investigated (e.g., by more precise measurements of the cracked surface area). The similar volumes and different base diameters of the NSC and HSC rear craters also infer that the HSC craters were more shallow than the NSC craters. Thinner rear face spalling implies steeper front face projectile impulse (Rinehart 1960) that may have been caused by the stiffer HSC specimens.

The limited tests conducted on concrete specimens with 0.76% of steel fibers had a perforation resistance that was similar to that of the plates that were reinforced with a mesh of steel rebars. These preliminary tests also indicated that in the perforated plates the fibers tend to decrease the rear face crater dimensions.

References

Barr P. (1990). *Guidelines for the Design and Assessment of Concrete Structures Subjected to Impact*, UK Atomic Energy Authority Safety and Reliability Directorate, London.

Haldar A. and Hamieh H. (1984). "Local Effect of Solid Missiles on Concrete Structures", *Journal of the Structural Division*, ASCE, Vol 110, No 5.

Holand I. (1989). "State of the Art of Design Aspects and Research Needs in the Future", *Design Aspects of High Strength Concrete, CEB, Bulletin D'Information*, No 193, 145-163.

Hughes G. (1984). "Hard Missile Impact On Reinforced Concrete", *Nuclear Engineering and Design* 77, 23-35, North-Holland, Amsterdam.

Kennedy R. P. (1976). "A Review of Procedures for the Analysis and Design of Concrete Structures To Resist Missile Impact Effects", *Nuclear Engineering and Design* 37, 183-203, North-Holland, Amsterdam.

Sliter G. E. (1980). " Assessment of Empirical Concrete Impact Formulas", *Journal of the Structural Division*, ASCE, Vol 106, No ST5, 1023-1045.

Rinehart J.S. (1960). *On Fractures Caused by Explosions and Impacts*, Quarterly of the Colorado School of Mines, Volume 55, Number 4, Department of Mining Engineering, Colorado School of Mines.

Williams M. S. (1994). "Modeling of Local Impact Effects on Plain and Reinforced Concrete", *ACI Structural Journal*, V. 91, No 2, March-April.

Zielinski A. J. and Reinhardt H. W. (1982). "Impact Stress-Strain Behaviour of Concrete in Tension", *Concrete Structures Under Impact and Impulsive Loading: RILEM-CEB-IABSE-IASS-Interassociation Symposium*, BAM, Berlin, 112-124.

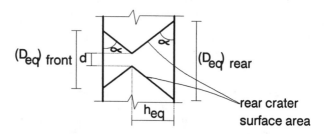

Figure 4: Equivalent measures of the craters

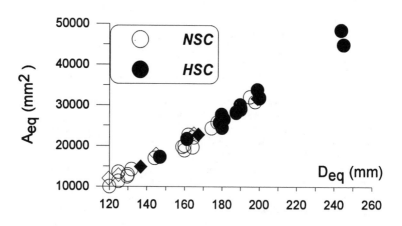

Figure 5: Rear craters equivalent surface areas, A_{eq} vs. D_{eq}

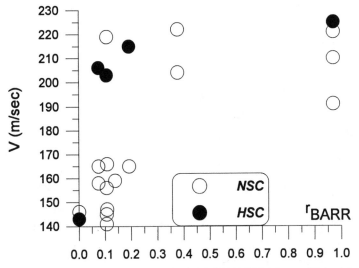

Figure 6: Projectile velocities of the PERFORATED 50 mm thick plates

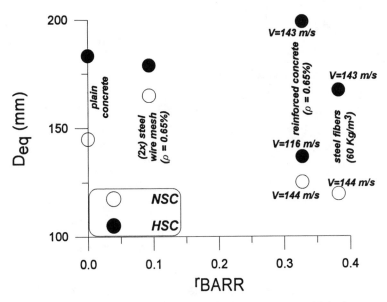

Figure 7: Rear craters equivalent diameter of the 40 mm thick plates

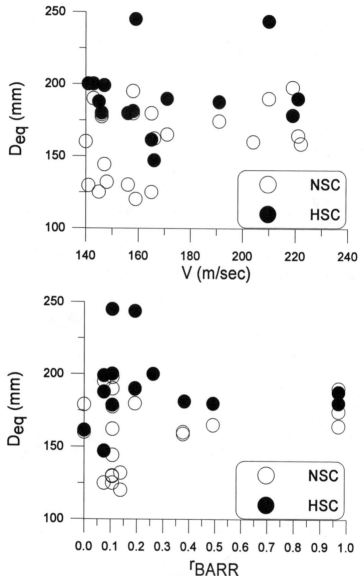

Figure 8: Rear craters equivalent diameter
of the 50 mm thick plates

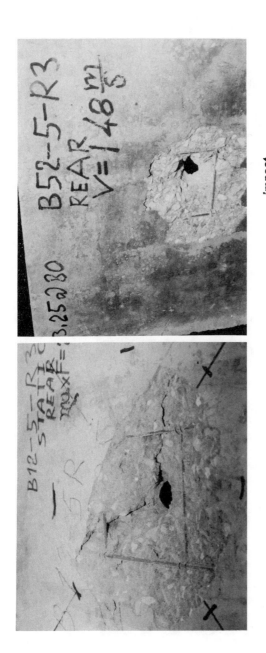

Figure 9: Surface of NSC rear face craters

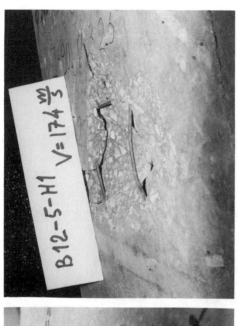

impact

static

Figure 10: Surface of HSC rear face craters

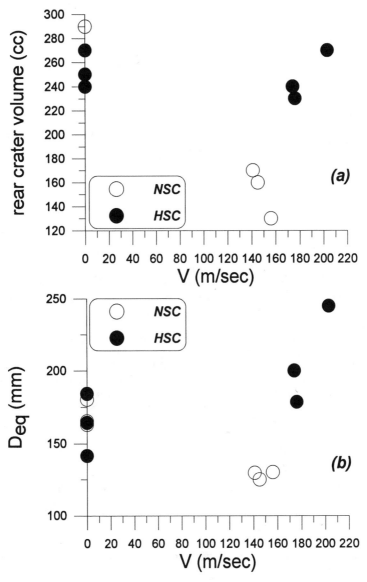

Figure 11: Static (V = 0) vs. impact penetration tests
(smooth bars, Φ3.25@104 mm e.w., ρ = 0.20%)

Influence of the Yield Strength of Steel Fibres on the Toughness of Fibre Reinforced
High Strength Concrete

Lucie Vandewalle[1]

Abstract

High strength concrete is known to be a brittle material. To overcome this
drawback discontinuous steel fibres are added to convert it into a ductile material and
further improve its mechanical properties. However, because of the better bond between
steel fibres and concrete it is possible that at appearance of the first crack the steel fibres
which cross the crack, fracture rather than being pulled out slowly.

At the Department of Civil Engineering of the K.U.Leuven, toughness of steel
fibre reinforced high strength concrete has been investigated by means of the four-point
bending test specified in the Belgian Standard NBN B15-238. The investigated
parameters were : compressive strength of concrete (50 MPa - 78 MPa - 92 MPa) and
yield strength of steel fibre (LC : 1000 MPa - HC : 2000 MPa). From these tests it
follows that from a compressive strength equal to 78 MPa, the toughness of concrete is
increased considerably by the addition of HC-fibres instead of LC-fibres.

Introduction

Toughness is a measure of the energy absorption capacity of a material and is used
to characterize the material's ability to resist fracture when subjected to static strains or
to dynamic and impact loading.

Behavior under flexure is the most important aspect for fibre reinforced concrete
because in most practical applications the composite is subjected to some kind of
bending load.

----

[1]Assistant Professor, Katholieke Universiteit Leuven, Department of Civil Engineering,
de Croylaan 2, 3001 Heverlee, Belgium.

The load-deflection curve of a concrete beam specimen, submitted to a displacement controlled flexural test, shows an almost vertical descending branch after the peak stress. Hence, concrete is often called a "brittle material". The flexural toughness of a material can be quantified as the area under the load-deflection curve up to a specific deflection. Adding steel fibres to a plain concrete matrix has little effect on its precracking behavior but does substantially enhance its post-cracking response, which leads to a greatly improved ductility and toughness.

The toughness of steel fibre normal strength concrete has been investigated extensively as mentioned in Reference 1 [Balaguru and Shah, 1992]. From this research it is known that fibre volume fraction, aspect ratio of the fibre and fibre geometry play an important role. Only very few data are available in literature concerning the flexural toughness of steel fibre high strength concrete. Usually it is concluded that when the matrix strength increases, concrete becomes more brittle and, hence, more steel fibres are needed to achieve the same amount of ductility. Increase of concrete compressive strength provides a better bond between fibre and matrix. From this consideration, it follows that after the appearance of the first crack the crossing steel fibres could fracture instead of being pulled out slowly. To investigate this aspect, an extensive test program was performed. The main parameters were : concrete compressive strength and yield strength of steel fibre.

Experimental Program

Complete details of the experimental work and results are given in Reference 2 [Van Deun and Vandewalle, 1995]. An abbreviated description is given in this paper. Eleven series of 6 concrete beam specimens (150 x 150 x 600 mm) were tested using the four-point flexural configuration specified in the Belgian Standard NBN B15-238. The prisms were simply supported with 450 mm clear span and subjected to two displacement controlled point loads (see Figure 1). The distance between the two loads was 150 mm and the total load (P) was measured using a load-transducer. The net-deflection of the specimen (δ_{net}) was measured at midspan by means of an electrical transducer. Load-deflection response was recorded continuously and stored in a computer file.

Three different compressive strength classes of concrete were investigated, i.e. one medium (MS) and two high strength (HS1 - HS2) concretes. The concrete compositions are given in Table 1. The fibre reinforced concretes have the same composition as the corresponding reference concrete without steel fibre. The used coarse aggregate was crushed stone. The steel fibres were hook-ended with length 60 mm and diameter 0.8 mm (aspect ratio = 75). Two types of fibres were tested, i.e. LC-fibres (=Low Carbon) with a yield strength of at least 1000 MPa and HC-fibres (=High Carbon) with a minimum yield strength of 2000 MPa.

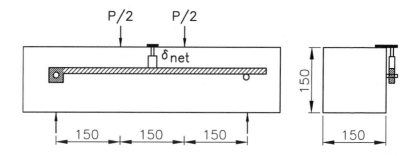

Figure 1. Schematic picture of test set-up

Constituents (kg/m³)	MS	HS1	HS2
limestone 7/14	560	-	-
limestone 2/7	560	-	-
porphyry 7/10	-	956	833
porphyry 2/7	-	478	416
sand 0/4	700	478	416
cement type CEM I 42.5	350	-	-
cement type CEM I 52.5	-	400	500
silicafume	-	-	50
superplasticizer	-	10	20.8
water	192.5	145	154

Table 1. Concrete compositions

The dosages of both fibre types were 30 and 60 kg/m³ of HS2-concrete whereas for the MS-concrete, HS1-concrete respectively, it was chosen equal to 30 kg/m³. The volume fraction V_f was equal to 0.375 or 0.75 percent which may seem quite low. However these volume fractions are encountered frequently in current applications and appear to be economically feasible. Moreover, for the type of fibres used, a volume fraction of 1 percent might limit workability. A general survey of the test program is given in Table 2. The concrete compressive strength has been determined on unflattened cubes with side 150 mm.

Series	Type of concrete	Fibre content in kg/m³ of concrete and fibre type	Mean cube strength in MPa	Age at loading in days
1	MS	0	51	14
2	MS	30 - LC	54	14
3	MS	30 - HC	52	14
4	HS1	0	72	14
5	HS1	30 - LC	78	14
6	HS1	30 - HC	85	14
7	HS2	0	92	28
8	HS2	30 - LC	92	28
9	HS2	30 - HC	96	28
10	HS2	60 - LC	92	28
11	HS2	60 - HC	92	28

Table 2. General survey of test program

Test Results

In the Belgian Standard NBN B15-238 the toughness B_n (Nm) is defined as the area under the load-deflection curve up to a midpoint deflection of ℓ/n (mm) with ℓ equal to 450 mm and n equal to 300, 150 respectively. The Belgian specification uses in addition to a toughness definition based on absolute energy, an equivalent flexural tensile strength (N/mm²), i.e. :

$$f_{f,n} = (B_n\, n)/(b\, h^2) \tag{1}$$
with   b : width of beam
          h : height of beam.

Table 3 gives the mean value of f_r, f_u, B_{300}, B_{150}, $f_{f,300}$ and $f_{f,150}$ for each of the 11 test series. f_r is equal to the flexural tensile stress at first-crack and f_u to the maximum tensile stress attained during the test. For the reference concretes without steel fibre, f_r is equal to f_u and no "toughness" and "equivalent flexural strength" are mentioned since the beam specimens failed at first-crack appearance.

Series	f_r (N/mm²)	f_u (N/mm²)	B_{300} (Nm)	B_{150} (Nm)	$f_{f,300}$ (N/mm²)	$f_{f,150}$ (N/mm²)
1	3.7	3.7	-	-	-	-
2	4.4	4.4	35.4	65.9	3.1	2.9
3	3.6	3.6	32.8	62.2	2.8	2.7
4	6.4	6.4	-	-	-	-
5	7.1	7.1	45.7	69.8	4.1	3.1
6	7.2	7.3	70.0	135.3	6.2	6.0
7	8.0	8.0	-	-	-	-
8	7.9	7.9	36.2	42.9	3.2	1.9
9	8.3	8.4	80.9	156.4	7.2	7.0
10	8.8	9.3	69.8	85.3	6.2	3.8
11	8.4	10.4	104.5	191.6	9.3	8.5

Table 3. Test results

For each series of fibre reinforced concrete only "one" representative load-deflection curve is shown in this paper. As mentioned previously more details can be found in Reference 2 [Van Deun and Vandewalle, 1995].

Figure 2 compares the representative load-deflection curves for prisms of series 2 (MS-LC-30) and series 3 (MS-HC-30). The toughness values presented in Table 3 and the load-deflection curves in Figure 2 confirm that the influence of yield strength of steel fibre is not significant if used in medium strength concrete. The planes of fracture of the specimens looked identical, i.e. the fibres were pulled out of the concrete and were not broken.

In Figure 3 the representative load-deflection curves of prisms of series 5 (HS1-LC-30), series 6 (HS1-HC-30) respectively are shown. For concrete reinforced with HC-steel fibres, there is a drop soon after the development of first-crack. However this drop is followed by a significant increase in load. This increase provides a stable postcrack behavior. The drop after the first peak is much more pronounced for LC-steel fibres and no increase in load afterwards is observed. The difference in behavior among the two fibre types can be seen even more clearly in Table 3 which shows the toughness values of B_{150} and B_{300}. HC-fibres perform better in both cases and the difference in absolute energy absorption capacity is more significant at larger deflections. From the observation of the planes of fracture of the specimens it follows that for HC-fibres

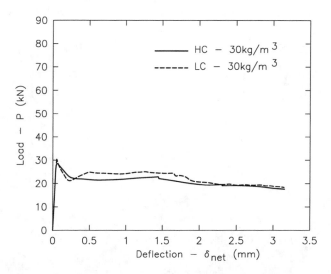

Figure 2. Load-deflection curves of series 2 (MS-LC-30) and 3 (MS-HC-30)

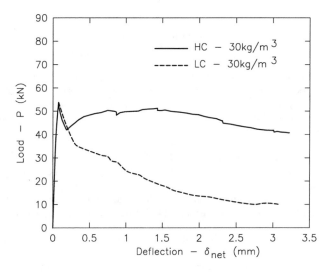

Figure 3. Load-deflection curves of series 5 (HS1-LC-30) and 6 (HS1-HC-30)

failure still occured as a result of fibre pullout although the better bond between high strength concrete and fibre (see Photograph 1 - left specimen). However for LC-fibres, this better bond produced fibre fracture rather than fibre pullout resulting in much less ductility (see Photograph 1 - right specimen).

Photograph 1. Planes of fracture

The representative load-deflection curves of specimens of series 8 (HS2-LC-30), series 9 (HS2-HC-30) respectively are shown in Figure 4. The same observations can be made as for concrete HS1. The influence of yield strength of steel fibre on ductility of fibre reinforced concrete becomes even more significant at higher concrete strength classes.

Figure 5 presents load-deflection curves of beams of series 10 (HS2-LC-60), series 11 (HS2-HC-60) respectively. For both fibre types the amount of drop soon after the development of first-crack is almost the same. For HC-fibres the postcrack increase in load is significant and from Table 3, it follows that the equivalent flexural tensile strengths ($f_{f,300}$ - $f_{f,150}$) are at least equal to the flexural tensile stress at first-crack. Also the maximum flexural tensile stress f_u increases with about 25 % in comparison with that of the reference concrete without steel fibre.

By comparing Figure 4 with Figure 5, one can see that the influence of fibre volume fraction on the energy absorption capacity is very significant, especially for LC-fibres. From Table 3, it indeed follows that an increase of the LC-fibre content with

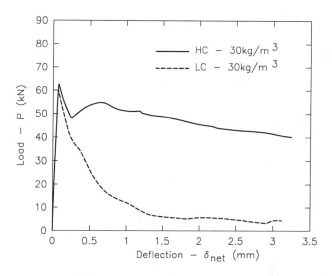

Figure 4. Load-deflection curves of series 8 (HS2-LC-30) and 9 (HS2-HC-30)

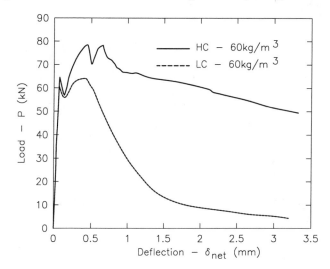

Figure 5. Load-deflection curves of series 10 (HS2-LC-60) and 11 (HS2-HC-60)

100 % roughly doubles the value of the toughness B_{300}, B_{150} respectively.

During the tests of series 10 (HS2-LC-60) it was observed that up to a deflection of about 0.5 mm the fibres were pulled out slowly. However, this behavior changed and was followed by fracture of fibres which resulted in a continuously decrease of the load as presented in Figure 5.

A comparison of the representative load-deflection curves of prisms of series 10 (HS2-LC-60) with those of series 9 (HS2-HC-30) is illustrated in Figure 6. This Figure 6 shows, just like Table 3 that the influence of yield strength of steel fibres when they are used in high strength concrete is at least as significant as that of the fibre content. At larger deflections the increase in yield strength of steel fibre from 1000 MPa to 2000 MPa results in an even higher improvement in energy absorption capacity than a change of the fibre content from 30 to 60 kg/m³ does. An additional benefit is that a smaller fibre volume fraction provides also a better workability of the concrete.

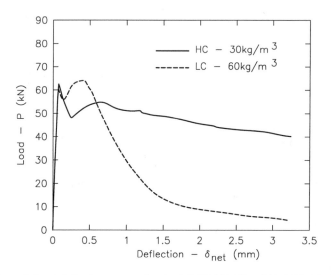

Figure 6. Load-deflection curves of series 9 (HS2-HC-30) and 10 (HS2-LC-60)

Conclusions

The influence of yield strength of steel fibre on toughness of fibre reinforced concrete is not significant when used in medium strength concrete (cube strength ≈ 50 MPa).

However, as the matrix strength increases, bond between concrete and fibre becomes better and, hence, for LC-fibres failure occurs as a result of fibre fracture rather than fibre pullout even at high volume fractions (0.75 vol.-%). This problem can be solved, as observed in this test program, by increasing the yield strength of steel fibre. The influence of yield strength of steel fibre on the energy absorption capacity of steel fibre high strength concrete is indeed very significant.

Appendix

1. Balaguru P.N., Shah S.P., Fiber Reinforced Cement Composites, Mc Graw-Hill, Inc., 1992, 530 pp.

2. Van Deun V., Vandewalle P., Mechanische eigenschappen van hoge sterkte beton versterkt met staalvezels, Thesis K.U.Leuven, june 1995 (in Dutch - English title: Mechanical properties of steel fibre reinforced high strength concrete).

Some Phenomenological Aspects of High Performance Concretes, and their Consequences for Numerical Analysis

Ravindra Gettu[1], Ignacio Carol[2] and Pere C. Prat[2]

Abstract

The microstructures of high performance concretes, such as high strength silica fume concrete and steel fiber reinforced concretes, give rise to some mechanisms that are significantly different from normal concrete. The phenomenological aspects that affect the long-term development, shrinkage, fracture and failure of these materials, and their implications for numerical analysis are discussed.

Introduction

The composition and quality of concrete have been improved constantly over the past few decades. New types of concretes are developed for specific applications and environments. Some of these are classified as high performance concretes since they have been engineered to exhibit significantly enhanced characteristics. Considerable research effort has been focussed on the study and characterization of these concretes. Nevertheless, several important aspects, including their numerical modeling, are not yet well established.

The implications of the difference in behavior between new and conventional concretes on material modeling have been discussed since the late 1970s (Bažant, 1979). However, the development of structural analysis tools for new concretes has been impeded by the lack of sufficient experimental evidence, as well as the lack of communication between material scientists and computational mechanists. With the further utilization of advanced materials and the consolidation of the fundamental knowledge of their mechanisms, it is expected that significant progress will be made

[1] Structural Technology Laboratory, Department of Construction Engineering

[2] Department of Geotechnical Engineering

Technical University of Catalunya, School of Civil Engineering (ETSECCPB-UPC).
Gran Capità, s/n — E-08034 Barcelona, Spain.

in the numerical modeling of these materials. The present work reviews some phenomenological aspects that are peculiar to new cement-based materials such as high strength and fiber-reinforced concretes, and examines some of the direct consequences for numerical modeling.

What is HPC?

There is no consensus on the precise definition of high performance concrete (HPC), which to some is synonymous to high strength concretes (with compressive strengths greater than 40–60 MPa). Others have defined HPC as concrete with high durability, in addition to high strength. It can be generally stated that HPC is a new class of concretes in which mineral and chemical admixtures, fibers and other unconventional materials may be incorporated, and which may require special processing techniques. Several objective definitions have been proposed, which have to be updated as technology and structural requirements advance. One set of criteria that summarizes the requirements of present-day HPCs is that of Zia et al. (1991):

- The compressive strength shall be at least 69 MPa at 28 days or 34 MPa at 24 hours or 21 MPa at 4 hours
- The freeze–thaw durability factor as determined by the ASTM C 666 method A shall be greater than 80%
- The water-cementitious material ratio shall not be more than 0.35

However, in practice, HPC is not a unique concrete that exhibits superior performance in all possible applications but one that performs best in the application under consideration. This concept is reflected in the HPC definition used by Carino and Clifton (1990): "Concrete having desired properties and uniformity which cannot be obtained routinely using only traditional constituents and normal mixing, placing and curing practices. As examples, these properties may include: ease of placement and compaction without segregation, enhanced long-term mechanical properties, high early-age strength, high toughness, volume stability, and long life in severe environments." Properties that satisfy the requisites of several applications are listed in their report. In general, HPC varies with the application; it is the concrete that best satisfies all the critical requisites, including cost, which affect the fabrication, utilization and long-term integrity of the material in the intended application (Gettu et al., 1993).

The HPCs considered in this work are silica fume concretes and fiber reinforced concretes. Silica fume concretes (FIP-CEB, 1990) are currently the most common type of high strength concrete (HSC). They are usually made with a low water-cement ratio ($\sim$0.35), high cement content (400–550 kg/m^3), silica fume (about 5–20% by weight of cement) and superplasticizer. Their dense high-strength matrices and matrix-aggregate interfaces provide high strength, stiffness and impermeability. The fiber reinforced concretes (FRCs) discussed here are mainly those that incorporate low volume-fractions (up to 2%) of short (up to about 60 mm) steel fibers. The principal benefit of such fibers is the increase in ductility, characterized by higher toughness, energy dissipation during failure and crack resistance.

What makes numerical analysis of HPC different?

Most constitutive models and modeling techniques used in simulating the behavior of concrete and concrete structures have been developed on the basis of the response observed in normal concrete. Such concretes are fabricated with high water-cement ratios and without special admixtures. These factors, as well as the mixing, placing and curing processes conventionally employed, give normal concrete a microstructure with weak matrix-aggregate interfaces, pores and flaws. The matrix is normally weaker, less stiff, more permeable and undergoes much larger time-dependent deformations than the aggregates. These aspects produce the characteristic composite behavior, including phenomena such as strain-softening, fracture toughening, shrinkage, creep and fluid transport. The HPCs are significantly different from normal concrete. HSC has a compact microstructure, which is free from major flaws, large interconnected pores and weak interfaces. Also, the matrix can be as strong as the aggregates. In FRC, the distributed fibers introduce additional micromechanisms such as fiber pullout and bridging, dowel effects, fiber rupture and distributed microcracking. Due to these aspects, conventional modeling techniques may sometimes be inadequate for HPCs. In the following sections, the main aspects of the behavior of HSC and FRC that differ from normal concretes are discussed, and some of their consequences for numerical analysis are outlined.

Time-dependent effects in high strength concretes

The differences in long-term behavior between HSC and normal concrete are due to significant dissimilarities in their water contents and permeabilities. In normal concrete, sufficient water is available in the mix to completely hydrate all the cement. When completely sealed or maintained in an environment with 100% relative humidity (RH), it remains practically saturated with an internal RH close to 100% and undergoes negligible shrinkage.

HSC has a much lower water content, and a more compact and impermeable microstructure than normal concrete (Nilsson, 1994). Therefore, the hydration reaction considerably depletes the pore water, even when the surface is exposed to 100% RH or is sealed. This process, known as self-desiccation, drastically decreases the internal RH, even to 75% within just the first week, leading to a reverse moisture gradient and internal residual stresses (Atlassi, 1993). The lack of water can also stop the hydration process before a significant part of the cement can be hydrated (FIP-CEB, 1990). This is not reflected adequately by the maturation curves used for normal concrete.

The decrease in the internal RH leads to autogenous shrinkage of the concrete, a phenomenon that is beyond the scope of many existing models (e.g., Bažant and Baweja, 1994). This is the shrinkage observed in HSCs that are sealed or maintained at 100% (external) RH, which can attain values up to 1000–2000 microstrains (Hansen and Jensen, 1989; Tazawa and Miyazawa, 1993; Tazawa et al., 1995). Shrinkage models that account for this phenomenon have been proposed (De Larrard et al., 1994).

Analysis of time-dependent deformations in concrete structures requires a model for the average shrinkage at each cross-section. More accurately, the evolution of the humidity is simulated and used as the input for the shrinkage model, which would give the intrinsic shrinkage of the material. In normal concrete, solution of the diffusion problem is simpler since it only involves the flow of moisture from the saturated pores to the drier free surfaces. This is sometimes handled through the use of empirical relations (McDonald and Roper, 1993). In contrast, for HSC the diffusion problem also involves the water depletion per unit volume due to the hydration reactions.

The consequences of self-desiccation on creep are more difficult to understand and quantify. The creep in a sealed specimen, which is conventionally denoted as the basic creep, includes the effect of self-desiccation of HPC, and therefore, could be higher than in normal concrete. Since the effect of self-desiccation cannot be separated from the creep of the saturated material, it should be noted that basic creep in this case does not have the same significance (i.e., it is not the intrinsic material behavior). It has also been observed that drying creep in HSC is significantly lower and the total creep is less than or at least equal to that of normal concrete, at the same relative stress levels (Brooks, 1994). These factors have to be considered in the application of existing creep models to HSC.

Cracking in high strength silica fume concretes

Due to its enhanced microstructure, HSC has very few initial flaws and weak interfaces. Under loading, cracks propagate through the aggregates instead of avoiding them and passing through the matrix-aggregate interfaces, as in normal concrete. This generally causes straighter cracks to develop with little tortuosity and aggregate-bridging, which decreases toughening and makes the concrete more brittle (Gettu and Shah, 1994). Consequently, the effective size of the fracture process zone will be less than that of normal concrete. Moreover, the tensile stress-separation response of the crack will be significantly different, as seen in Fig. 1. Other failure mechanisms are also affected. For example, in shear failure, the failure of the coarse aggregates practically eliminates aggregate interlock; Walraven and Stroband (1994) reported that the shear capacity of a crack in a HSC with a compressive strength of 110 MPa could be about the same as that of a 17 MPa concrete, where no aggregate fracture occurs. Additionally, compressive stresses generated across the crack faces due to dilation will be much smaller in HSC. This effect has been observed in compression tests of HSC (Ahmad, 1994), where the increase in lateral strains is much lower than in normal

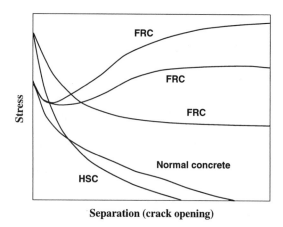

Separation (crack opening)

Figure 1. Tensile stress-separation curves of cracks.

concrete. This implies that the effects of triaxial stresses and confinement will vary disproportionately with an increase in strength. Another difference in the compressive behavior is in the failure mechanism itself. It has been observed that in HSC there is a greater tendency for splitting cracks along the direction of compressive loading than for inclined shear cracking, which predominates in normal concrete.

All these aspects have to be accounted for in the numerical analysis. Some of them may just require changing the values of the parameters of the models used for normal concrete, for example, the values of f_t and G_f, and the shape of the stress-separation curve. However, in other cases such as the change in crack orientation under compression or the decreased dilatancy, some models may require more fundamental modifications in order to be applicable to HSC.

Cracking in reinforced concrete is influenced further by the bond between the steel rebars and concrete, which can be characterized by the bond stress versus slip relation. In HSC, it has been shown that the maximum bond stress is much higher due to the compact microstructure, but the stress-slip relation is more brittle (*cf.* FIP-CEB, 1990). These aspects, as well as the rupture of the aggregates, may be the cause of the steeper and more concentrated secondary cracks that occur at bond failure in HSC structures. It has also been observed that there is decreased hysteretic bond action, and consequently, lower damping under cyclic loading. The different bond behavior also leads to a larger number of more closely-spaced cracks under flexure, as demonstrated by Jaccoud et al. (1993) in tests of reinforced HSC slabs. In the numerical modeling of normal concrete, these phenomena are often taken into account in a simplified way through the tension stiffening effect. This approach seems to be also valid for

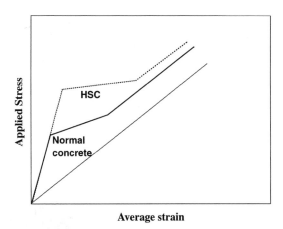

Average strain

Figure 2. Idealized tension stiffening in reinforced concrete tie (Jaccoud et al., 1993).

HSC with the appropriate parameters that ensure the higher tension stiffening observed experimentally (Fig. 2).

Failure of concretes with low volume-fractions of steel fibers

The main objective for the incorporation of steel fibers in concrete is to enhance its response against cracking. Usually a maximum volume fraction of 2–4% of fibers, with a maximum length of about 50 mm, is incorporated in the FRCs. The pullout behavior of the fiber depends on the properties of the steel and the matrix, as well as the shape, length and diameter of the fiber. In high strength FRCs, the matrix-fiber bond is much stronger, which increases the possibility of fiber rupture as opposed to fiber pullout. This has lead to the use of high-strength steel fibers, with tensile strengths up to 2000 MPa (e.g., Gopalaratnam et al., 1995).

The tensile stress-separation curve of a crack in FRC can differ considerably from that of normal concrete. Assuming that the fibers do not rupture, the stress-separation curve of FRC may exhibit a plastic plateau or hardening (Fig. 1), instead of monotonic softening (Li et al., 1993). It is not generally possible to obtain the fracture energy or the entire stress-separation curve since complete separation occurs at very large displacements (in the order of half the fiber length), which are beyond practical interest or the range of the tests. Hillerborg (1985) proposed the use of an idealized bilinear softening-plastic curve for the finite element analysis of FRC, with the fictitious crack model. On the basis of such analyses of FRC beams, he concluded that low fiber volumes do not affect the bending strengths of smaller beams. However, the flexural strength is nearly constant for larger beams, which corresponds to plastic failure. Therefore, there is a transition in the mode of failure and size effect from

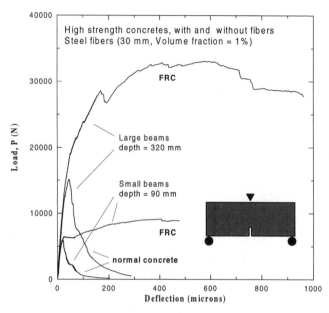

Figure 3. Load-displacement curves of two sizes of notched HSC beams, with and without steel fibers.

fracture mechanics type to plastic, as the structural size and fiber volume increase. This is in sharp contrast to (plain) concretes without fibers. This response has been observed experimentally (Bryars et al., 1994; Gopalaratnam et al., 1995) in notched beams of high strength concrete reinforced with steel fibers (Fig. 3).

There are important phenomenological differences between the fracture processes of FRC and plain concrete. The governing mechanisms in FRC are those related to the fiber-matrix interaction. The width and length of the fracture process zone are much larger in FRC than in the plain concrete. This is due to fiber-bridging, fiber debonding and pullout, crack deflection and multiple cracking. All these features increase the energy dissipation and ductility of the material. The benefits in flexural failure are obvious from Fig. 3. In the tensile failure of split-cylinders, the fibers lead to multiple cracks and greater energy dissipation, as seen in Fig. 4 where the load versus crack opening curves of high strength concrete with and without fibers are compared (Carmona et al., 1995). The response of the plain HSC exhibits softening after the peak followed by a plateau, which corresponds to the formation of the wedge cracks under the loading platens. In FRCs, the softening (corresponding to the matrix-dominated cracking) is followed by hardening-type behavior due to fiber-bridging across the main crack and distributed cracking.

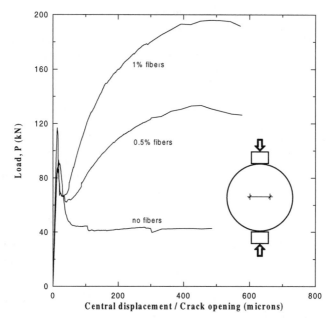

Figure 4. Response of HSC cylinders under splitting tension.

Other differences between FRC and plain concrete are also observed in compression, where the fibers cause internal confinement leading to failure mechanisms, such as shear bands, which are similar to those in plain concrete under external confinement. In shear and punching failures, there is additional shear resistance due to the dowel action of the fibers. In general, there is practically no improvement in compressive strength but there are significant increases in tensile and shear strengths. This implies that the empirical relations used for determining the latter strengths from the conventional compressive strengths are not valid for FRCs (Li et al., 1992).

In reinforced concrete elements, the incorporation of distributed fibers leads to changes in the failure mechanisms. In tests of reinforced concrete beams subjected to shear failure, Li et al. (1992) observed that the diagonal shear cracking of plain concrete changed to bending-shear failure with multiple flexural and shear cracks. In short beams, they observed splitting cracks along the line of arching compression in plain concrete while the FRC failed due to crushing or shear-compression cracking. They also concluded that fibers increase the effectiveness of the rebars by preventing splitting cracks and debonding along the rebar.

Important changes are, therefore, observed in the material and structural behavior due to the incorporation of fibers in concrete. This suggests that new or fundamentally different models are needed for the numerical modeling of FRCs. Some new approaches have recently been proposed (e.g., Carpinteri and Massabó, 1994) with specific representation of the fiber action across a localized crack. Further research is needed to develop more general analytical tools for FRCs.

Pseudo-ductility of high volume-fraction fiber composites

Cement-based composites made with high volumes of fibers are fabricated using special proprietary techniques, such as casting with vibration, compaction, extrusion and injection molding. The main characteristic of these composites is the formation of distributed cracking without localized macrocracks. Under tensile loading, multiple short microcracks develop in the fiber reinforced composite leading to plastic or hardening type behavior.

The Compact Reinforced Composite (CRC) of Aalborg Portland (Bache, 1987) is made with silica fume and 6% of steel fibers, and reinforced with conventional rebars. In test beams of CRC, bending stresses of 160 MPa and strains of 3mm/m have been observed before cracking. This implies an increase in flexural strength of about 5–10 times and an increase in deformation capacity of 15–30 times. The high amount of reinforcement prevents softening and localization until the deformations are one or two orders of magnitude higher than normal concrete.

The extruded composites developed at the ACBM Center (Marikunte and Shah, 1994) contain up to 15% of fibers. Dramatic increases in tensile strain capacities have been observed in these composites. Stabilized homogeneous microcracking has been obtained even at strains of 1%, that is, about 100 times the strain at which localization occurs in plain concrete.

In general, it can be stated that conventional fracture analysis is not applicable to composites where energy dissipation occurs due to considerable distributed microcracking instead of independent localized macrocracks.

Conclusions

The behavior of high performance concretes, such as high strength silica fume concretes and fiber reinforced concretes, depends on some mechanisms that are significantly different from those governing normal concretes. Some of these phenomenological aspects can be represented with conventional models using appropriate parameters. Other aspects, however, often necessitate the modification of existing tools and the formulation of new techniques. This is especially true when fibers are incorporated in the concrete.

Acknowledgements

Partial support from DGICYT research grants MAT93-0293 (P.I.: A. Aguado) and PB93-0955 (P.I.: I. Carol) to UPC is gratefully acknowledged.

References

Aguado, A., Gettu, R., and Shah, S., editors (1994). *Concrete Technology: New Trends, Industrial Applications*, Barcelona. E&FN Spon, London, U.K.

Ahmad, S. (1994). Short term mechanical properties. In (Shah and Ahmad, 1994), pages 27–64.

Atlassi, E. (1993). Effect of moisture gradients on the compressive strength of high performance concrete. In (Holand and Sellevold, 1993), pages 646–653.

Bache, H. (1987). Compact reinforced composite — Basic principles. CBL Rapport 41, Aalborg Portland, Aalborg, Denmark.

Bažant, Z. (1979). Discussion of "Material behavior under various types of loading". In Shah, S., editor, *High Strength Concrete (Proc. NSF Workshop, Chicago)*, pages 79–92.

Bažant, Z. and Baweja, S. (1994). Creep and shrinkage prediction model for analysis and design of concrete structures. Structural Engineering Report 94-10/603c, Dept. of Civil Engineering, Northwestern University, Evanston, USA. Summarized by RILEM Committee TC107 (1995), *Mater. Struct.*, 28, 357–365.

Bažant, Z. and Carol, I., editors (1993). *Creep and Shrinkage of Concrete (ConCreep 5)*. Chapman & Hall.

Brooks, J. (1994). State-of-the-art report: Elasticity, creep and shrinkage of concretes containing admixtures, slag, fly ash and silica fume. Draft report, ACI Committee 209.

Bryars, L., Gettu, R., Barr, B., and Ariño, A. (1994). Size effect in the fracture of fiber-reinforced high-strength concrete. In Bažant, Z., Bittnar, Z., Jirásek, M., and Mazars, J., editors, *Fracture and damage in quasi-brittle structures*, pages 319–326. E & FN SPON.

Carino, N. and Clifton, J. (1990). Outline of a national plan on high-performance concrete: Report on the NIST/ACI Workshop, May 16–18, 1990. Technical Report NISTIR 4465, U.S. Department of Commerce, National Institute of Standards and Technology, Gaithersburg, USA.

Carmona, S., Gettu, R., Hurtado, C., and Martín, M. (1995). Use of the splitting-tension test to characterize the toughness of fiber-reinforced concrete. *Anales de Mecánica de la Fractura*, 12:233–238.

Carpinteri, A. and Massabó, R. (1994). Continuous versus discontinuous bridged crack in the description of reinforced material flexural collapse. In Mang, H., Bićanić, N., and de Borst, R., editors, *Computational modelling of concrete structures*, pages 233–242, Innsbruck. Pineridge Press.

De Larrard, F., Acker, P., and Le Roy, R. (1994). Shrinkage, creep and thermal properties. In (Shah and Ahmad, 1994), pages 65–114.

FIB-CEB Working Group on High Strength Concrete (1990). High strength concrete, State-of-the-art Report. CEB Bulletin d'Information 197, Fédération Internationale de la Précontrainte, London, U.K.

Gettu, R., Aguado, A., and Pacios, A. (1993). Need for application-oriented high-performance concretes. In Oxford & IBH, New Delhi, India, editor, *International Symposium on Innovative World of Concrete (ICI-IWC-93)*, volume 3, pages 299–303, Bangalore, India.

Gettu, R. and Shah, S. (1994). Fracture mechanics. In (Shah and Ahmad, 1994), pages 161–212.

Gopalaratnam, V., Gettu, R., Carmona, S., and Jamet, D. (1995). Characterization of the toughness of fiber reinforced concretes using the load-CMOD response. In Wittman, F., editor, *Fracture Mechanics of Concrete Structures (FraMCoS 2)*, pages 769–782, Zürich, Switzerland.

Hansen, P. and Jensen, O. (1989). Selfdesiccation shrinkage of low porosity cement-silica mortar. Nordic Concrete Research 8, Norske Betongforening. Nordic Concrete Federation (eds.).

Hillerborg, A. (1986). Determination and significance of the fracture toughness of steel fibre concrete. In Shah, S. and Skarendahl, Å., editors, *Steel fiber concrete (Proc. US-Sweden joint seminar)*, pages 257–271. Elsevier Applied Science, London, U.K.

Holand, I. and Sellevold, E., editors (1993). *Utilization of high strength concrete*, Lillehammer, Norway. Norwegian Concrete Association, Oslo, Norway.

Jaccoud, J., Charif, H., and Farra, B. (1993). Cracking behaviour of HSC structures and practical consequences for design. In (Holand and Sellevold, 1993), pages 225–232.

Li, V., Stang, H., and Krenchel, H. (1993). Micromechanics of crack bridging in fibre-reinforced concrete. *Materials & Structures RILEM*, 26:486–494.

Li, V., Ward, R., and Hamza, A. (1992). Effect of fiber modified fracture properties on shear resistance of reinforced mortar and concrete beams. In Carpinteri, A., editor, *Applications of fracture mechanics to reinforced concrete*, pages 503–522, Torino, Italy. Elsevier Applied Science, London, U.K.

Marikunte, S. and Shah, S. (1994). Enginnering of cement-based composites. In (Aguado et al., 1994), pages 83–102.

McDonald, D. and Roper, H. (1993). Prediction of drying shrinkage of concrete from internal humidities and finite element techniques. In (Bažant and Carol, 1993), pages 259–264.

Nilsson, L. (1994). The relation between the composition, moisture transport and durability of conventional and new concretes. In (Aguado et al., 1994), pages 63–82.

Shah, S. and Ahmad, S., editors (1994). *High Performance Concretes and Applications*. Edward Arnold, London, U.K.

Tazawa, E. and Miyazawa, S. (1993). Autogenous shrinkage of concrete and its importance in concrete technology. In (Bažant and Carol, 1993), pages 159–168.

Tazawa, E., Miyazawa, S., and Kasai, T. (1995). Chemical shrinkage and autogenous shrinkage of hydrating cement paste. *Cement and Concrete Research*, 25(2):288–292.

Walrawen, J. and Stroband, J. (1994). *Shear friction in high-strength concrete*, pages 311–330. Special Publication 149. Malhotra, V.M. (ed.), American Concrete Institute.

Zia, P., Leming, M., and Ahmad, S. (1991). High performance concretes: A state-of-the-art report. Report SHRP-C/FR-91-103, Strategic Highway Research Program, Washington, D.C.

Non Linear Computation Of Fiber Reinforced Micro-concrete Structures

Mouloud Behloul[1], Régis Adeline[2], Gérard Bernier[1]

Abstract

It is shown that the modeling of structures made of fiber-reinforced micro-concrete is well suited to linear calculation using only equivalent behavioral laws for the constituent materials. The equivalent tensile behavioral law, in the post-peak part, is deduced from direct tensile strength tests by dividing the crack opening by a characteristic length.

This length, which depends on the geometry of the structure, is taken as being equal to the depth of the section loaded. Results of tests on thin plates and on a prestressed Reactive Powder Concrete (RPC) beam have been compared to the results of the calculation. The size effect induced by this approach is demonstrated; it is nevertheless slight in the case of Reactive Powder Concrete.

Introduction

Composites with fiber-reinforced cementitious matrices have been developed over the last thirty years or so without the methods of calculation of structures made with these materials being entirely defined and unified: this is certainly due to the distinction that should be made between two types of composites:
- fiber-reinforced ordinary concrete and high-performance concrete,
- fiber-reinforced micro-concrete.

----

[1] ENS Cachan/CNRS/Université Paris VI Laboratoire de mécanique et technologie. 61, avenue du Président Wilson, 94235 CACHAN CEDEX, FRANCE.
[2] Direction Scientifique de BOUYGUES.Challenger, 1, avenue Eugène Freyssinet-78061 Saint-Quentin-Yvelines Cedex.

In the first type, fiber contents do not exceed 40 kg/m^3, this limitation being due to placement difficulties [11]. This low content has little effect on the flexural strength: the increase is between 0 and 50% as compared to the flexural strength of the matrix [2-10]. The fibers do not distribute cracking.
This location of the cracking in the case of a beam in flexure requires a calculation in 2 phases: a first phase until the appearance of cracking (tensile yield point), where all materials are in their elastic area, followed by a second phase involving pseudo-mechanism behavior [1-3].

For micro-concretes, such as reactive powder concretes (RPC), and fiber-reinforced cement mortars or slurries, the fiber contents are higher, and can reach as much as 12% by volume [4]. Furthermore, the fibers are large relative to the constituents of the matrix.

Fibers function in similar fashion to steel rebars in reinforced concrete, and thus allow multi-cracking of the material [8]. The flexural strength of the matrix is then substantially increased [9-12]: (by 100 to 200%).
In the case of a high-performance matrix such as that of RPC, low fiber contents (from 1 to 4%) make for a considerable improvement to mechanical performance [7].

The multi-cracking aptitude of these micro-concretes means they can be modeled as a homogeneous material governed by a stress-strain behavioral law. The tensile post-peak part is deduced from the "stress-crack opening" curve obtained with a direct tensile strength test where the opening of the crack is divided by a characteristic length.
This type of modeling can be used to predict the ultimate bending moment, the equivalent rigidity, and the evolution of deflection in accordance with loading.

The behavior of Reactive Powder Concrete (RPC)

Reactive Powder Concrete (RPC) constitutes a new range of materials developed by the Scientific Division of Bouygues. Studies have led in particular to compressive strengths between 200 and 800 MPa being achieved.
The principles that have guided the elaboration of Reactive Powder Concrete are:
- elimination of the coarse aggregate in order to increase the homogeneity of the material,
- improvement of the voids ratio by optimizing the grading and pressure during setting,
- improvement of the microstructure by heat treatment after setting,
- addition of steel fibers to improve ductility.

The use of superplasticizers and silica fume has made it possible to reduce the water/cementitious materials to less than 0.15. The result is a reduction in the average

pore diameter and the total porosity of the cementitious paste.
In the rest of this article, only RPC200 (200 MPa compressive strength) is studied.

The typical composition of 1 m³ of RPC200 comprises 685 Kg of cement and 222 Kg of silica fume. The ratio Water/(Cement+SF) reaches 0.19 .
The steel fibers are 12 mm long and 0.15 mm in diameter. They represent 2% of the total volume.

Compressive behavior.
The compressive strength has been measured on 7 cm diameter and 14 cm high cylindrical test specimens.
RPC200 has a compressive strength between 170 and 230 MPa, and a Young's modulus between 50 and 60 GPa.
Its behavior can be approximated with a bilinear law: linear behavior until 50% of the ultimate strength, with the initial modulus, followed by a linear curve until failure with a tangent modulus with a value of 85% of the initial modulus.

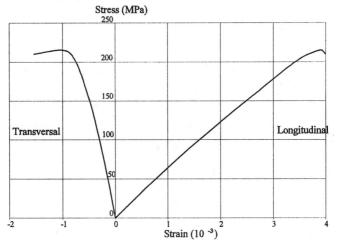

Figure 1. Stress-strain curve of RPC in compression

Tensile behavior.
To be representative of real behavior in structures, tensile strength tests were made on unnotched test specimens (Figure 2). The particular shape was chosen by applying 2 criteria:
-    to obtain failure in the central zone where the stress pattern is uniform, whence a reduction by half of the sectional area loaded;

- to prevent the concentration of stresses by choosing a sufficient radius of curvature.

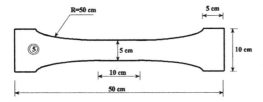

Figure 2. Direct tensile test specimen

The tests were carried out on an INSTRON machine, the 2 ends the specimen being glued to the frame of the machine. The test was conducted with displacement at a speed of 0.01 cm/minute.

In most cases, the residual force was canceled out for a crack opening of 6 mm, which corresponds to half the length of the fibers (Figure 3).

The post-peak part is a polynomial curve that can be determined if the points of intersection with the lines and tangents at those points are known [3].

In the case of RPC, the stress-deflection curve can be described by the following equation:

$$\sigma(w) = f_t \left(-1.6 \left(\frac{w}{6} \right)^3 + 3.6 \left(\frac{w}{6} \right)^2 - 3 \left(\frac{w}{6} \right) + 1 \right)$$

Where w is the opening of the crack in mm, and $\sigma(w)$ is the residual stress.

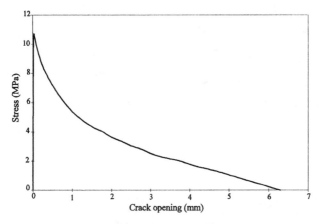

Figure 3. A stress-elongation curve of RPC200 in tension

Flexural behavior.
Test specimens of dimensions 15x100x400 mm were loaded in
4-point flexure (Figure 4).

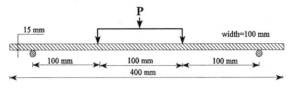

Figure 4. Flexural test loading arrangement.

Test specimens exhibited extensive multi-cracking in the
bottom part of the central zone. The results of Figure 5
show the increase of flexural strength and ductility in
relation to the fiber content.

A comparison was made between 3-point flexure on
40x40x160 mm specimens, 4-point flexure on 15 mm thick
specimens, and direct tensile tests.
This study revealed a direct connection between the
tensile strength and 3-point flexural strength:

$$ft=f3p/(2.7x1.6)$$

The coefficient 2.7 corresponds to 4-point-flexural-
strength-to-tensile-strength ratio for a perfect plastic
elastic material [6].

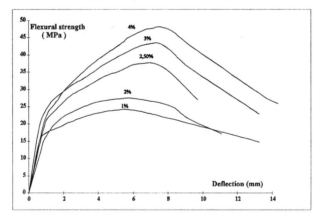

Figure 5. Flexural strength-midpoint displacement
response for Vf=0 to 4 %

The coefficient 1.6 observed on several 3- and 4-point flexural tests may be explained by the particular mechanism of the 3-point flexural test.

Equivalent behavioral law of Reactive Powder Concrete

Multi-cracking makes it possible to go from the stress-crack opening law to a stress-strain law by dividing by a characteristic length.
Choosing this length is a delicate matter: two authors - Casanova [1] and Naaman [5] - have taken the depth of the cracked section, h, as the characteristic length. Two justifications have been given for this:
- beyond a distance h/2 from the crack, the stress pattern is not disturbed by the crack;
- the pitch of cracking is of similar magnitude to the depth of the beam (Naaman).

This definition has been adopted for RPC, even if observations made on the prestressed beam, which will be dealt with below, reveal a pitch of cracking of the order of one third of the depth of the beam.
In compression, the behavioral law is that derived directly from the tests.

Non linear analysis procedure

For modeling the behavior of structures in Reinforced Fiber Concrete, a non-linear multi-layer calculation program has been developed. The basic principle is the balance of cracked sections. The section is discretized into layers, which allows modeling of complex forms.

Passive or prestressed reinforcement can be represented.

The calculation assumptions are those of beams (plane sections remain plane); the bond between reinforcement and concrete is assumed to be perfect.
Figure 6 presents the algorithm for calculation of section balance where it is sought to determine the distribution of strains that balances external efforts by taking account of the behavioral laws of each material.

This algorithm can be integrated into a general structural calculation program, Figure 7 (under development).

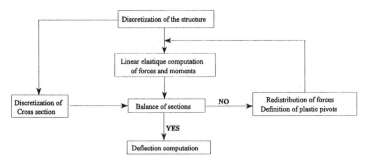

Figure 6. Flow chart of the non linear analysis of sections

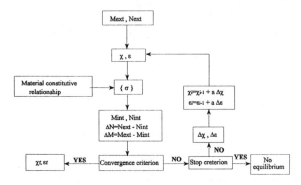

Fig 7. General flow chart of the computation

Size effect

Consideration of one dimension of the structure in the tensile behavioral law induces a size effect. The theoretical evolution of the flexural strength as a function of depth is represented on Figure 8 which shows that the size effect expected for RPC remains slight; the gain in flexure as compared to direct tensile stress remains greater than 2 to a depth of 40 cm.

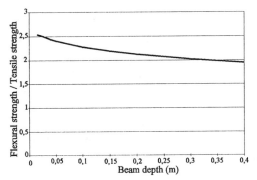

Figure 8. Size effect

Comparison of test results and predictions
Two types of structures were studied: a highly prestressed
beam and thin plates without main reinforcement.

Prestressed beam.
The beam tested is 15 meters long, 37 cm deep, and 32 cm
wide. Its cross-section is roughly X-shaped, forming the
central part of an "hour-glass" in which the curved
sides have a radius of 14 cm (Figure 9).

The beam, without passive reinforcement, is prestressed by
eight 15 mm strands which were tensioned before casting.
The tendons have a sectional area of 140 mm, an elastic
limit of 1540 MPa, and a breaking strength of 1770 MPa.

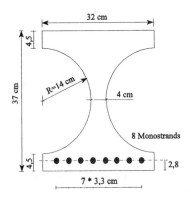

Figure 9 Cross section

The load is applied by means of 8 regularly spaced jacks
(Figure 10).

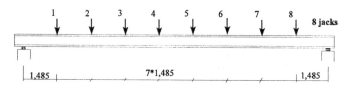

Figure 10. Testing device

The behavioral law was defined in full, on the basis of the direct tensile stress (deduced from the 3-point flexural tests) compressive strength tests, and measurements of Young's moduli using the Grindosonic method.

The moment-deflection curve derived from the test is represented in Figure 11. It is compared to the results of calculation.
There is seen to be a good match between the 2 curves, and that the mode of failure, in particular - by compressive crushing of the concrete - was predicted by the calculation.

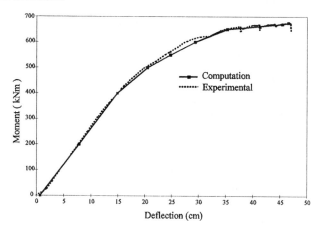

Figure 11. Moment deflecion curve, prestressed beam.

During this test, the strain on the upper and lower face of the beam was recorded (Figure 12). The strains calculated are a good match to those measured, which confirms the good representativity of the approach used.

Thin plates.
Thin plates (25 mm thick) were tested for 4-point flexure in 2 loading configurations: 1 m distance between loads, and spans of 3.80 and 2.20.

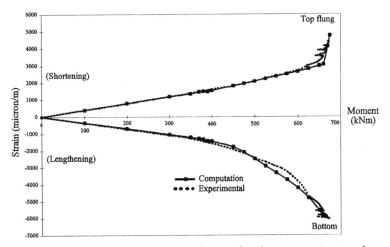

Figure 12. Mid-section strain evolution, prestressed beam.

In this type of structure, the ultimate strength is largely governed by the peak tensile strength of the RPC. Considering the relative scatter of flexural test results, two laws bracketing the average law by +-5% were used for the modeling.

The results presented in Figures 13 and 14 show a good match with the test as regards the following two aspects:
- value of the ultimate flexural strength.
- deflection/imposed load ratio.

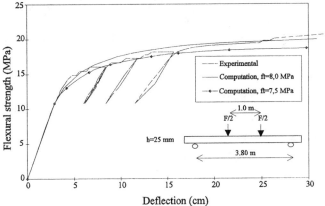

Figure 13. Midpoint displacement response, plate 1

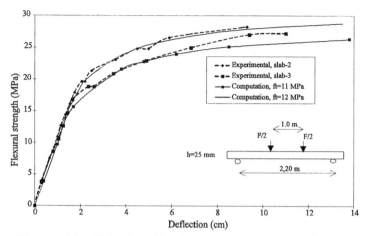

Figure 14. Midpoint displacement response, plates 2-3

Conclusion
Fiber-reinforced micro-concretes have high flexural
strength and their pre-failure behavior is characterized
by multi-cracking. The ultra-strong matrix of RPC benefits
fully from these two effects, even with low fiber
contents.
This ductile behavior means the material can be
represented as a homogeneous material; the equivalent
strain of the behavioral law for tensile stress is deduced
from the opening of the crack in a direct tensile strength
test by dividing the opening by a characteristic length .
This characteristic length can be assumed to be equal to
the depth of the cracked section.
The application of this method to a prestressed beam and
to thin plates without main reinforcement enabled good
prediction of the overall behavior of these structures.

References

[1]   Casanova.P, Rossi.P, LeMaou.F.
      "Calcul en Flexion et caractérisation des bétons de
      fibres métalliques".
      Annales ITBTP N°528, novembre 1994. pp 81-91.
[2]   Gopalaratnam.V.S and Cie.
      "Fracture Toughness of fiber reinforced concrete"
      A.C.I. materials Journal, july-august 1991. pp339-
      353.

[3] Maalej.M, Li.V.C.
"Flexural strength of fibre cementitious composites"
Journal of materials in civil engineering,
vol.6, N°3, august 1994. pp 390-406.

[4] Naaman.A.E.
"SIFCON:Tailor properties for structural performance"
High Performance Fiber Reinforced Cement Composites.
Edited by H.W Reinhardt and A.E.Naaman. pp 39-56.
1992 RILEM.ISBN 0 419 17630 6.

[5] Naaman.A.E, Reinhardt.H.W, Fritz.C and Alwan.J.
"Non-linear analysis of RC beams using a SIFCON
matrix"
Materials ans structures, 1993, 26, 522-531.

[6] Olsen.P.C.
"Some comments on the bending strength of concrete
beams". Magazine of concrete research, 1994, 46,
N°168, 209-214.

[7] Richard.P, Cheyrezy.M.
"Reactive Powder Concretes with high ductility
and 200-800 Mpa compressive strength"
San Fransisco, 1994.

[8] Shao.Y, Li.Z and Shah.S.P.
"Matrix cracking and interface debonding in fiber-
reinforced cement-matrix composites".
Adv Cem Bas Mat 1993;1:55-66.

[9] Tjitobroto.P and Hansen.W.
"Tensile Strain hardening and multiple cracking in
high performance cement based composites containing
discontinuous fibers".
A.C.I. materials journal/ january-February 1993.
pp134-142.

[10] Trottier.J.F and Banthia.N.
"Toughness characterization of steel-fiber reinforced
concrete".
Journal of materials in civil engineering,vol.6, N°2,
may 1994. pp 264-288.

[11] Projet national, voies nouvelles du béton, béton de
fibres.
"Le béton de fibres métalliques, état actuel
des connaissances"
Annales ITBTP, N°515 juillet-aout 1993. pp 39-67.

[12] Ward.R.J and Li.V.C.
"Dependance of flexural behavior of fiber-reinforced
mortar on material fracture resistance and beam
size".
ACI Materials journal, November-december 1990. pp
627-637.

Seismic Analysis of Concrete Towers of the San Diego-Coronado Bridge and
Evaluation of a High Performance Concrete Retrofit

Robert A. Dameron, MS, PE[1], Daniel R. Parker, MS[2]

Abstract

As part of a comprehensive seismic vulnerability assessment sponsored by the California Department of Transportation, the authors performed detailed analysis of three concrete towers of the San Diego-Coronado Bay Bridge. The first objective of the work was to determine the behavior, strength and ductility capacities, and failure modes of the towers when subjected to strong ground motion postulated to occur within one kilometer of the bridge site. Upon determining the vulnerabilities, the second objective was to develop potential retrofit solutions that would prevent collapse or long-term closure of the bridge after occurrence of the postulated event. While the 6 month study also involved many other components and structural details, this paper presents the procedures used and findings related to the detailed evaluation of the concrete towers. The towers are tall and slender and have an arched bent cap that gives the bridge aesthetic distinction that must be preserved in any retrofit solution. Potential shear failure of the bent caps in the tallest towers must therefore be addressed with an innovative retrofit consisting of use of prestressing in conjunction with high performance concrete to maintain slenderness and aesthetic balance. This paper describes the bent cap vulnerability study, the retrofit options being considered, and the analysis procedures in use to evaluate a high performance concrete retrofit strategy.

Introduction

In response to the damage to California bridges sustained in the 1989 Loma Prieta Earthquake, the California Department of Transportation has taken extensive action to seismically retrofit the state's bridges. This has included not only retrofit of over 1,000 bridges, but also large scale research of earthquake induced damage phenomena and how best to prevent it. Some of these activities have been influenced by the recommendations of the Governor's Board of Inquiry, most notably, the steps taken to seismically evaluate the state's toll bridges, a subset of the state's most important transportation structures. The Governor's Board Report [1] calls for the performance of "comprehensive earthquake vulnerability analyses and evaluation of important transportation structures, . . . using state-of-the-art

[1]Principal Engineer, ANATECH Corp., Member ASCE, 5435 Oberlin Drive, San Diego, CA 92121
[2]Engineer, ANATECH Corp., 5435 Oberlin Drive, San Diego, CA 92121

methods in earthquake engineering." Caltrans has now conducted or contracted for such seismic analyses or vulnerability assessments of all of the state's toll bridges.

The San Diego-Coronado Bay Bridge connects San Diego with Coronado by carrying State Route 75 across San Diego Bay. Plan and elevation views of the 8,000 foot long bridge are shown in Figure 1. The bridge's design was controlled by constraints on approach alignments and by channel navigation requirements. Thus, the bridge consists of three long channel spans of orthotropic box girder construction, twenty-six shorter spans of steel plate girder and composite lightweight concrete deck construction, and a 90-degree, 1800-foot radius curve on the western approach spans. The bridge supports for Piers 2 through 24 consist of pile supported reinforced concrete towers and, east of Pier 24, of towers on spread footings. The long length, height, flexibility, and curve of the bridge are all issues that contribute to unique seismic performance.

ANATECH performed a study for the State of California that consisted of local and global analyses to make a seismic vulnerability assessment of the bridge in its current condition, development of conceptual retrofit schemes, and an approximate retrofit cost estimate [2]. The study required that the bridge's vulnerability to two earthquake events (the Safety Earthquake and the Functional Earthquake) be evaluated using both "Minimum Performance Criteria" and "Important Bridge Performance Criteria." Minimum Criteria allows significant damage for the safety event but does not allow collapse, while Important Bridge Criteria requires damage to be repairable under traffic for the safety event, i.e. no extended bridge closure. To evaluate the bridge's ability to meet a "no collapse" criteria under the larger earthquake event required extensive use of nonlinear analysis for predicting inelastic behavior and ductility.

In the concrete components of the bridge, the study revealed that the bridge in its current condition had vulnerable foundations and towers. In the foundations, the pile-to-pile cap moment connections are not ductile. The rebar cage connecting the piles to the pilecaps was found to be significantly susceptible to bond-slip and anchorage loss, thus causing relatively brittle failure. The twenty-nine batter-leg towers simply supporting the superstructure consist of reinforced concrete shafts and bent caps. The tower shafts have inadequate confinement steel, and the vertical steel extending from the footings is spliced at the base. This reduces the ability of the towers to form full-strength plastic hinges. However, detailed analysis described in this paper shows that the plastic hinges are fairly long, so damage is widely distributed. Thus reasonable displacement ductility (~5) is achievable, so most tower columns do not require any retrofit. However, large shear forces must be transmitted through the bent caps, and in the critical shaft/cap joint region there is very little shear reinforcement. This appears to be of particular concern on the two main channel piers (19 and 20) which is the subject of this paper.

Analysis Methodology

Actual member response to combined bending, shear, and axial force is very complex especially for the Coronado tower geometries, with tapered, battered columns, and bent caps with curved haunches. Linear analysis techniques cannot determine the true stresses in the reinforcement or concrete under these combined conditions. Starting in the early 80s, the authors and others at ANATECH developed a concrete constitutive modeling approach for the nonlinear continuum analysis of reinforced concrete. This work is based on the work by Rashid, the original developer of smeared cracking finite element technology in the 1960's [3]. In the past several years, the methodology has been used to successfully predict laboratory tests of several bridge joints (e.g. [4,5]). During the 80's it also successfully predicted large nuclear containment structure behavior (e.g. [6]) where continuum response was critical to the solution. In the tests of scale bridge joint models, the methods accurately predicted cracking, crushing, compressive plasticity (softening), cyclic hysteresis/stiffness degradation, rebar fracture, and ultimate structure failure. This nonlinear modeling methodology for concrete was used in the current work with the nonlinear finite element program, ANACAP [7], for conducting detailed predictions of three representative towers.

Within the concrete constitutive model, cracking and all other forms of nonlinearity are treated at the element integration points; thus, the crack status and stress-strain state can vary within an element. Cracks are assumed to form perpendicular to the principal directions that exceed the cracking criteria $\left(\varepsilon_{cr} > 0.0001\text{in/in}\right)$, and multiple cracks are allowed to form, but they are constrained to be mutually orthogonal. If cracking occurs, the normal stress is reduced to zero gradually through a tension stiffening algorithm, and the distribution of stresses around the crack is recalculated. The loading must be incrementally applied to allow for this redistribution to rebars or to neighboring elements, thus these analyses are computationally intensive. The shear stress is also reduced upon cracking and further decays as the crack opens through detailed treatment of post-cracking shear stiffness. This effect is known as "shear retention" and is attributed to crack roughness and aggregate interlock [8]. Experimental data shows that upon cracking, the shear modulus is reduced to a value of about 40% of the uncracked values and is subsequently reduced in proportion to the crack opening strain. Once a crack forms, the direction of the crack remains fixed, and it can never "heal." However, cracks do close and re-open under load reversals.

The reinforcement in the models is represented as individual sub-elements within the 3D concrete continuum elements. The rebar sub-element stiffnesses are superimposed onto the concrete element stiffness in which the bar resides. The rebar material behavior is handled with a separate constitutive model that treats the steel plasticity, strain hardening, and bond-slip behavior. The concrete and rebar formulations can handle arbitrary strain reversals at any point in the response, whether in tension or compression.

A modeling parameter important to predicting bridge structure failure is rebar bond-slip and anchorage failure modeling. The bond-slip effects are important for characterization of both the shape of a structure's hysteresis loops and of the ultimate capacity (and ductility) a structure can reach. The former is manifested by the characteristic pinched loops of joints with significant bond-slip as compared with the non-pinched loops exhibited by joints with strong confinement and little bond-slip action. The bond-slip model used in the current work incorporated both the loop pinching effect and the ultimate anchorage loss effect. The loop pinching is captured by modifying the elastic unloading behavior of the steel constitutive model in conjunction with the implementation of Bauschinger Effect. The anchorage loss is modeled by monitoring the concrete strain normal to the bar (the dilatational strain) and degrading the bar's effectiveness (degrading its strength) after this strain reaches 0.001. This critical level of dilatational strain was specified by Priestley, et al., as a result of extensive column testing that included rebar lap-splices (see, for example, [9]). Similar conclusions have also been reached by other researchers. The loop pinching associated with bond-slip was found to be very important in the Coronado Pier model behavior predictions.

Computational Models

Detailed continuum models of three piers were developed: Piers 2, 16, and 20. This selection was based on the need to study, in detail, conventional short pier behavior, tall pier behavior, and the unique behavior exhibited by the tall, hollow column, Channel Span towers. Time and budget constraints of the project did not allow any further continuum modeling, except those used on pile-pilecap models described in [2].

Each model was developed with 8-node brick concrete elements and rebar subelements as previously described. The models and results are illustrated in Figures 2, 3, and 4. Each model used an assumption of a fixed boundary condition at the location of each pile head and applied displacements to drive the model at the points of attachment of the bearings. Each model was also pre-loaded with the superstructure dead load and with the pier's self weight. Separate analyses were run for transverse and longitudinal motion, and in each case, a monotonic push was applied first to evaluate the general response envelope, gage the unit-ductility, and select the loading peaks to prescribe for the cyclic analysis. Cyclic displacement pushover analysis was then performed at increasing amplitudes much like the quasi-static cyclic testing to failure normally used in testing laboratories. Analysis were performed in both the transverse and longitudinal directions, but only the transverse results are shown here since this is the direction that creates large bent cap shear demands.

The calculated pier response and damage patterns are illustrated in Figures 2, 3, and 4. Each set of plots provides Force-deflection behavior, Deformed Shape and Crack Patterns, and strain contour information. In all cases, plastic hinges form in the columns (generally top first) and not in the cap beams and the plastic hinges are fairly long, on the order of 2xd for the tall piers and approximately d for the short piers. In the case of Pier 20 (identical to Pier 19) the large strains associated with plastic hinging extend far up into the cap beam joint. For this reason, this cap beam has little residual shear capacity. For the tall piers, cycles of increasing amplitude lead to fairly broad hysteresis loops and, ultimately, failure by fracture of column reinforcement. For the short pier (Pier 2), hysteresis loops are more pinched and failure is in shear in the columns. The ramifications of this play a major role in the vulnerability evaluation and retrofit strategy.

Summarized interpretation of these analysis results is shown in Table 1.

The continuum analyses of the towers demonstrated overall system ductility to be relatively uniform at about 5. The failure modes, however, were significantly different. The cap beams for the approach span pier, though cracked in the cyclic exercise, are not prone to shear failure. Damage is more significant in the Channel Span Cap beams as shown in Figure 4. Even though the analysis shows ductility capacity of 5, because of the importance of the towers and because brittle shear failures must be avoided with absolute certainty, the allowable ductility to use in the retrofit design was reduced by 0.5 to a value of 2.5.

Table 1. Tower Continuum Analysis Results Summary (Transverse Direction)

Results Item	Pier 2	Pier 16	Pier 20
Unit Ductility	1.2"	11.0"	11"
Displ. Capacity	6.0"	50"	60"
Ductility Capacity	5	4.5"	5.5
Force Capacity	3000	1250	3900
Failure Mode	Column Shear	Column Flexure	Column flexure, but major bent cap shear damage

Once the ductility is known from the local tower analyses, a global model of the structure was analyzed with the postulated earthquake to obtain the demands on the structure components. In keeping with Caltrans retrofit design procedures, the ratio of the force demands to force capacity is then compared to the available ductility to determine the retrofit needs of the structure. The bent cap shear capacity was determined as follows.

$$Vn = Vc + Vt$$

$$V_t = \frac{Avfyd}{S} \quad V_c v_c A_e$$

A_n = Total cross-sectional area of transverse reinforcement within spacing S.

f_y = Probable yield strength of transverse reinforcement

d = Effective depth of cap beam

S = Spacing of transverse reinforcement

A_e = Effective shear area taken as .8Ag (Ag = gross section area)

$$v_c = 3.5\sqrt{f_{c'}}$$

$f_{c'}$ = concrete crushing strength

Using this force capacity calculation, the global demand-capacity ratios for the bent cap shear ranged from 2.5 on the shorter piers to about 4.5 on Piers 19 and 20, well in excess of the allowable ductility.

Retrofit Solutions

The basic concept for retrofit of the Coronado Bridge is to strengthen certain elements to guard against such severe damage that would lead to collapse or long-term failure of the bridge. These strengthened elements include the superstructure hinges, the bridge abutment on the Coronado side, some of the pile systems, the columns of the shorter piers, and most importantly for topic of this paper, the bent caps of Piers 19 and 20. In addition to strengthening, seismic isolation of the piers is being considered, but since the Coronado Bridge is such a long period structure to begin with, the reduction in force demands imparted to the towers by the superstructure is not as advantageous as in other applications of seismic isolation. For these reasons and due to the critical importance of the main Channel Span towers (supporting the central 600 foot spans), strengthening of the Pier 19 and 20 bent caps is deemed necessary.

The preliminary concept for the bent cap retrofit is drawn in Figure 5. This concept strengthens the bent cap for carrying shear. The preliminary retrofit solution is being evaluated with a modification of the detailed continuum element model of Bent 20. The retrofitted cap beam shows significantly improved performance and much less damage to the cap beam than the as-built structure, so this retrofit strategy is acceptable on the basis of strength considerations. However, as previously mentioned the towers of the bridge are tall and slender and have an arched bent cap that gives the bridge aesthetic distinction that must be preserved. To do this the retrofitted cap beam must be made more slender and less obstructive than the preliminary retrofit concept. A modified retrofit solution has been proposed using prestressing in conjunction with high performance concrete to maintain slenderness. This modified solution is shown in Figure 6.

High performance concrete not only has the advantage of slenderness, but it also brings additional, much needed ductility to the cap beam. Increased ductility of beam-column joints for seismic applications through the use of fiber-reinforced concrete has been the subject of considerable research in recent years. For example, Ramey in [10] showed that the confining rebar requirements for concrete building joints could be considerably relaxed if fiber reinforced concrete were used. Applying these findings to Coronado would imply that the addition of a fairly slender fiber reinforced concrete retrofit would add considerable confinement and ductility to the joint. Concrete of 10 ksi compressive strength and reinforced with 1 inch long steel fibers, 2% by volume is proposed.

The proposed scheme is now being evaluated with a modified continuum element model of the more slender design and with a modified constitutive model. The modifications to the constitutive model are to the simulation of tensile behavior. The tensile strain at cracking is increased to 0.0003 based on observed fiber reinforced concrete behavior, but the tension stiffening algorithm is also modified to include slower strength attenuation after the onset of cracking, due to the enhanced crack distribution behavior provided by the fiber reinforcement. More details on the fiber reinforced constitutive model can be found in [7]. Preliminary analyses of the modified retrofit solution have demonstrated superior performance to the original retrofit concept, even though the proposed retrofit is much more slender than the original concept.

Conclusions

A detailed finite element analysis with cyclic loading was performed for three representative towers of the San Diego-Coronado Bay Bridge. While each tower exhibited transverse displacement ductilities of approximately 5, the shorter tower failed in column shear, the tall-solid-column tower failed in flexure, and the tall Channel Span tower failed in bent cap shear. A preliminary retrofit solution of thickening the bent cap and adding post-tensioning produced satisfactory strength enhancing performance, but was aesthetically undesirable. A more slender retrofit appears to be achievable through the use of high performance concrete. Evaluation of the more slender design is in progress using a modification of the ANACAP smeared cracking concrete constitutive model that accounts for the enhanced tensile and compression performance of the high performance, fiber reinforced concrete.

References

1.   Housner, G. W., et al., "Competing Against Time," Report to Governor George Deukmejian from the Governor's Board of Inquiry on the 1989 Loma Prieta Earthquake, State of California, Office of Planning and Research, May 31, 1990.

2.   Dameron, R. A., Sobash, V. P. "Seismic Analysis of the Existing San Diego-Coronado Bay Bridge," ANATECH report to Caltrans, October, 1995.

3.   Rashid, Y. R., "Ultimate Strength Analysis of Prestressed Concrete Pressure Vessels," Nuclear Engineering and Design, 7, 1968, pp. 334-344.

4.   Parker, D. R., Dameron, R. A., "China Basin Viaduct Foundation Blind, Pretest Prediction Analysis of 1/3 Scale UCSD Specimen," pre-test prediction report for CH2M HILL and Caltrans, April 1994.

5.   Dameron, R. A., Kurchubasche, I. R., Rashid, Y. R., "Predictive Analysis of Outrigger Knee-Joint Hysteresis Tests: A Torsion/Flexure Test at UCB, and two Flexure Tests at UC San Diego," Caltrans First Annual Seismic Research Workshop, November 1991.

6.   Dameron, R. A., Rashid, Y. R., Sullaway, M. F., "Pretest Prediction of a 1:10 Scale Model Test of the Sizewell-B Containment Building," ANATECH report to Sandia National Labs, Albuquerque NM, NUREG/CR-5671, 1990.

7.   ANATECH Corp., "ANACAP Users Manual, Version 2.0, April, 1994.

8.   Dameron, R. A., Dunham, R. S., Kurchubasche, I. R. Rashid, Y. R., "A Rough Crack Constitutive Model for Concrete," Proceedings of SMIRT-11, Tokyo, 1991.

9.   Dameron, R. A. and Dunham, R. S., "Continuum F. E. Analyses of Concrete Bridge Outrigger Cyclic Loading Tests with a New Rebar Bond Slip Model," Proceedures of the NSF Workshop on Bridge Engineering Research in Progress, November 16 & 17, 1992.

10.   Gefken, P. R. and Ramey, M. R., "Increased Joint Hoop Spacing in Type 2 Seismic Joints Using Fiber Reinforced Concrete," ACI Structural Journal, March - April, 1989, pp. 168-172.

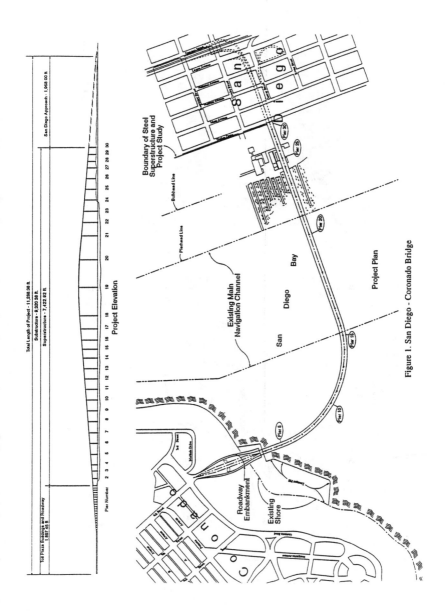

Figure 1. San Diego - Coronado Bridge

Vertical Strain Contour

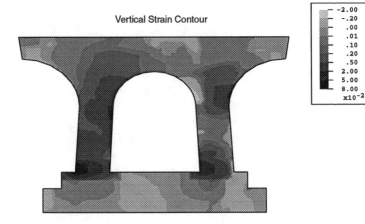

Top Displ. = +6.0"

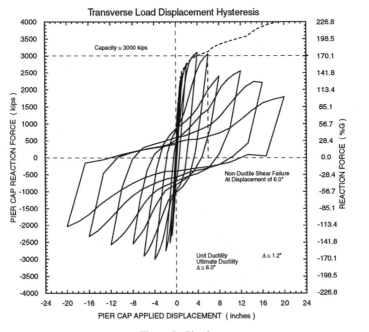

Figure 2. Pier 2

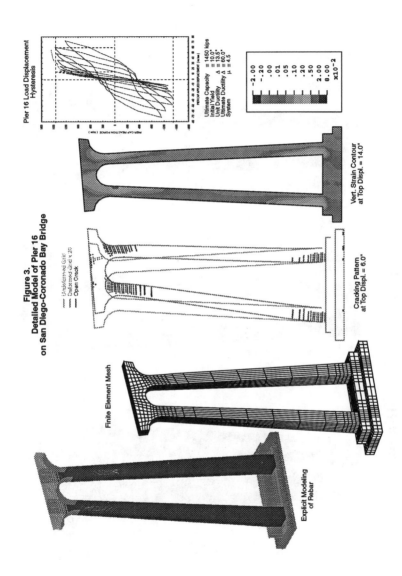

Figure 3.
Detailed Model of Pier 16
on San Diego-Coronado Bay Bridge

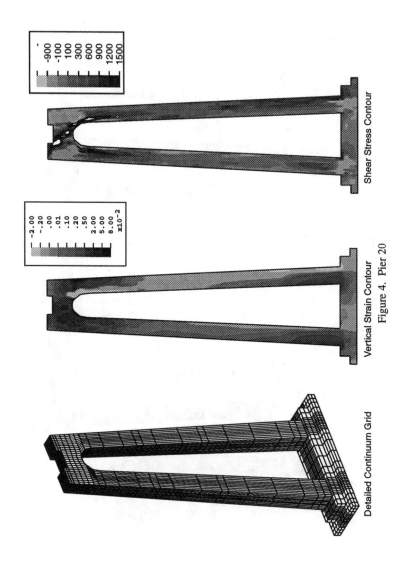

Shear Stress Contour

Vertical Strain Contour
Figure 4. Pier 20

Detailed Continuum Grid

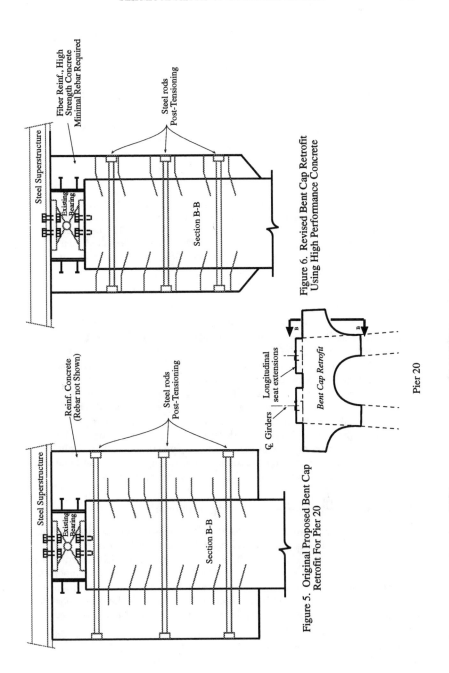

Figure 6. Revised Bent Cap Retrofit Using High Performance Concrete

Figure 5. Original Proposed Bent Cap Retrofit For Pier 20

Computational Modeling of Early Age HPC

Rob ter Steeg*   Jan Rots *   Ton van den Boogaard[†]

Abstract

In this paper, a discussion of the theoretical background of the computational model for early age HPC will be given. The application of the developed model to a structural element of a bridge will be discussed where the material properties of the young HPC have been obtained from experimental measurements of the "Stichtse Brug" test program in the Netherlands.

Introduction

Concrete can be regarded as a material that changes its mechanical properties drastically with the elapse of time, especially at early ages. The stiffness and strength will increase during the first days of the hydration process. The tensile stresses due to the hydration process can result in cracks because of the low tensile strength at early ages. In the last decade, high strength concrete has proved to be an extremely sensitive material regarding thermal cracking at early ages because of the increased hydration heat in these types of material. The simulation of thermal cracking in HPC structures at early ages is therefor important to asses the pre-damaged regions of the structure. The simulation can be performed with a numerical procedure similar to the numerical procedure for normal strength concrete, only with the appropriate adjustments of the relevant material properties which can be determined from small–scale experiments.

Computational model for aging materials

During the hardening process the strength and stiffness of concrete increase from a negligible value to the ultimate value for fully hardened concrete. Also the creep of concrete reduces substantially as concrete ages. A model to predict the mechanical behavior of young concrete should incorporate the most important factors determining the mechanical state of concrete i.e. the degree of reaction and the temperature. The degree of reaction depends on properties like time, moisture content and temperature. Because these properties are almost independent of the

---

*TNO Building and Construction Research, P.O. Box 49, 2600 AA Delft, The Netherlands
[†]currently at University of Twente, P.O. Box 217, 7500 AE Enschede, The Netherlands

mechanical load, the development of temperature and degree of reaction can be calculated before the structural analysis. For a more extensive discussion of this staggered procedure, see [5].

Thermal analysis

The mechanical properties of young concrete are mainly determined by the degree of reaction and the temperature. The degree of reaction $r(t)$ is defined as the ratio of accumulated heat production at time t to the total heat production at infinity Q_∞

$$r(t) = \frac{1}{Q_\infty} \int_{-\infty}^{t} q(\tau, T(\tau)) \, d\tau \tag{1}$$

where the heat production $q(\tau, T(\tau))$ depends on the temperature, T, and the type of concrete, see e.g. van Breugel[6]. The effect of the temperature on the total amount of hydration heat is relatively small but the rate of heat liberation is to a large extend influenced by the temperature. At higher temperatures the reaction processes are accelerated. The heat production as function of temperature and time is modeled by

$$q(r(t), T) = \gamma f(r(t)) e^{\dfrac{-b}{T}} \tag{2}$$

in which γ a material parameter, b the Arrhenius constant, T the absolute temperature, and $f(r(t))$ the normalized heat production. The function $f(r)$ and the constant γ are usually determined from an adiabatic calorimetric test. The heat production during an adiabatic test is determined from

$$q(r, T) = \rho(r, T) c(r, T) \dot{T} \tag{3}$$

The specific heat c and the density ρ have to be known as function of temperature and degree of reaction. Subsequently eqs. (1) and (2) can be used to determine function $f(r)$.

As soon as the heat development as function of degree of reaction is known, the heat diffusion equation can be used to calculate temperature distribution and development:

$$\nabla(\lambda \nabla T) + q = \rho c \dot{T} \tag{4}$$

In this equation λ is the conductivity, c is the heat capacity and q the heat production given by (2). Eq.(4) is solved using a finite element method. The boundary conditions are usually described using heat convectivity

$$q = \alpha (T_{concrete} - T_{air}) \tag{5}$$

in which α is the convection coefficient. Most large concrete structures are casted outdoors and the boundary conditions are usually not clearly defined. The influence of variations in environmental properties like wind speed, precipitation or solar radiation on boundary conditions may be quite large.

Structural analysis

The structural problem is also solved using a finite element method. Besides the spatial discretization the problem is discretized in time to account for the effects of temperature and deformation history. Every time increment the parameters of the material model are updated to simulate the development of material properties within that time increment. To avoid oscillating stress fields the order of discretization in the structural analysis should be higher than the order of discretization in the thermal analysis. The effect of temperature on the development of mechanical properties is incorporated by the introduction of an equivalent age which is calculated at the integration points of the finite elements,

$$t_{eq}(t) = \int_0^t e^{\frac{b}{T_0} - \frac{b}{T(\tau)}} \, d\tau \tag{6}$$

with the Arrhenius constant b, the reference temperature T_0, and the temperature T which follows from the thermal analysis.

Age dependent constitutive model

A basic feature of the proposed constitutive model is a decomposition of the strain into an elastic, a crack, a creep, and an autogeneous strain component, according to

$$\varepsilon = \varepsilon^e + \varepsilon^{cr} + \varepsilon^c + \varepsilon^\theta \tag{7}$$

The autogeneous strain component depends solely on the time, and in this study it is assumed that this strain component represents the thermal strain component only, and consequently, depends on the temperature and expansion coefficient. The stress is related to the elastic strain component through a general stiffness relation, $\mathbf{D}^e$,

$$\sigma = \mathbf{D}^e \varepsilon^e \tag{8}$$

The constitutive model for concrete cracking will be assumed to be given by the multi–directional fixed smeared crack model with aging material parameters, e.g. [3], and the material model for concrete creep by an aging visco–elastic Kelvin model, e.g. [1]. The resulting algorithm will not be discussed here because of space limitations, see e.g. [3], and only a brief discussion of the material model for cracking and creep will be given in the following.

Crack model

The concept of a multi–directions fixed smeared crack model with strain decomposition is in essence given by assumption that for each crack i a stress rate $\dot{s}_i$ and strain rate $\dot{e}_i^{cr}$ exists. The crack strain rate vector for crack i is given by

$\dot{\mathbf{e}}_i^{cr} = \{\dot{e}_{nn}^{cr} \dot{\gamma}_{ns}^{cr}\}^T$ in a two-dimensional situation. The relation between the global strain rate and the vector $\dot{\mathbf{e}}^{cr}$ is given by the transformation

$$\dot{\boldsymbol{\varepsilon}}^{cr} = \mathbf{N}\dot{\mathbf{e}}^{cr} \tag{9}$$

with $\mathbf{N}$ the transformation matrix of crack i. The subscript i has been dropped for convenience. The crack stress rate vector is given by $\dot{\mathbf{s}}^{cr} = \{\dot{s}_{nn}^{cr} \dot{s}_{ns}^{cr}\}^T$. The relation between the global stress rate and the vector $\dot{\mathbf{s}}^{cr}$ can be derived as

$$\dot{\boldsymbol{\sigma}} = \mathbf{N}^T\dot{\mathbf{s}}^{cr} \tag{10}$$

The constitutive relation between the crack stress rate vector and the crack strain vector is given by

$$\dot{\mathbf{s}}^{cr} = \mathbf{D}^{cr}\dot{\mathbf{e}}^{cr} \tag{11}$$

with the matrix $\mathbf{D}^{cr}$

$$\mathbf{D}_i^{cr} = \begin{bmatrix} D^I & 0 \\ 0 & D^{II} \end{bmatrix} \tag{12}$$

where D^I and D^{II} the tangential stiffness moduli of the crack. The shear stiffness of the crack, D^{II}, is assumed to be determined by a fraction of the linear–elastic shear stiffness, βG, where β the shear retention factor is assumed to be constant. The normal stiffness of the crack, D^I, is determined by tensile strength $f_{ct,m}$ and the fracture energy G_f,

$$D^I = -\frac{1}{k(e_{nn}^{cr})} \frac{f_{ct,m}^2 h}{G_f} \tag{13}$$

in which the crack band width h has been introduced to avoid mesh–size dependency, [4]. The factor $k(e_{nn}^{cr})$ reflects the shape of the softening branch.

The time dependent behavior of concrete cracking can be modeled by modification of eq.(13) in order to simulate the aging of concrete. It is assumed that during aging the shape of the softening curve remains similar to the initial curve regardless of changes in ultimate stress or changes in fracture energy. This seems a fair approximation, since the influence of the exact shape of the curves on computational results in engineering calculations are small, as long as the correct fracture energy is dissipated. If the shape of the softening curve remains similar, the development of the softening curve is completely determined by the development of the tensile strength and the fracture energy during the hydratation process. The development of the tensile strength and fracture energy is assumed to be similar to the development of the compressive strength, f_{cm}, which is given by a modification of the CEB-FIP recommendation

$$f_{cm}(t) = f_{cm}(28) \exp\left(s\left[1 - \sqrt{\frac{28t_1}{t - t_0}}\right]\right) \tag{14}$$

where s is a constant depending on the type of cement which varies from 0.2 for rapid hardening high strength concretes to 0.38 for slowly hardening concrete, t_1 a constant equal to 1 day, and $f_{cm}(28)$ the mean compressive strength after 28 days. The modification compared to the CEB-FIP relationship is the introduction of the constant t_0 to account for a dormant period. During the dormant period stiffness and strength of the material can be neglected.

Creep Model

The creep behavior of concrete can to a certain extend be described by viscoelasticity. The creep of concrete strongly depends on properties like temperature, humidity and degree of reaction. When following the creep of concrete during a lifetime cycle of a structure, the influence of these properties has to be accounted for in order to obtain reliable deformation predictions.

The Kelvin chain model is an example of a creep function $J(t, \tau)$. This creep function can be used to calculate the relation between stresses and strain-history

$$\varepsilon(t) = \int_{-\infty}^{t} J(t, \tau) \bar{\mathbf{C}} \dot{\boldsymbol{\sigma}}(\tau) \, d\tau \tag{15}$$

The dimension-less compliance matrix $\bar{\mathbf{C}}$ is used to transform a 3-dimensional stress situation to a 1-dimensional relation, just as in the generalized Hooke law. The creep function $J(t, \tau)$ can be written as a Dirichlet series

$$J(t, \tau) = \sum_{\alpha=0}^{\infty} \frac{1}{E_\alpha(\tau)} \left[1 - e^{-\frac{t - \tau}{\lambda_\alpha}} \right] \tag{16}$$

which can be interpreted as a finite chain of Kelvin elements. The stiffness and viscosity of the spring and damper in each Kelvin element determine the retardation time λ_α

$$\lambda_\alpha = \frac{\eta_\alpha}{E_\alpha} \tag{17}$$

The stiffness of a viscoelastic material can vary due to external influences like temperature and humidity or due to an aging process. For young concrete the stiffness increases with age. Material that is loaded with a prescribed stress will immediately deform (elastically) to a certain strain. This stress-strain behavior is described in the Kelvin model by a chain-element without damper and only a stiffness, say E_0. As a first approximation we could assume that all chain-elements with a damper vary in time with the same time-dependent multiplier. The time dependent stiffness of the chain-element without damper is now determined by the instantaneous stiffness at different loading times and the compliance (strain/stress) at $t \to \infty$ is determined by the summation of all chain-element compliances. In this case creep curves at different loading ages are similar. This class of creep functions is given by

$$J(t,\tau) = \frac{1}{E(\tau)} + f(\tau)\Phi(t-\tau) \tag{18}$$

where τ is the time at load application, $\Phi(t,\tau)$ the normalized creep coefficient function and $f(\tau)$ a scaling factor. The age dependent Young's modulus $E(\tau)$ will be approximated using the same modification as the modification of the compressive strength, eq.(14),

$$E(\tau) = E_c(28)\sqrt{\exp\left(s\left[1 - \sqrt{\frac{28t_1}{t-t_0}}\right]\right)} \tag{19}$$

were $E_c(28)$ the Young's modulus at 28 days, s a constant depending on the type of cement, and t_1 equals 1 day. The constant t_0 accounts for the dormant period. The functions $\Phi(t,\tau)$ and $f(\tau)$ are assumed according to the standard CEB-FIP recommendations, see [2], in which the influence of temperature on the development of the mechanical properties is accounted for with an equivalent time.

Application

The example calculation is derived from an experimental investigation of the applicability of high strength concrete for the segmentally casted 'Stichtse Brug' [7]. First experimental data from small experiments will be used to derive a material model, subsequently this material model will be used for a calculation of temperature and stress distribution in the test structure.

Material parameters

In order to derive the material properties, we will first determine the heat development as function of time. This heat development is derived from an adiabatic calorimetric test. In Figure 1 the measured and the calculated adiabatic lines, and the calculated normalized heat production as function of degree of reaction are depicted. Experiments were performed to obtain the development of the instantaneous stiffness as function of time. The development of Young's modulus in time can be approximated using the modified CEB-FIP relationship, eq.(19). The dormant period, t_0, was approximately 17 hours. The factor s indicating the reaction speed has been assumed equal to 0.25, the regular value for normal and rapid hardening cements. The s value was relatively high, because the HPC used in the experiments was of a special compound characterized by a slower course of reaction than ordinary high strength concrete.

In order to gain insight into the creep during the hardening of concrete, creep tests were performed on the hardening concrete. The creep strain was measured

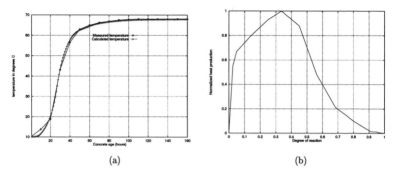

(a)   (b)

Figure 1: Stichtse brug. (a) Temperature development during an adiabatic test.
(b) Normalized heat production as function of degree of reaction.

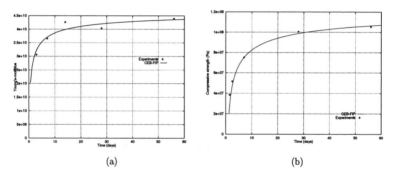

(a)   (b)

Figure 2: Stichtse brug. (a) Development of Young's modulus in time. (b)
Development of compressive strength in time.

after 19, 24, 34 and 48 hours. After 48 hours the largest part of the development
of mechanical properties has occured. The compilation of loading times allows
for a reasonably well recovering of the influence of the continuing hydration on
the creep behavior. The creep experiments are performed at two different stress
levels. In a first series of tests the imposed load was 20% of the actual strength
at load application. In a second series of tests the imposed load was 40% of the
actual strength at load application.

In Figure 2 the measured development of the Young's modulus is compared
to the approximations obtained with the CEB-FIP model. The scattering of the
experimental values is large, but the CEB-FIP model seems to predict the Young's
modulus quite accurate. Taken into consideration this reasonable description
of the development of instantaneous stiffness by the CEB-FIP model code, the

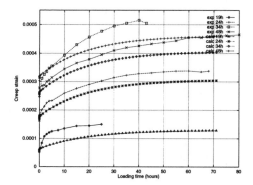

Figure 3: Stichtse brug. Measured creep curves at a stress level of 20% and approximations.

creep curves were first modeled using the standard CEB-FIP models. The results were however disappointing. Therefore the creep part of the strain (total minus instantaneous strain) of the 34 h creep curve was fitted to a Kelvin chain directly, resulting in the following creep formulation

$$J(t, t_0) = \frac{1}{E_c(t_0)} \left[1 + f(t_0)\phi\left(t - t_0\right) \right] \tag{20}$$

with $\phi(t - t_0)$ a function determining the shape of the creep function, $E_c(t_0)$ is the instantaneous stiffness and $f(t_0)$ a scaling factor for the creep deformation. The function $f(t_0)$ is described using a power law:

$$f(t_0) = (t_0)^{-n} \tag{21}$$

Combination of the equations for describing the instantaneous stiffness $E_c(t_0)$ the fitted creep function and the scaling factor $f(t_0)$ gives a function for describing creep. In Figure 3 it can be seen that the accuracy of the total deformation is reasonable, the accuracy of the prediction of creep deformation however is only small. Taken into consideration the complexity of the processes within the concrete and the scattering of experimental results, the obtained results are satisfying for practical applications. It is admitted that a globally scaled creep curve is only a rough approximation, as in reality the shape of the creep curves changes in time. The developed material model tends to underestimate material deformation, this means that autogeneous stresses due to temperature effects will probably be overestimated. This is a save approximation when considering temperature induced cracking. Other experiments were performed to determine the development of compressive strength in time. In Figure 2 the experimental values and development of compressive strength according to the CEB-FIP are shown. The ultimate tensile strength is estimated from the tensile strength according to CEB-FIP model code,

$$f_{ct,m}(t) = 1.4 \left(\frac{f_{cm}(t)}{10} \right)^{2/3} \tag{22}$$

Analysis of a bridge segment

The material model derived in the previous section will be applied to an analysis of a part of a segmentally casted cantilever bridge, currently under construction near Rotterdam in the Netherlands. The experiment has been carried out under supervision of the Dutch Ministry of Traffic, Public Works, and Water Management, in order to study the risk of crack formation during the construction.

The segment used in the experiment, is a floor-wall construction part of a tubular shaped cantilever bridge. The cross-section of the construction part is depicted in Figure 4. The length of the specimen is 5 meter, the thickness of the floor is 550 mm the thickness of the wall is 350 mm. The height of the left flange varies between 4900 mm and 6100 mm, the height of the right flange is 1550 mm. Since the cross section was nearly constant and the length of the specimen was large compared to its thickness, a simplified geometry could be used in the calculations. The heat flow was assumed to be two dimensional. The calculation of autogeneous stresses was a quasi three dimensional calculation. The construction was modeled using one layer of brick elements. Because of the three dimensional stress state, brick elements are required in the structural analysis and it is convenient to use brick elements in the thermal analysis as well. The mesh used in the analysis is depicted in Figure 4.

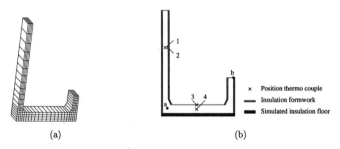

Figure 4: Stichtse brug. (a) Mesh used in simulations. (b) Boundaries in the simulation.

For the thermal calculations boundary conditions were simplified. During the test the outer temperature varied between approximately 3°C and 7°C, in the calculations a uniform external temperature of 5°C was assumed. The heat exchange with the environment was simulated using a convection eq.(5). The convection coefficient α incorporates a constant wind speed of 14 m/s and the effect of the form work on the heat conduction. The concrete was poured onto a specially manufactured floor. The floor was simulated by an insulation at the bottom of the specimen. The temperature of the subsoil was assumed to be uniform at a temperature of 5°C. In Figure 4 the boundary conditions are shown, and the material and boundary parameters given in Table 1.

Table 1: Material and boundary data.

Convection coefficient free surface	6.0	$Wm^{-2}K^{-1}$
Convection coefficient surface with formwork	2.3	$Wm^{-2}K^{-1}$
Convection coefficient to subsoil	3.0	$kWm^{-2}K^{-1}$
Thermal capacity of concrete c	2.075	$MJm^{-3}K^{-1}$
Thermal conductivity of concrete λ	1.8	$Wm^{-1}K^{-1}$
Thermal expansion coefficient α	$1.1 \cdot 10^{-5}$	K^{-1}
Arrhenius constant	5000	K
Total heat production	119.5	MJm^{-3}
Initial degree of reaction	0.0001	

The temperature development during the first 160 hours was calculated. After this period, the effects of temperature changes on structural behavior are marginal. During the hardening of concrete, temperatures in the floor and in

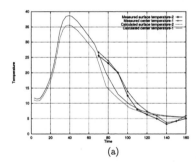
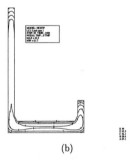

(a)   (b)

Figure 5: Stichtse brug. (a) Measured and calculated temperatures in the wall. (b) Temperature distribution after 39 hours.

the wall were measured, the position of the used thermo couples is depicted in Figure 4. These measurements can be used to verify the results of the thermal calculation. In Figure 5 the calculated and measured development of temperature at two points in the wall is depicted. It is seen that a reasonable agreement between calculated and measured results is obtained. The differences between calculation and results may be caused by the simplifications of the boundary conditions and by deviations in material parameters or pouring sequence. In Figure 5, temperatures after 39 hours are depicted. After 39 hours temperatures in the calculation have reached their maximum value. It is seen that temperatures within the bottom slab are high compared to temperatures in the wall. This is to be expected, because of the large concrete mass in the lower construction part and the good insulation by the subsoil at the bottom of the floor slab. The highest temperatures occur in the corners where wall and floor meet.

The temperatures calculated in the thermal analysis where used in a structural analysis to calculate stresses and strains. The displacements in z-direction of the back plane were suppressed. The front plane of the mesh was assumed to remain flat, this was accomplished by means of tyings of nodes. This way homogeneous expansion as well as warping of the construction can be simulated.

In an early stage of the hardening process it was found that the surfaces without formwork are submitted to tensile stresses. These tensile stresses are caused by large thermal expansion of the center of the structure, where temperatures are much higher than at the surface. However the stresses in the structure are low compared to the tensile strength at this age and no cracks did arise. In the calculations an initial strength of 0.01 MPa was used. In a later stage of the

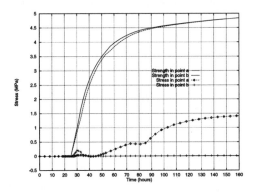

Figure 6: Stichtse brug. Strength and stress development for point a and b.

hardening process tensile stresses were found in the middle of the structure while compressive stresses were found at the surface of the structure. At this age the structure is cooling down and temperatures are leveling. After 70 hours, the maximum tensile stress in the structure is approximately 2.6 MPa, whereas the tensile strength has increased to approximately 4 MPa. In Figure 6 the development of stresses and strains in two points of the structure are depicted. The position of point a and point b is indicated in Figure 4. It is obvious that in this problem the influence of temperature differences within the structure on the development of tensile strength is only small. The effect of temperature on specific age is most clearly indicated by the increase of the dormant period, which was 17 hour at 20°C. In point a tensile stresses early in the hardening process are seen, whereas point b in the center of the structure is submitted to tensile stresses during a later stage in the hardening process. During the entire period the tensile strength of the concrete is considerably larger than the tensile stresses, so the risk of thermal cracking in structure is negligible.

Conclusions

The increasing performance of young concrete requires finite element algorithms that incorporate the development of mechanical properties. For an adequate adaptation of mechanical properties the development of temperature and degree of reaction in time have to be calculated. This information has been used to adapt the material properties in a subsequent structural analysis. Well established numerical algorithms for normal strength concrete can be modified to account for aging effects and can be applied to HPC with the appropriate material properties.

Acknowledgment

This research has been sponsored by the Civil Engineering Division of the Dutch Ministry of Traffic, Public Works and Water Management. The simulations have been performed with a research version of the DIANA finite element code.

References

[1] CAROL, I., AND BAŽANT, Z. P. Solidification theory, a rational and effective framework for constitutive modeling of aging viscoelasticity. In *Proc. 5th Int. RILEM Symposium* (1993).

[2] CEB-FIP Model Code 1990. Comité Euro-International du Béton, 1993. Thomas Telford, London.

[3] DE BORST, R., AND VAN DEN BOOGAARD, A. H. Finite Element Modeling of Deformation and Cracking in early-age Concrete. *J. Eng. Mech. 120*, 12 (1994), 2519–2533.

[4] ROTS, J. G. *Computational Modeling of Concrete Fracture*. PhD thesis, Delft University of Technology, 1988.

[5] TER STEEG, R. T. M. Computational modelling of young concrete. Tech. Rep. 95-NM-R999, TNO Bouw, Rijswijk, 1995.

[6] VAN BREUGEL, K. Numerical simulation of hydration and microstructural development in hardening cement-based materials. *Heron 37* (1994).

[7] Toepasbaarheid van beton met hoge sterkte in de 2e Stichtse Brug, 1994. Testprogram by order of: Ministry of Traffic, Public Works and Water Management.

Embedded Crack Approach to Regularize Finite Element Solutions of Concrete Structures

E. Pramono, J. C. Mould, Jr. and H. S. Levine
Weidlinger Associates, Inc.
Los Altos, CA

Abstract

An embedded crack approach is used to regularize strain softening that is contained within a fictitious crack band at the finite element centroids. A comprehensive concrete model is used to represent the material inside the crack band. With this regularization, the spurious mesh-size sensitivity of the finite element solutions can be eliminated.

Introduction

In this paper, we report development of a concrete model that is able to capture strong localization phenomenon through the inclusion of regularized displacement discontinuity in the displacement field of the finite element formulation. To obtain a displacement jump (discontinuity) during cracking, we adopt the approach outlined in [Dvorkin et al., 1990], where the elements are enriched to permit a displacement jump at the element centroids. From a physical point of view, the inclusion of a discontinuous displacement formulation is a natural way to model cracking in concrete type materials. Similar approaches using an embedded displacement discontinuity in the formulation have been reported by various investigators [Klisinski et al., 1991, Simo et al., 1993, Lotfi et al., 1994]. Other investigators [Larsson and Runnesson, 1994, Xu and Needleman, 1994] formulated the displacement discontinuity at element boundaries.

By enforcing the cracking to occur within a finite band width and by using the kinematic properties of the classical definition of a shear band, a constitutive relation between the traction and the displacement across the crack is developed from an existing concrete model [Mould and Levine, 1994]. This concrete model is able to represent complex deformation modes of the crack band from mode I to complex mixed-mode failures.

Formulation of Discontinuous Displacement

The principle of virtual work of a solid body, that is subjected to external pressure **p** along the external boundary surface S_p, with boundary condition $\mathbf{u}_b$ along the external boundary surface S_u, and a crack in the form of the singular surface , can be written as

$$\int_V \boldsymbol{\sigma}\delta\boldsymbol{\epsilon}\,dV - \int_{S_p} \mathbf{p}\delta\mathbf{u}_p\,dS_p - \int_{S_{int}} \mathbf{t}\delta\mathbf{u}_j\,dS_{int} + \int_V \rho\ddot{\mathbf{u}}\delta\mathbf{u}\,dV = 0 \quad (1)$$

where $\boldsymbol{\sigma}$ and $\boldsymbol{\epsilon}$ are the stress and strain tensors in the continuum, $\mathbf{u}_j$ is the displacement jump across the singular surface, and V is the volume of the solid body. The first and third terms of Eq. 1 represent the work done in the continuum and crack band, respectively. In this problem, as long as the constitutive relation in the continuum does not exhibit strain softening or loss of normality, a unique solution can be obtained.

To include the displacement discontinuity in the finite element formulation, a discrete displacement jump is introduced where $\mathbf{u}_j$ are the internal degrees of freedom at the finite element centroids. Then, the displacement field approximation, that is decomposed into two parts, can be written as

$$\mathbf{u} = \mathbf{N}\mathbf{u}_n + \mathbf{M}\mathbf{u}_j \qquad (2)$$

where $\mathbf{N}\mathbf{u}_n$ represents the continuous part, the regular classical finite element approximation, and $\mathbf{M}\mathbf{u}_j$ represents the discontinuous part of the displacement field due to the jump. The strain field in the continuum is, then, expressed as

$$\boldsymbol{\epsilon} = \mathbf{B}\mathbf{u}_n - \mathbf{B}\mathbf{P}\mathbf{u}_j \qquad (3)$$

The second term represents the strain unloading due to crack band deformation. Matrix is an operator to represent the change of the displacement field due to the displacement jump.

Crack Band Formulation

Assuming that the localized deformation resulting from micro cracking occurs within a fictitious thin band δ, then the crack band is defined as a region located between two singular parallel surfaces with the orientation defined by a unit normal **n**. Following the classical kinematics of a shear band of finite width, the deformation is defined by a discrete displacement vector $\mathbf{u}_j$ that consists of one normal and two shear displacements of the crack surface. With this definition then, the strain rate inside the crack band is calculated as

$$\dot{\boldsymbol{\epsilon}} = \frac{1}{2\delta}(\dot{\mathbf{u}}_j\mathbf{n} + \mathbf{n}\dot{\mathbf{u}}_j) \qquad (4)$$

Once an element experiences cracking or shear-band formation, the element is modeled to behave in a heterogeneous manner where the stress degradation, driven by the displacement jump across the crack or shear band, follows a comprehensive constitutive law while the remaining regions in the element can still behave elastically or inelastically.

Constitutive Model Inside Crack Band

The stress strain relation inside the crack band is represented by a comprehensive concrete model [Mould and Levine, 1994] that provides the constitutive relation of the stress - strain increments inside the band. The model has been calibrated using a wide-range of experimental data for various concrete types. The model consists of a three-invariant failure surface and a three-invariant strain hardening cap. To account for the increasing/decreasing strength during hardening/softening, we adopt kinematic shifting of the failure surface along the pressure axis. The same model is used for the continuum, but without softening.

Equilibrium Across Crack Band

The displacement jump $\dot{u}_j$ can be evaluated or condensed out using the equilibrium condition at the crack surface or on the finite element level. The equilibrium along the singular surface is written as

$$n\dot{\sigma}_b = n\dot{\sigma}_c \tag{5}$$

where $\dot{\sigma}_b$ is the stress tensor in the crack band and $\dot{\sigma}_c$ is the stress tensor in the continuum ahead of the crack. Using the previously defined crack band in Eq. 4 and evaluating $\dot{\sigma}_b$ for displacement jump $\dot{u}_j$, the incremental traction $\dot{t}_j$ can be written as

$$\dot{t}_j = \frac{1}{\delta}nE_{ep}n\dot{u}_j = \frac{1}{\delta}Q\dot{u}_j \tag{6}$$

where E_{ep} denotes the elasto-plastic moduli, that are functions of , and Q is the classical definition of the acoustic tensor for the material. Substituting Eq. 3 and Eq. 6 into Eq. 5 and evaluating equilibrium at the element centroid, the equation 5 can be written as

$$\frac{1}{\delta}Q\dot{u}_j + nEBP\dot{u}_j = nEB\dot{u}_n \tag{7}$$

The increment of displacement jump $\dot{u}_j$ can be obtained using a robust iterative technique to satisfy the equilibrium condition stated in Eq. 7.

If an element stiffness matrix needs to be formed explicitly, then $\dot{u}_j$ needs to be condensed out. Using the free body of the crack element, the equilibrium equation on the element level can be written as

$$-P^T \int_V B^T\dot{\sigma}_c \, dV + \int_{Sint} \dot{t}_j \, dS_{int} = 0 \tag{8}$$

Substituting Eq. 3 and Eq. 6 into Eq. 8, the equilibrium equation can be written as

$$-P^T \int_V B^TEB \, dV\dot{u} + \int_V P^TB^TEBP \, dV\dot{u}_j + \int_{Sint} \frac{1}{\delta}nE_{ep}n \, dS_{int} \dot{u}_j = 0 \tag{9}$$

Now, the element stiffness matrix can be constructed by condensing out $\dot{u}_j$ from Eq. 9.

Numerical Simulation:

The constitutive model was implemented in the FLEX [Vaughan 1991] finite element code to simulate the following example problem. A numerical study of a single-edge-notched beam was performed to compare with results obtained from the experiment [Schlangen, 1993]. The beam was discretized using three finite element meshes. Figure 1 compares force vs. crack mouth sliding displacement (CMSD) of the numerical and the experiment results. Figure 2 shows the mesh deformations after the crack is completely open. Although the proposed model does not use a discrete crack approach, it is able to predict closely the trajectory of the crack pattern observed in the experiment.

Concluding Remarks:

Results from the numerical example show that the model is able to eliminate the spurious mesh sensitivity when the structure undergoes intense localized deformations in the form of a crack band. The model is efficient enough to be used for practical problems and meets all theoretical requirements with regard to uniqueness, stability and convergence.

References:

Dvorkin, E N., Cuitino, A. M., and Gioia, G.,"Finite Element with Displacement Interpolated Embedded Localization Lines insensitive to Mesh Size and Distortions," I. J. for Num. Meth. in Engr., vol 30, 1990.

Klisinski, M., Runesson, K, and Sture, S.,"Finite Element with Inner Softening Band," Journal of Engineering Mech., 3, 1991.

Larsson, R. and Runesson, K,"Discontinuous Displacement Approximation for Capturing Plastic Localization," Int. J. for Num. Meth. in Engr., Vol. 26, pp. 2087-2105, 1993.

Lotfi, H. R. and Shing, P. B.,"Analysis of Conrete Fracture with an Embedded Crack Approach," Proc. Int. Confer. on Computational Modelling of Concrete Structures, Vol. 1, 1994.

Mould, J. and Levine, H., "A Rate-Dependent Three Invariant Softening Model for Concrete," Mechanics of Materials and Structures, Edited by G. Z. Voyiajis, L. C. Bank, and L. J. Jocobs, Elsevier Science Publishers, Amsterdam, 1993.

Schlangen, E.," Experimental and Numerical Analysis of Fracture Processes in Concrete," Dissertation, Delft University of Technology, Delft, 1993.

Simo, J. C., Cervera, M., and Armero, F.,"An Analysis of Strong Discontinuities Induced by Strain-Softening in Rate Independent Inelastic Solids, Int. J. of Comp. Mech., Vol 12, pp 277-296, 1993.

Vaughan, D. K., "FLEX User's Guide," Report UG8298, Weidlinger Associates, Palo Alto, CA, May 1983 plus updates through 1991.

Xu, X-P. and Needleman, A.," Numerical Simulations of Fast Crack Growth in Brittle Solids," J. of Mech. and Phys. of Solids, 1994.

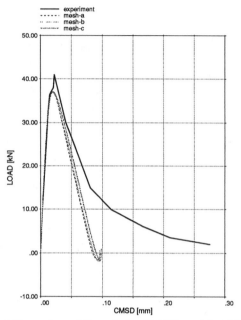

Figure 1 Comparison of Numerical, with Different Mesh Discretizations,
and Experimental Results

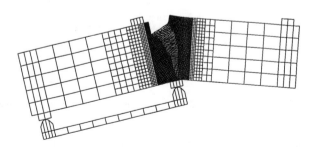

Figure 2 Deformed Mesh of SEN Beam

The Effect of Moisture on Spalling of Normal and High Strength Concretes

[1]N. Khoylou, [2]G.L. England, Member ASCE

ABSTRACT

Spalling of concrete in fire is an explosive mode of failure. It occurs during rapid rates of surface heating and is related to the moisture content at the time of heating. It is also dependent on the flow rate of moisture within and from the concrete during the heating period.

Hardened concrete, *as cast*, contains free water in its porous skeleton but is not physically saturated. The presence of this free moisture is in one sense beneficial to the performance of concrete under fire because water absorbs large quantities of latent heat during vaporisation. When an exposed surface is heated this absorption of heat delays the increase of temperature within the concrete mass. However, as temperatures rise, the free pore water will expand simultaneously with sustaining an increasing saturated vapour pressure. Eventually the continued expansion of the water together with moisture flow away from the heated zone, where chemically bound water is also progressively released, will lead to physical saturation of the pore structure elsewhere. If the expanding moisture there cannot escape from the saturated pores during further increases of temperature, additional large tensile strains will be imposed on the skeleton due to the low compressibility of water. In turn, these strains create tensile stresses in, and possible fracture of, the solid skeleton of the concrete. High strength concrete is particularly vulnerable to this behaviour because of its inherent, low porosity and low permeability to water flow.

This paper introduces a simple theoretical/mathematical model used to describe the spalling behaviour of concrete while exposed to fire. A crucial parameter relating to spalling is the rate of water flow through and from the saturated pores.

1 Research Engineer, PhD Candidate, Civil Engineering Department Imperial College, London, UK

2 Professor of Mechanics and Structures, Civil Engineering Department, Imperial College, London, UK

INTRODUCTION

Concrete has a low thermal conductivity and is often used for fire protection. Its ability to perform well in this capacity, depends on its ability to withstand high temperatures and rapid rates of heating, without spalling.

Spalling is a feature commonly observed in concrete structures exposed to fire. It is frequently explosive and leads to the detachment of pieces of concrete from the heated surface. In extreme cases spalling may expose the steel reinforcement, in RC structures, directly to fire temperatures. It is, therefore, important to understand the mechanism of spalling in concrete if good fire protection and structural serviceability are to be guaranteed.

Although the spalling mechanism is poorly understood, it is clearly related to the free moisture content of concrete and its distribution at the time of heating. Test results have indicated that "when the moisture content is low (7% v/v) which can be expected in centrally heated buildings, about five years after completion, the risk of spalling diminishes" [Copier, 1980]. Other parameters are also important; for example, permeability, the degree of pore saturation (i.e. ratio of volume of free water in the pores to total pore volume), and length of migration path. Experimental evidence generally suggests that high strength concrete is more vulnerable to spalling in fire than normal strength concrete [Sanjayan and Stocks, 1993] because of its low permeability concomitant slower rate of drying. A good understanding of the moisture flow behaviour of concrete at high temperatures is therefore necessary in order to investigate spalling.

MOISTURE MIGRATION AT NON-UNIFORM TEMPERATURES

Drying of concrete usually starts at a very early age, after striking the formwork. At ambient temperatures, drying occurs because humidity inside the pore structure of concrete is higher than the external humidity. Moisture evaporates from the exposed surfaces into the surrounding environment. As the concrete dries, a drying front moves progressively into the concrete mass, where water then evaporates *in situ* and the resulting vapour diffuses to the exposed surfaces and escapes.

Heating concrete, affects its drying behaviour significantly. Moisture migration in non-uniformly heated concrete is governed by many inter-related factors: mainly, volume of free water, V_{fw}, initial unfilled pore volume, V_{uf}; permeability, k, and temperature, T. As heat penetrates into a concrete member from an exposed surface, pore vapour pressures build up. The initial unfilled free pore volume of the hardened *as cast* concrete allows the development of Saturation Vapour Pressures of water, P_{svp}, in the partially water-filled pores. The high pressures of the hottest regions then force moisture to move down the pressure gradient towards neighbouring cooler regions. During this migration, the initial free pore volume acts as reservoirs to permit rapid initial flow rates to occur. These flow rates depend upon the pressure gradients and permeability of concrete, which themselves are strongly influenced by temperature. Thus, the rate of flow will vary from point to point in the concrete mass. The lower rates associated with cooler regions will lead to physical water saturation and the formation of a *saturation plug*. As heating continues, the saturation front moves further inward. The direction of flow in the drying region is controlled by the boundary conditions. If the heated face is sealed the pore pressure gradients will force moisture to move away from the heated face and into the concrete. This will lead to progressive water condensation at the upstream face of the saturation plug. Thus, the saturated region will lengthen in time as it moves down the pressure gradient. However, when the heated surface is unsealed, as is usually the case in a fire situation, the moisture flow from the dry region will be, at least in part towards the exposed face also.

The formation of the saturation plug influences the rate of movement of mass into the concrete. Within the saturation plug, water moves under the action of a hydraulic gradient governed by the saturation vapour pressures at its two ends. The pressures in the saturated zone, are higher than those prior to physical saturation. With time the permeability of concrete in these regions reduces. This, however, is a long-term process and is unlikely to influence the spalling behaviour in fire. The permeability in the hottest regions is however likely to increase significantly as drying progresses [Greathead, 1986].

The proximity of the downstream saturation front of the *plug* normally will not influence spalling. However, in thin sections the plug can migrate to the unheated face and eject water there. This loss of water will reduce the risk of spalling.

Experimental evidence and numerical predictions of this type of migrational behaviour in non-uniformly heated concrete have been discussed in detail elsewhere [England and Khoylou, 1995].

SPALLING OF CONCRETE

The free moisture content together with the internal moisture flow properties of concrete appear to be the key parameters in the determination of spalling during fire. The moisture content diminishes in time as drying progresses at ambient temperatures. This is usually beneficial. However, when concrete is heated, the free water always expands at a greater rate than its solid skeleton and this action increases the degree of pore saturation. Also, the chemically bound water is progressively released as temperatures rise. Thus, the degree of pore saturation (i.e. percentage filling of the pores) increases with increasing temperature and can lead to physical saturation of the pores. Then if heating continues additional strains will be induced in the solid skeleton surrounding the pore, due to the low compressibility of the expanding water, in this strain controlled situation. If fracture does not occur, these strains can theoretically become rapidly very large and create high tensile stresses and high pore water pressures. Meanwhile, because the pores are interconnected, water can escape from any zone of high pressure to the neighbouring pores. The amount of water which can migrate in this way depends primarily upon the permeability of concrete. When the flow rate is insufficient to prevent the build up of tensile stresses in the skeleton these stresses may quickly exceed the fracture strength of concrete. This is an example of hydraulic fracture and is believed to be one mechanism which is responsible for spalling. The explosive spalling behaviour which is often observed can be related to this mechanism as a post-fracture phenomenon created by the rapid vaporisation of the superheated water following the creation of a free water surface, similar to a boiling liquid expanding vapour explosion [Birk and Cunnigham, 1995].

A simple spherical model is used later to describe the onset of fracture in the concrete skeleton of a saturated porous structure, from which water is not allowed to escape during heating. The behaviour of water in a partially saturated pore, under steadily increasing temperature is now described.

Heating of a Partially Saturated Pore:

During heating, the differential thermal expansion between the contained water and the pore skeleton results in an increasing degree of pore saturation. As long as the pore is only partially filled with water, free volume is available to allow saturation vapour pressures to increase. Eventually, with further increase of temperature the pore will become physically saturated. This is a critical stage and defines a discontinuity of behaviour with respect to continued heating. The related temperature, pore volume and specific volume of water are denoted by T^*, V^* and ϑ^* respectively. T^* can be evaluated from the knowledge that at first saturation:

$$\frac{V_{fw}^*}{V^*} = 1.0 \tag{1}$$

The volume of free water, V_{fw}^*, and the pore volume, V^*, can be calculated using the volume of free water, pore volume, temperature and specific volume of water before start of heating. These latter quantities are denoted by $\overline{V}_{fw}$, $\overline{V}$, $\overline{T}$ and $\overline{\vartheta}$ respectively. Thus,

$$V_{fw}^* = \overline{V}_{fw} \, \frac{\vartheta^*}{\overline{\vartheta}} \tag{2}$$

and $\qquad\qquad V^* = \overline{V} \, [1 + \beta \, (T^* - \overline{T})] \tag{3}$

In Eq.(3) β is the volumetric coefficient of thermal expansion of the skeleton. Substituting for V_{fw}^* and V^* in Eq.(1) leads to,

$$\frac{\vartheta^*}{\vartheta} = \frac{\overline{V}}{\overline{V}_{fw}} [1 + \beta (T^* - \overline{T})] \tag{4}$$

T^* can thus be found from Eq.(4), by iteration. ϑ^* is a function of T^*.

From this stage onwards, further heating will generate significant tensile strains in the pore skeleton.

Heating of Saturated Pores:

(a) Skeleton The pore skeleton is subjected to two components of volumetric expansion during heating at a rate, $\frac{dT}{dt}$. Firstly, there is the free rate of thermal expansion, $\frac{dV_{th}}{dt}$; and secondly, there exists a component due to the straining of the skeleton by the expanding water, $\frac{dV_{el}}{dt}$. Thus the total rate of change of pore volume is,

$$\frac{dV_s}{dt} = \frac{dV_{th}}{dt} + \frac{dV_{el}}{dt} \tag{5}$$

(b) Water The expansion of the water, volume V_w, during heating at the rate, $\frac{dT}{dt}$, is

$$\frac{dV_w}{dt} = \frac{d(m\,\vartheta)}{dt} = m\frac{d\vartheta}{dt} + \vartheta\frac{dm}{dt} \tag{6}$$

m and ϑ^* are respectively the mass and specific volume at temperature T. The first term on the right hand side represents the changes in the volume of water due to thermal expansion, while the second term represents the changes in volume of water due to mass transfer (*into* the pore as defined here).

Volumetric compatibility requires that the rate of increase in volume of shell, $\frac{dV_s}{dt}$, be equal to the rate of volume expansion of water, $\frac{dV_w}{dt}$, at the water/skeleton interface of the pore. Hence, from Eqs. (5) and (6),

$$\frac{dV_{el}}{dt} = m \frac{d\vartheta}{dT} \frac{dT}{dt} + \vartheta \frac{dm}{dt} - \frac{dV_{th}}{dt} \tag{7}$$

Noting that,

$$\frac{dV_{th}}{dt} = \beta \, V \frac{dT}{dt} \tag{8}$$

Eq.(7) then becomes,

$$\frac{dV_{el}}{dt} = m \frac{d\vartheta}{dT} \frac{dT}{dt} + \vartheta \frac{dm}{dt} - \beta \, V \frac{dT}{dt} \tag{9}$$

For tensile fracture of the skeleton to be avoided, $\frac{dV_{el}}{dt} = 0$. This implies that an escape of water at a rate $\frac{dm}{dt}$ must be permitted. Thus,

$$\frac{dm}{dt} \geq -m \frac{dT}{dt} \left[\frac{1}{\vartheta} \frac{d\vartheta}{dT} - \beta \right] \tag{10}$$

The negative sign prefixing the term on the right hand side of this inequality means that the mass flow is out of the pore.

NUMERICAL MODEL

The spherical model of Figure 1 has been used to evaluate tensile stresses in the pore skeleton during expansion of the water in the physically saturated condition, and with no loss of water (i.e. $\frac{dm}{dt} = 0$ in Eq.(7)). The stresses calculated refer to a hypothetical elastic material of high tensile strength and are derived simply to demonstrate that concrete can be expected to undergo tensile fracture soon after pore saturation is achieved in a rising temperature field.

The initial conditions for the sphere (i.e. thickness of shell and water content) are determined from the mix proportions and the initial free pore volume of the concrete. For the spherical pore of Figure 1,

$$\frac{r^3}{R^3} = \frac{V}{V_{tot}} \tag{11}$$

Here, V and V_{tot} are respectively the *total* pore volume and the *bulk* volume of concrete; and r and R define the inner and the outer radii of the pore. The total pore volume is,

$$V = V_{uf} + V_{fw} \qquad (12)$$

where, V_{uf} is the initial *as cast* unfilled pore volume and V_{fw} if the volume of free water.

 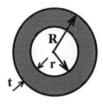

Figure 1 - Spherical pore model for concrete

The model is used firstly to determine the temperature at which pore saturation first occurs and then to evaluate skeletal stresses during heating and thermal expansion of the water in the saturated state, on the assumption that water remains incompressible. Figures 2 and 3 give some results. From the temperature, T^*, and time, t^*, of first saturation, Eq.(5) is used, with $\dfrac{dm}{dt} = 0$, to determine the elastic volumetric changes ΔV_{el} corresponding to specified temperatures, $T > T^*$.

Thus, during an interval $\Delta t = (t - t^*)$, and temperature change, $\Delta T = (T - T^*)$, the volume change of the pore due to thermal expansion of the skeleton is from Eq.(8),

$$\frac{dV_{th}}{dt} = \beta\, V \frac{dT}{dt}$$

i.e. $\qquad\qquad \Delta V_{th} = \beta\, V^* \Delta T \qquad (13)$

V^* is the volume at first saturation. The corresponding expansion of the water during the interval Δt is obtained from Eq.(6) as,

$$\frac{\Delta V_w}{\Delta t} = m \frac{\Delta \vartheta}{\Delta t} \tag{14}$$

Thus, $\Delta V_w = m \Delta \vartheta$ (15)

It then follows that the component of volume change in the pore, due to the elastic strains in the skeleton, may be derived from Eqs.(5) and (6), as

$$\Delta V_{el} = \Delta V_w - \Delta V_{th} \tag{16}$$

The change of water temperature, ΔT, creates a resultant outward radial displacement, u, at the pore boundary of the spherical elastic model. The following equations then apply,

From Eq.(15), $\Delta V_w = m \Delta \vartheta$

From Eq.(16), $\Delta V_{th} = \beta V^* \Delta T$

The elastic volumetric change component of the skeleton is simply,

$$\Delta V_{el} = 3 \frac{u}{r} \tag{17}$$

At a constant water content $\Delta V_w = \Delta V_{th} + \Delta V_{el}$. It then follows that the radial displacement, u, is given by,

$$u = \frac{r}{3} \left[\frac{m \Delta \vartheta}{V^*} - \beta \Delta T \right] \tag{18}$$

The water pressure caused by the radial displacement, u, and thus, the stresses in the wall of the sphere can then be calculated using the theory of thick shells. Cracking will be initiated when the calculated stresses, exceed the tensile strength of concrete.

NUMERICAL EXAMPLE

The method described has been used to investigate the spalling behaviour of the two concretes A and B, while heated at a rate equal to 2°C/s. The following additional data help to define the problem:

Concrete A. Normal strength concrete (42 MPa)

(a) Concrete mix proportions by weight: 1 : 2.67 : 4: 0.60 :: cement : fine aggregate : coarse aggregate: water.

(b) Free pore volume: 3.0% of bulk volume.

(c) Water in hydration, 0.2; free water, 0.4 (as a water-cement ratio).

(d) Permeability of concrete $10^{-17}\,m^2$

Concrete B. High strength concrete (90 MPa)

(a) Concrete mix proportions by weight: 1: 1.57 : 2.54 : 0.1 : 0.348 :: cement : fine aggregate : coarse aggregate : micro silica : water plus plasticizer.

(b) Free pore volume: 0.95% of bulk volume.

(c) Water in hydration, 0.2; free water, 0.148 (as a water-cement ratio).

(d) Permeability of concrete $10^{-20}\,m^2$.

Table 1 summarises those volume ratios required for the analysis.

The spherical model with external radius, R, internal radius, r, and thickness, t, is now used to represent the concretes, A and B, as defined in Table 1.

Table 1 - Volume ratios for numerical analysis

Type of concrete	w/c	V_{uf}/V_{tot} %	V_{fw}/V_p %	V_p/V_{tot} %
Normal strength: Concrete A	0.6	3.0	79.1	14.35
High strength: Concrete B	0.358	0.95	87.8	7.86

Note : V_p = Pore volume of concrete (denoted as V in the text and defined in Eq.(12))

V_{uf} = Initial unfilled pore volume of concrete V_{fw} = Volume of free water in concrete

V_{tot} = Bulk volume of concrete

The following ratios then apply;

$$(\frac{r^3}{R^3})_A = (\frac{V_p}{V_{tot}})_A = 0.1435 \qquad \frac{r_A}{R_A} = 0.524 \qquad \frac{t_A}{R_A} = 0.476$$

$$(\frac{r^3}{R^3})_B = (\frac{V_p}{V_{tot}})_B = 0.0786 \qquad \frac{r_B}{R_B} = 0.428 \qquad \frac{t_B}{R_B} = 0.572$$

For no mass transfer from the model pores, Eq.(4) indicates that physical saturation occurs as follows;

Concrete A. After 2 minutes of heating and at a temperature of 260°C.
Concrete B. After 1.45 minutes and at a temperature of 194°C.

Additional heating then causes elastic tensile strains in the shells. Figures 2 and 3 illustrate, (for descriptive purposes only), the development of tensile stresses with temperature for a range of initial pore water contents, as might occur during natural drying preceding exposure to fire temperatures.

Both figures indicate that physical saturation occurs later for the dryer concretes. However, drier concretes show higher rates of stress development immediately after becoming physically saturated. Because the values of all these stresses are significantly higher than the tensile strength of concrete, (Figs 2 and 3), and change rapidly after the pores first become saturated, tensile fracture can be anticipated whenever physical saturation occurs whilst temperatures continue to rise.

CONCLUDING REMARKS

Explosive spalling of concrete in fire is strongly related to the amount free water in concrete and its distribution during heating.   At high temperatures, chemically bonded water is also released [3].   This then acts to sustain high pore vapour pressures in a similar manner to free water.   Because the quantity of chemically bound water retained at high temperatures ($300°C$ to $350°C$) may be as low as 0.06 (as an equivalent water/cement ratio) the implications are that high strength concretes of low water/cement ratios are not free from the risk of spalling at high temperatures.

Differential thermal expansion of water and concrete, at the time of heating, will normally, lead to physical saturation of pores. Further heating of the physically saturated pore will rapidly, generate additional tensile strains in the solid skeleton surrounding the pores, due to the low compressibility of the expanding water. The resultant stresses in the solid skeleton can quickly exceed the tensile fracture strength of concrete and initiate cracking/spalling.

Generally, mass transfer in concrete will delay the onset of physical saturation and the occurrence of hydraulic fracture. The extent of this delay will depend on the rate of mass flow and the initial unfilled pore volume in the as cast hardened concrete. For a given rate of heating, $\dfrac{dT}{dt}$, fracture can be avoided only if the rate of mass flow from the saturated pores exceeds a minimum value as defined in Eq.(10), for the local conditions. When the rate of flow is lower than this minimum value, spalling is predicted. For predicting spalling and its location it is necessary to identify the position of the saturation plug. This can be achieved from a separate moisture migration analysis, which incorporates the following influential parameters; permeability and its time-, temperature-, and moisture content-dependent variations; pore pressure distribution; temperature distribution; geometry and boundary conditions [England and Khoylou, 1995].

The risk of spalling in normal strength concrete is greatest at early ages. During natural drying, the loss of water reduces the effective water content of the pores close to the boundary. Thus, physical saturation of these partially dry pores will only occur after longer periods of fire exposure, i.e. higher temperatures, or not at all. Spalling of high strength concrete appears not to be affected by age in a similar manner, due primarily, to its low inherent permeability.

Future work will aim to integrate the spalling model described here with the moisture migration analysis [England and Khoylou, 1995] for the purpose obtaining more accurate predictions of the location of spalling in concrete during fire.

REFERENCES

Birk, A.M., and Cunnigham, M.H. (1995). "The boiling liquid expanding vapour explosion". *Proceedings of 15th Canadian Conference of Applied Mechanics, CANCAM 95*, May-June, Victoria

Copier, W.J. (1980). "The spalling of normal weight and lightweight concrete exposed to fire". *ACI Special Publication, SP 80-7*, pp 219-236, Puerto Rico

England, G.L. and Khoylou, N. (1995). "Moisture flow in concrete under steady-state non-uniform temperature states - experimental observation and theoretical modelling". *Journal of Nuclear Engineering and Design* Vol.156 pp 83-107

Greathead, R.J. (1986). "Permeability of concrete containing blast furnace slag as affected by temperature, moisture and time." *Ph.D. Thesis, University of London.*

Sanjayan, G. and Stocks, L.J. (1993). "Spalling of high-strength silica fume concrete in fire." *ACI Materials Journal*, Vol. 90, No.2, pp 170-173.

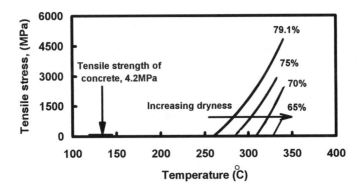

Figure 2.   Tensile stresses in the skeleton of the spherical model for different initial pore fillings (Concrete A).

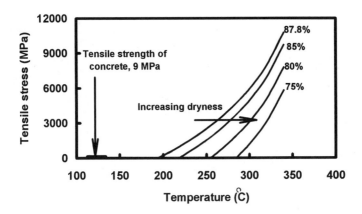

Figure (3).   Tensile stresses in the skeleton of the spherical model for different initial pore fillings (Concrete B)

SUBJECT INDEX
Page number refers to first page of paper

AUTHOR INDEX
Page number refers to first page of paper

575